T0398613

A Laboratory Manual in Biophotonics

A Laboratory Manual in Biophotonics

Vadim Backman
Adam Wax
Hao F. Zhang

CRC Press
Taylor & Francis Group
Boca Raton London New York

CRC Press is an imprint of the
Taylor & Francis Group, an **informa** business

CRC Press
Taylor & Francis Group
6000 Broken Sound Parkway NW, Suite 300
Boca Raton, FL 33487-2742

© 2018 by Taylor & Francis Group, LLC
CRC Press is an imprint of Taylor & Francis Group, an Informa business

No claim to original U.S. Government works

Printed on acid-free paper

International Standard Book Number-13: 978-1-4398-1051-4 (Hardback)

Library of Congress Cataloging-in-Publication Data

Names: Backman, Vadim, author. | Wax, Adam, author. | Zhang, Hao (Hao F.), author.
Title: A laboratory manual in biophotonics / Vadim Backman, Adam Wax, and Hao Zhang.
Description: Boca Raton : Taylor & Francis, 2017. | Includes bibliographical references and index.
Identifiers: LCCN 2017037476| ISBN 9781439810514 (hardback : alk. paper) |
ISBN 9781498744331 (ebook)
Subjects: | MESH: Light | Tissues--physiology | Phototrophic Processes |
Optics and Photonics--methods | Optics and Photonics--instrumentation | Laboratory Manuals
Classification: LCC R857.T55 | NLM QT 162.L5 | DDC 616.07/548--dc23
LC record available at https://lccn.loc.gov/2017037476

Visit the Taylor & Francis Web site at
http://www.taylorandfrancis.com

and the CRC Press Web site at
http://www.crcpress.com

eResource material is available for this title at https://www.crcpress.com/9781439810514

Contents

Preface

Optical imaging and sensing technologies played key roles in almost all fronts of fundamental biomedical investigations and clinical patient care. In the past two decades, the breadth and depth of technology development in biomedical optics have grown exponentially and unremittingly. New technologies enabled physicians to, for example, examine human retinas for better management of multiple blinding diseases, saving billions of dollars of the healthcare system. New technologies also allowed researchers to visualize complex biological processes at the most fundamental individual molecule scale, which were recognized by the 2014 Nobel Prize in Chemistry. As a result, more students are becoming interested in biophotonics careers and an increasing number of institutions are starting to offer biomedical optics courses. There are already many outstanding textbooks in biomedical optics authored by leading researchers around the world. However, almost all of them focus on theoretical foundations of tissue optics and principles of optical imaging. After teaching biomedical optics courses in our home institutions for over 10 years, we found that there still lacks a suitable textbook that covers the essential experimental skills to, for example, measure tissue optical properties and construct basic optical instruments. We observed that students can reach the best understandings of optical technologies after they finish both well-guided and exploratory hands-on laboratories. To serve this need, we developed this new book called *A Laboratory Manual for Biophotonics*. This book starts with brief descriptions of fundamental mathematics, biology, and basic optical components and electrical instruments. Then we introduce fundamental optical properties of tissues with a focus on light–tissue interactions. The following three chapters guide students through key microscopic imaging, macroscopic imaging, and spectroscopic analysis technologies. The last chapter explains computational tools either to facilitate imaging and sensing or to achieve comprehensive data analysis. This book is suitable for classes taken by senior undergraduate students and graduate students. It can also be used as a reference book for biophotonics researchers.

   We sincerely acknowledge the prodigious contributions from all of our talented group members, without whom this book would be impossible. We also feel extremely grateful to our family members. They sacrificed tremendously in order to support our careers.

Vadim Backman, Adam Wax, and Hao F. Zhang

MATLAB® is a registered trademark of The MathWorks, Inc. For product information, please contact:

The MathWorks, Inc.

3 Apple Hill Drive

Natick, MA 01760-2098 USA Tel: 508 647 7000

Fax: 508-647-7001

E-mail: info@mathworks.com

Web: www.mathworks.com

Authors

Dr. Vadim Backman is the Walter Dill Scott Professor of Biomedical Engineering at Northwestern University in Evanston, IL, and the program leader of the Program in Cancer and Physical Sciences, Lurie Comprehensive Cancer Center in Chicago. He directs one of the highest NIH-funded laboratories in the United States. He received his PhD in medical engineering and medical physics from Harvard University and the Massachusetts Institute of Technology in 2001. His research is focused on bridging advances in biophotonics into biology and clinical medicine. The Backman laboratory develops novel optics technologies for characterization and imaging of biological tissue, with a focus on the nanoscale and molecular levels. The research spans cancer biophysics to novel optical diagnostic and imaging techniques to multicenter clinical trials. Dr. Backman has received numerous awards, including being selected as one of the top 100 young innovators in the world by *MIT Technology Review Magazine*, being elected as a Fellow of the American Institute for Medical and Biological Engineering, and receiving the National Science Foundation CAREER award. He has published more than 214 papers in peer-reviewed journals, including some of the most prestigious, such as *Nature*, *Nature Medicine*, *PNAS*, and *Physical Review Letters*, and he holds over 20 patents. He serves as a member of the Steering Committee of the National Cancer Institute Early Detection Research Network and is a frequent participant in NIH and NSF review panels and study sections. Dr. Backman is the cofounder of two biotech companies.

Dr. Adam Wax received dual BSs in 1993, one in electrical engineering from Rensselaer Polytechnic Institute, Troy, NY, and one in physics from the State University of New York at Albany, and a PhD in physics from Duke University, Durham, NC, in 1999. He joined the George R. Harrison Spectroscopy Laboratory at the Massachusetts Institute of Technology as a postdoctoral fellow of the National Institutes of Health immediately after his doctorate. Dr. Wax joined the faculty of the Department of Biomedical Engineering at Duke University in the fall of 2002. In 2006, Dr. Wax founded Oncoscope, Inc., to commercialize early cancer detection technology developed in his laboratory. In 2010, he was named a Fellow of the Optical Society of America and SPIE, the international society for optics and photonics. He is currently the Theodore Kennedy Professor of Biomedical Engineering at Duke University. His research interests are the use of light scattering and interferometry to probe the biophysical properties of cells for both diagnosis of disease and fundamental cell biology studies.

Dr. Hao F. Zhang received his bachelor's and master's degrees from Shanghai Jiao Tong University in 1997 and 2000, respectively, and his PhD from Texas A&M University in 2006, all in biomedical engineering. From 2006 to 2007, he was a postdoctoral fellow at Washington University in St. Louis. From 2007 to 2010, he was an Assistant Professor in the Department of Electrical Engineering and Computer Science at the University of Wisconsin-Milwaukee. In 2011, he moved to the Department of Biomedical Engineering at Northwestern University and was tenured in 2013. Dr. Zhang has received multiple awards, including the Shaw Scientist Award by the Greater Milwaukee Foundation in 2009, the Director's Challenging Award by the National Institutes of Health in 2010, the CAREER Award by the National Science Foundation in 2011, and the Frontiers of Engineering Young Investigator Award by the Georgia Institute of Technology in 2013. His research interests include optical imaging technologies and their applications in biomedicine. For more information, please visit http://foil.northwestern.edu.

1

General Introductory Topics

Part 1: Fundamental Mathematics

Here, we introduce basic concepts in mathematics and statistics that will be useful in the coming chapters and experiments. The content is very limited and serves only to remind readers what should be known prior to continuing. For more detailed explanations and derivations, please refer to subject matter textbooks that focus on those topics.

Dirac Delta Pulse

There is often a need to consider the effect on a system by a forcing function that acts for a very short time period, such as a "kick" or an "impulse." Such an impulse is called the **Dirac delta (δ) pulse**, shown in Figure 1.1, and is defined as

$$\delta(x) = \begin{cases} +\infty, & x = 0 \\ 0, & x \neq 0 \end{cases} \tag{1.1}$$

Furthermore, the delta pulse satisfies an identity constraint:

$$\int_{-\infty}^{+\infty} \delta(x)dx = 1 \tag{1.2}$$

The Dirac delta is not a function, as any extended-real function that is equal to zero everywhere except one single point cannot have a total integral of 1. While it is more convenient to define the Dirac delta as a distribution, it may be manipulated as though it were a function, thus conferring its many properties and characteristics.

The Dirac delta can be scaled by a nonzero scalar α:

$$\int_{-\infty}^{+\infty} \delta(\alpha x)dx = \int_{-\infty}^{+\infty} \delta(u)\frac{du}{|\alpha|} = \frac{1}{|\alpha|} \tag{1.3}$$

$$\text{so that} \quad \delta(\alpha x) = \frac{\delta(x)}{|\alpha|} \tag{1.4}$$

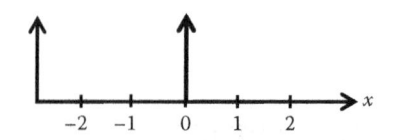

FIGURE 1.1
Dirac delta pulse.

The Dirac delta is symmetrical and follows an even distribution such that

$$\delta(-x) = \delta(x) \tag{1.5}$$

The Dirac delta exhibits a translation property where the integral of a pulse delayed by d returns the original function evaluated at d shown by

$$\int_{-\infty}^{+\infty} f(x)\delta(x-d)dx = f(d) \tag{1.6}$$

This is also called the **sifting property** of the delta pulse, as it "sifts out" the value $f(d)$ from the function $f(x)$. Although the integral here ranges $x = \pm\infty$, the same result is obtained for any integral bounds α and β provided that $\alpha \le d \le \beta$.

Convolving a function $f(x)$ with a delta pulse delayed by d has the effect of also time-delaying $f(x)$ by d:

$$(f(x)*\delta(x-d)) = \int_{-\infty}^{+\infty} f(\tau)\cdot\delta(x-d-\tau)d\tau, \tag{1.7}$$

which gives

$$\int_{-\infty}^{+\infty} f(\tau)\cdot\delta(\tau-(x-d))d\tau = f(x-d) \tag{1.8}$$

If $f(x)$ is a continuous function that vanishes at infinity, the following integral evaluates to 0:

$$\int_{-\infty}^{+\infty} f(x)\delta(x)dx = [f(x)H(x)]_{-\infty}^{+\infty} - \int_{-\infty}^{+\infty} f'(x)H(x)dx, \tag{1.9}$$

which evaluates to

$$-\int_{0}^{+\infty} f'(x)dx = [-f(x)]_{0}^{+\infty} = 0. \tag{1.10}$$

Here, $H(x)$ is the Heaviside step function whose derivative is the Dirac delta pulse.

Kronecker Delta Pulse

Similar to the Dirac delta, the Kronecker delta is represented by an impulse that equals 1 at zero and zero everywhere else. The Kronecker delta is frequently denoted by δ_i to distinguish it from the Dirac delta pulse.

Kronecker Comb

If a series of equally spaced impulses of amplitude 1 were delivered to a system (such as in the case of discrete sampling), we can see a periodic construction of Kronecker delta pulses known as a **Kronecker comb**.

$$\Delta_T(t) \stackrel{\text{def}}{=} \sum_{k=-\infty}^{\infty} \delta_i(t - kT) \tag{1.11}$$

where $\delta_i(x)$ is the Kronecker delta, and T represents a given period between successive pulses. Figure 1.2 shows a graphical representation of the Kronecker comb.

Notice that the Kronecker comb multiplied by any function $f(t)$ would depend on $f(t)$ only at the locations of the Kronecker pulses; hence, $f(t)$ is "sampled" and is commonly referred to as the **sampling function** in cases of signal processing and electrical engineering disciplines. In Figure 1.3, this sampling process is shown for an arbitrary curve.

sinc Function

Since the Dirac delta pulse is not defined as a function, it is typically constructed by taking a limit on a function. An example of this is by using the **sinc** function. This function, shown in Figure 1.4, is defined as

$$\text{sinc}(x) = \frac{\sin x}{x} \tag{1.12}$$

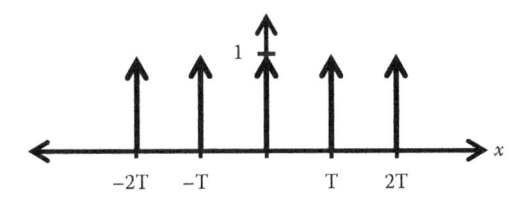

FIGURE 1.2
Kronecker comb.

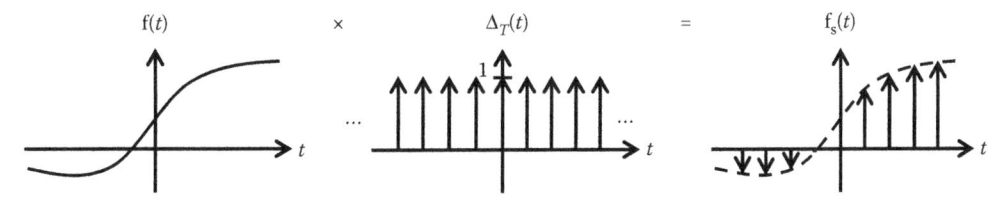

FIGURE 1.3
Piecewise sampling of *f(t)* via combination with Kronecker comb $\Delta_T(t)$.

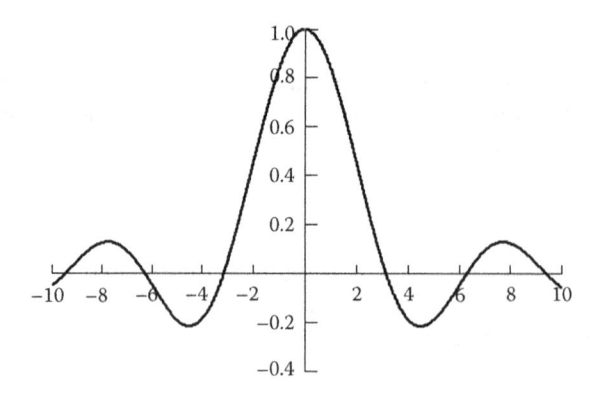

FIGURE 1.4
Sinc function.

To construct the pulse, let

$$F(x) = \frac{\text{sinc}(x/a)}{a}, \tag{1.13}$$

$$\text{as } a \to 0 \quad \text{and} \quad F(x) \to \delta(x) \tag{1.14}$$

Exercise: Show that $F(x) = (\text{sinc}(x/a))/a \to \delta(x)$ when $a \to 0$ in MATLAB®. Let $x = -10{:}0.001{:}10$ and $a = 10, 1, 0.1, 0.01$ and note how $F(x)$ approaches the Dirac delta pulse.

Convolution

Convolution is a mathematical operation on two functions that gives the amount of overlap of the first function as it translates through the second. This operation is denoted with an asterisk:

$$(f * g)(t) \overset{\text{def}}{=} \int_{-\infty}^{\infty} f(\tau)g(t - \tau)d\tau \tag{1.15}$$

Note that convolution is commutative, so $(f * g)(t) = (g * f)(t)$. Convolution is useful for calculating a moving average and is therefore used extensively in image and signal filtering. A visual representation of convolution is shown in Figure 1.5.

Cross-Correlation

Cross-correlation represents an important signal processing tool, which provides a measure of similarity between two signals. This similarity is expressed as a function of a supposed time-lag between the two waveforms.

For continuous functions $f(t)$ and $g(t)$, the cross-correlation $(f * g)(t)$ is defined as follows:

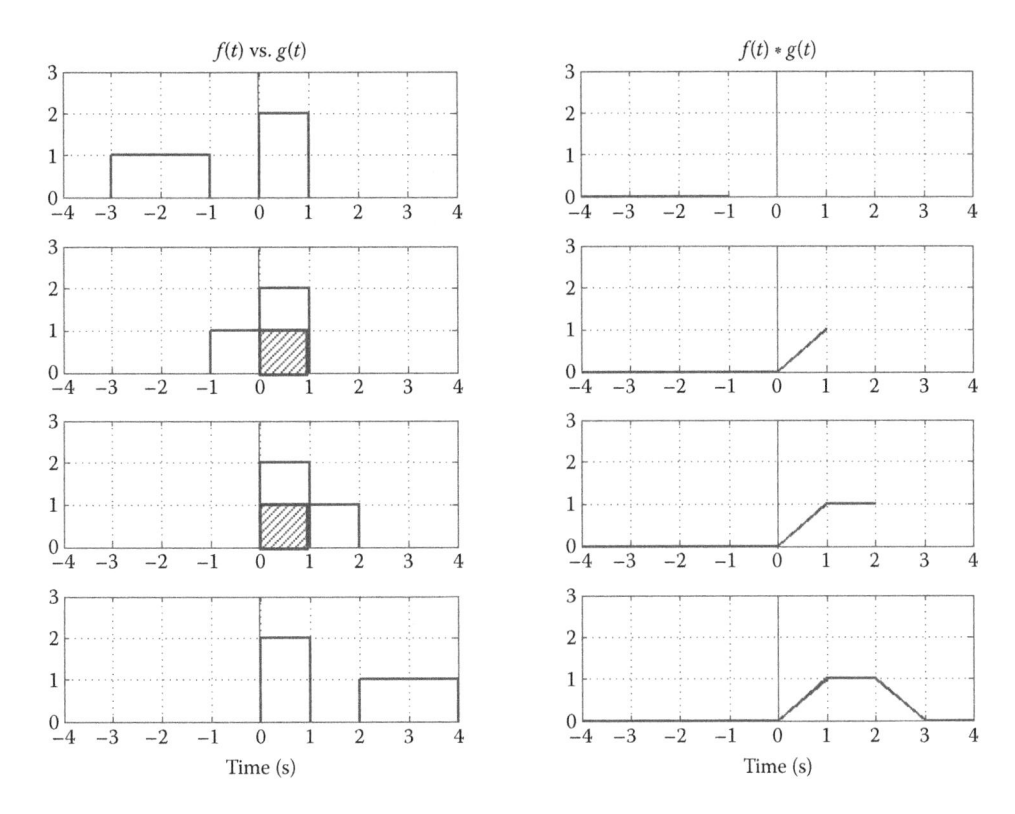

FIGURE 1.5
Convolution of two square impulses, one stationary and one moving along the positive *x*-direction. The shaded area is correspondingly plotted as the convolution result.

$$(f * g)(t) \stackrel{\text{def}}{=} \int\limits_{-\infty}^{+\infty} f^*(\tau)g(t + \tau)d\tau \tag{1.16}$$

Recall that $f^*(t)$ denotes the complex conjugate of $f(t)$.

The usefulness of cross-correlation is apparent in signal processing where temporal similarities between two waveforms can be isolated and examined. The cross-correlation of two sinusoidal curves is shown in Figure 1.6.

The cross-correlation of two continuous functions f and g is, in fact, related to its convolution where $(f * g)(t)$ is equivalent to the cross-correlation of $f^*(-t)$ and $g(t)$.

Autocorrelation

Autocorrelation represents a special case of cross-correlation where the temporal similarity of a signal is examined against itself. Informally, it presents a measure of similarity between observations as a function of the time separation between them.

Given a signal $f(t)$, the continuous autocorrelation $R_{ff}(\tau)$ at lag τ is defined as

$$R_{ff}(\tau) = (f(t) * f^*(-t))(\tau) \tag{1.17}$$

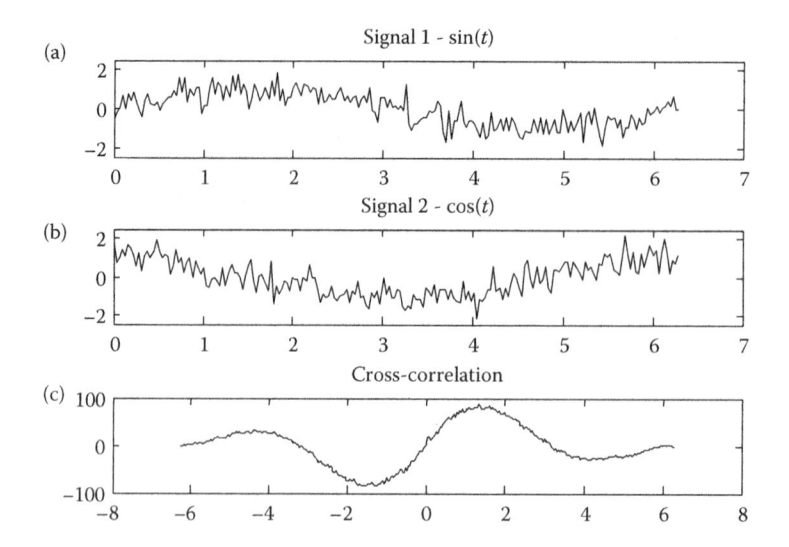

FIGURE 1.6
Two sinusoidal signals (a) and (b) corrupted by white noise are displaced by a factor of $\pi/2$. The resulting cross-correlation sequence (c) displays as a percentage the temporal correlation of (a) and (b). Since $\sin(t)$ leads $\cos(t)$ by $\pi/2$, the cross-correlation sequence shows a maximum at $\pi/2 = 1.5708$.

which is further defined as

$$\int_{-\infty}^{+\infty} f^*(t)f(t+\tau)dt = \int_{-\infty}^{+\infty} f(t)f^*(t-\tau)dt \tag{1.18}$$

In signal-processing disciplines, autocorrelation acts as a useful tool for identifying periodicities and repeating patterns within time domain signals, as well as identifying fundamental frequencies in a signal implied by its harmonic frequencies.

Fourier Transform

In mathematics, Fourier series are decompositions of periodic functions into an infinite set of simple oscillating functions of sines and cosines. In engineering, however, the forward Fourier transform is typically used. This extension of the Fourier series transforms a function of time into a function of frequency. There is also an inverse Fourier transform that converts a function of frequency back into a function of time. To simplify the math involved, the transform is expressed with complex exponentials instead of sines and cosines.

$$F(\omega) = \int_{-\infty}^{\infty} f(t)e^{-j\omega t}\, dt, \text{ forward Fourier transform} \tag{1.19}$$

$$f(t) = \frac{1}{2\pi} \int_{-\infty}^{\infty} F(\omega)e^{j\omega t}\, d\omega, \text{ inverse Fourier transform} \tag{1.20}$$

Pay special attention to the fact that the forward Fourier transform has the $e^{-j\omega t}$ term in the integral, while the inverse Fourier transform has the $e^{j\omega t}$ term. There is also a $1/2\pi$ term for the inverse Fourier transform because the integral is with respect to the angular frequency instead of frequency.

Fourier transforms are widely used in signal processing to calculate the frequency spectrum. Given a periodic signal in the time domain, the absolute value of its Fourier transform gives the frequency information of the signal. In Figure 1.7, the Fourier transform of a compound sinusoidal signal shows the frequency component.

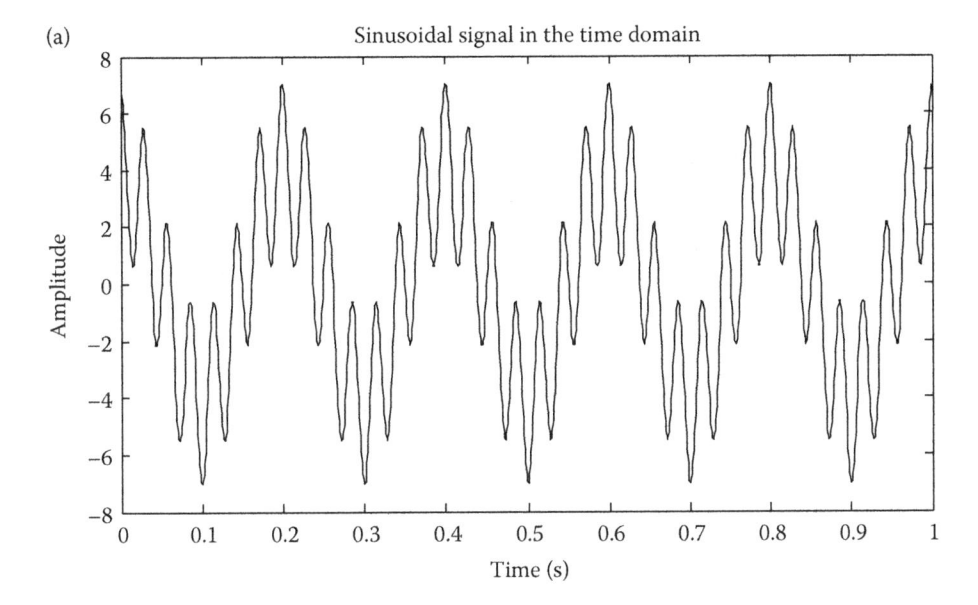

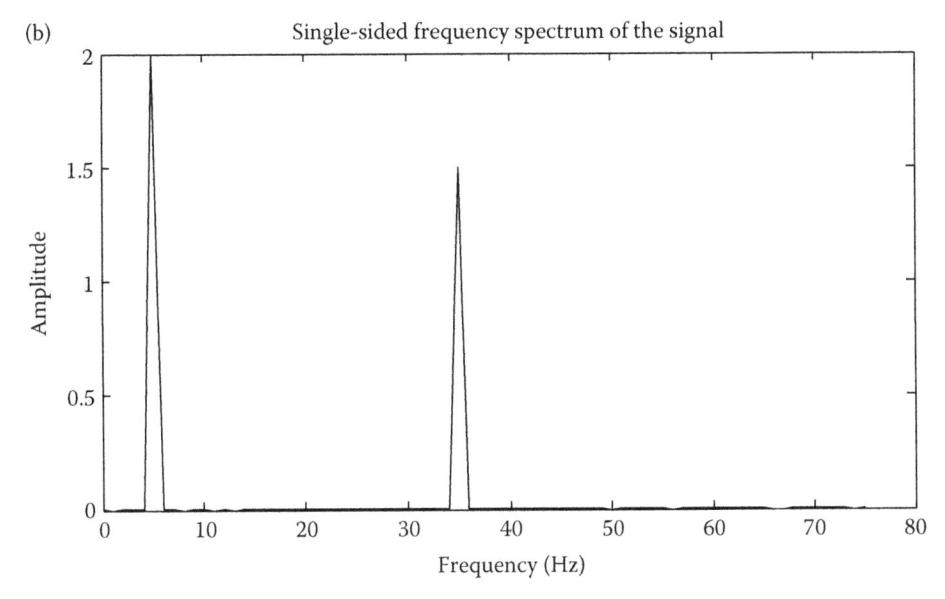

FIGURE 1.7
Demonstration of Fourier transform pair. (a) Sum of two sinusoidal signals: $4\cos(2\pi * 5t) + 3\cos(2\pi * 35t)$. (b) Frequency spectrum of the signal sampled at 1000 Hz: two peaks at 5 and 35 Hz.

The Fourier transform also has several important properties:
Linearity

$$af(t) + bg(t) \Leftrightarrow aF(\omega) + bF(\omega) \tag{1.21}$$

$$\text{Multiplication} \Leftrightarrow \text{Convolution} \quad f(t) \cdot g(t) \Leftrightarrow (F * G)(\omega) \tag{1.22}$$

$$\text{Convolution} \Leftrightarrow \text{Multiplication} \quad (f * g)(t) \Leftrightarrow F(\omega) \cdot G(\omega) \tag{1.23}$$

$$\text{Time shift} \Leftrightarrow \text{Phase shift} \quad f(t - t_0) \Leftrightarrow e^{-j\omega t_0} F(\omega) \tag{1.24}$$

Discrete Fourier Transform

In signal processing, it is common to use the discrete Fourier transform (DFT) instead of the continuous version. However, to use the discrete transform, the signal must first be sampled. After converting the continuous signal into a discrete signal of N points, the DFT can be calculated using the formula:

$$X_k = \sum_{n=0}^{N-1} x_n \cdot e^{-j2\pi(k/N)n} \quad k = 0, \ldots, N-1 : \text{Discrete Fourier transform} \tag{1.25}$$

$$x_n = \frac{1}{N} \sum_{k=0}^{N-1} X_k \cdot e^{j2\pi(k/N)n} \quad n = 0, \ldots, N-1 : \text{Inverse discrete Fourier transform (IDFT)} \tag{1.26}$$

Note that DFT returns a discrete sequence of N points, which is the sampled signal in the frequency domain. The resultant DFT is sampled at k/N cycles per sample, and the amplitude is equal to $|X_k|/N$.

In MATLAB, there are function commands that calculate the DFT and IDFT of signals called fft() and ifft(). These functions use an algorithm called the fast Fourier transform (FFT) that efficiently solves for the DFT of signals. These functions in MATLAB calculate the DFT over the frequency range of $(f_s/2, f_s)$, where f_s is the sampling frequency. Standard two-sided frequency spectrums, however, are over the frequency range of $(-f_s/2, f_s/2)$. Thus, in order to shift the frequency to the correct range, students should use fftshift(). This function sets the direct current (DC) frequency ($f = 0$ Hz) as the center frequency. The following shows sample code in MATLAB that calculates the frequency spectrum of the signal in part (a) of Figure 1.7 and produces (b).

```
% Time specifications:
F_s = 100;                    % Sampling frequency
dt = 1/F_s;                   % Time step
Time_stop = 3;
t = (0:dt:Time_stop-dt)';
N = size(t,1);                % Number of points

% Signal:
x = 4*cos(2*pi*15*t)+6*cos(2*pi*25*t)+3*cos(2*pi*35*t);
```

```
% Discrete Fourier transform
X = fftshift(fft(x));           % fftshift used to center the DC frequency

% Frequency specification
df = F_s/N;                     % Frequency step
f = (-F_s/2:df:F_s/2-df);

% Plot
plot(f,abs(X)/N)
```

Least-Squares Fit

Often in the real world, one expects to find linear relationships between variables. Experimental data in the form (x_n, y_n) for $n \in \{1, \ldots, N\}$ serve to examine these proposed relationships. The method of **least-squares fit** is a procedure to determine the best-fit linear regression line $y = ax + b$ to such a data set.

When it comes to quantifying "best fit," if we believe $y = ax + b$, then $y - (ax + b) = 0$. Thus, given observations:

$$\{(x_1, y_1), \ldots, (x_N, y_N)\}$$

We examine the error in our $y = ax + b$ prediction by looking at

$$\{y_1 - (ax_1 + b), \ldots, y_N - (ax_N + b)\}$$

This is illustrated in Figure 1.8 as the vertical distance a particular data point lies from a "prediction" or regression line and is marked for several points.

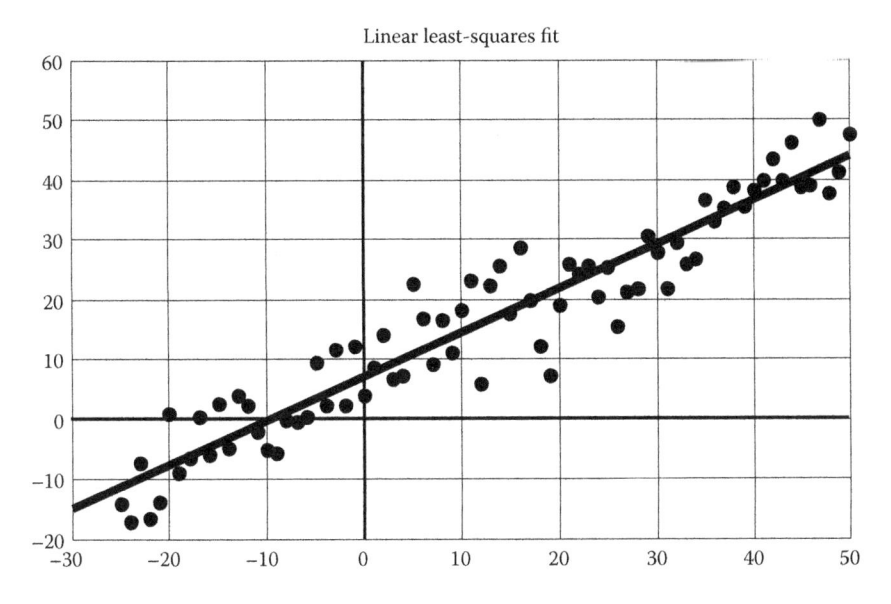

FIGURE 1.8
Scatterplot with regression line.

Given data $\{(x_1,y_1), \ldots, (x_N,y_N)\}$, we can define an error function E associated with our prediction $y = ax + b$ by

$$E(a,b) = \sum_{n=1}^{N} (y_n - (ax_n + b))^2 \tag{1.27}$$

The error function E represents the sum of squares of the error between predicted and actual data. Squaring the error gives larger errors a higher weight than smaller errors, as well as eliminating the sign of the error, allowing scrutiny of only its distance from our prediction.

The goal of the least-squares method is to find values a and b to minimize E, requiring solutions (a, b) to satisfy

$$\frac{\partial E}{\partial a} = 0, \quad \frac{\partial E}{\partial b} = 0 \tag{1.28}$$

Differentiating $E(a, b)$ and dividing by 2 yields:

$$\sum_{n=1}^{N} (y_n - (ax_n + b)) \cdot x_n = 0, \quad \text{and} \quad \sum_{n=1}^{N} (y_n - (ax_n + b)) = 0 \tag{1.29}$$

which may be rewritten as

$$\left(\sum_{n=1}^{N} x_n^2 \right) a + \left(\sum_{n=1}^{N} x_n \right) b = \sum_{n=1}^{N} x_n y_n, \quad \text{and} \quad \left(\sum_{n=1}^{N} x_n \right) a + \left(\sum_{n=1}^{N} 1 \right) b = \sum_{n=1}^{N} y_n \tag{1.30}$$

The values of a and b are those that satisfy the system of linear equations, in matrix form below:

$$\begin{bmatrix} \sum_{n=1}^{N} x_n^2 & \sum_{n=1}^{N} x_n \\ \sum_{n=1}^{N} x_n & \sum_{n=1}^{N} 1 \end{bmatrix} \begin{bmatrix} a \\ b \end{bmatrix} = \begin{bmatrix} \sum_{n=1}^{N} x_n y_n \\ \sum_{n=1}^{N} y_n \end{bmatrix} \tag{1.31}$$

Method of Least-Squares Fit for Multiple Parameters

The introduction of multiple-parameter relationships is often conveniently expressed in a compact matrix form of the least-squares method. Suppose we have an overdetermined system of equations:

$$\begin{aligned} y_1 &= ax_{11} + bx_{21} + c \\ y_2 &= ax_{12} + bx_{22} + c \\ y_3 &= ax_{13} + bx_{23} + c \\ y_4 &= ax_{14}{}^2 + bx_{24} + c \end{aligned} \tag{1.32}$$

This can be rewritten into matrix form as

$$X\beta = Y \tag{1.33}$$

where

$$X = \begin{bmatrix} x_{11} & x_{21} & 1 \\ x_{12} & x_{22} & 1 \\ x_{13} & x_{23} & 1 \\ x_{14} & x_{24} & 1 \end{bmatrix}, \quad \beta = \begin{bmatrix} a \\ b \\ c \end{bmatrix}, \quad Y = \begin{bmatrix} y_1 \\ y_2 \\ y_3 \\ y_4 \end{bmatrix} \tag{1.34}$$

The vector β is called a *least-squares* solution of the overdetermined system and can be solved by the **normal equation**:

$$(X^T X)^{-1} \cdot X^T Y = \beta \tag{1.35}$$

Nyquist–Shannon Sampling Theorem

The Nyquist–Shannon sampling theorem states that for a signal containing the highest frequency component f_{max}, the sampling rate should be at least $2f_{max}$ in order to avoid aliasing. This theorem is important because it provides a minimum condition on the sampling rate of the signal, so when you are trying to record an AC signal, the sampling rate should be at least twice that of the highest-frequency component. There is a critical case for this theorem: when the signal is a perfect sinusoid with frequency f_{max}, then the sampling frequency needs to be greater than $2f_{max}$.

This case is illustrated in Figure 1.9, where a 1-Hz sinusoidal signal is sampled at 2 Hz. If the data sampling starts at $\pi/2$ and continues to sample at every $n\pi/2$ where $n = 1, 3, 5, \ldots$, then the sampling will return the up-and-down peaks in part b. The shape of the sine curve is preserved, and reconstruction is possible. However, if the sampling starts at 0 and continues to sample at every $n\pi$ where $n = 0, 1, 2, \ldots$, then it will return 0 for each data sample. The shape is then completely destroyed, and reconstruction is impossible. Thus, to avoid this complication, people usually set the sampling rate to four to five times that of the highest-frequency component. There is no upper limit on the sampling rate, since oversampling will only help the reconstruction process. However, high sampling rates will result in more data points, so memory allocation and program runtime should still be considered in determining the optimal rate.

Exercise: Assume that a digital acquisition system has a sampling rate of 200 Hz. Please write MATLAB code to duplicate Figure 1.10. Note that when the signal frequency is 55 Hz, we cannot recognize a sine wave anymore; however, the sampling rate is still higher than the required minimal Nyquist sampling frequency. Why is this happening?

Basic Statistics

Measures of Central Tendency—Measures of central tendency relate to the aggregation of quantitative data around some central value.

- **Mode**—Number that appears most often in a set of numbers or distribution
- **Median**—Number separating the higher half of a distribution from the lower half

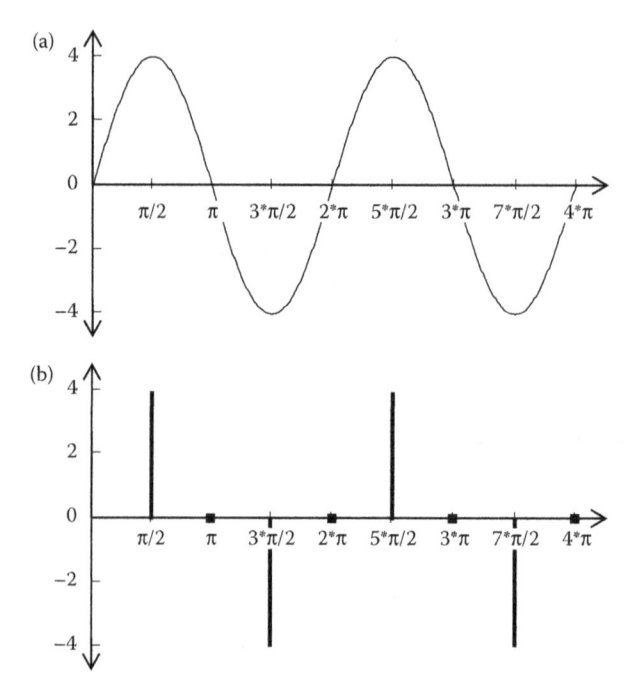

FIGURE 1.9
Demonstration of sampling theorem. (a) 1-Hz sine wave. (b) Two sampling attempts at 2 Hz with accurate reconstruction marked by the alternating impulses and failed reconstruction marked by the points on the origin.

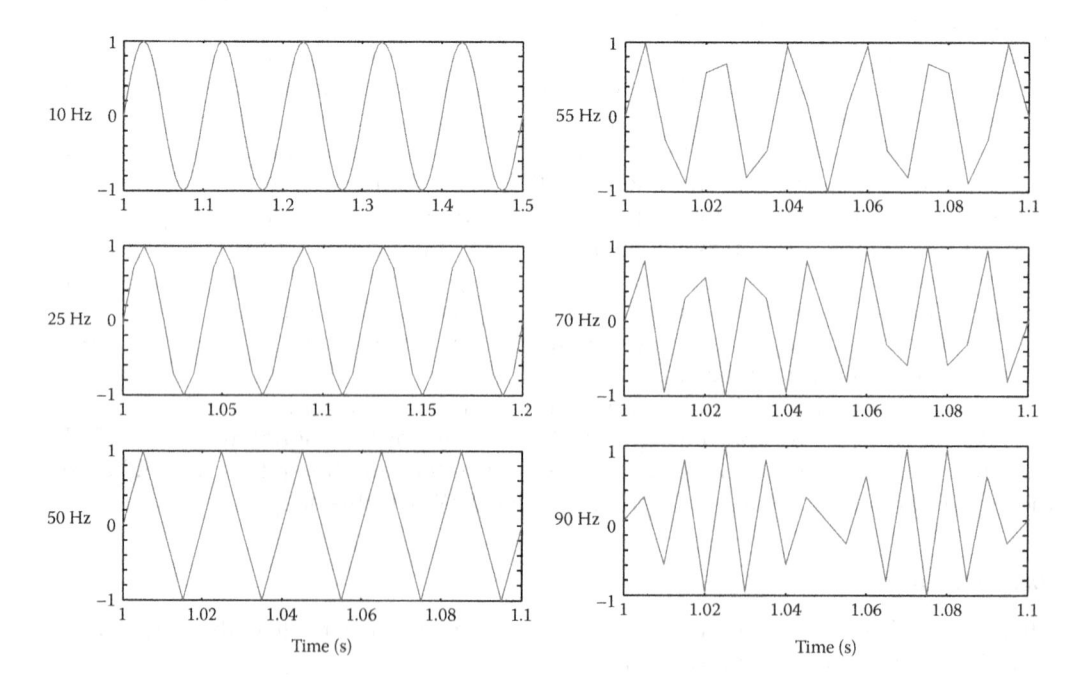

FIGURE 1.10
Examples of undersampling of sine waves at different frequencies.

- **Mean**—The mean $\bar{x}$ of a sample space $\{x_1, \ldots, x_n\}$ is defined as

$$\bar{x} = \frac{1}{n}\sum_{i=1}^{n} x_i \tag{1.36}$$

Measures of Dispersion—Dispersion measurements quantify the variability of a distribution.

- **Range**—The difference between a distribution's largest and smallest values.
- **Standard Deviation**—The standard deviation σ of a sample space $\{x_1, \ldots, x_n\}$ measures the variation of its values from its mean (or expected value) and is defined as

$$\sigma = \sqrt{\frac{1}{n}\sum_{i=1}^{n}(x_i - \bar{x})^2}, \quad \text{or} \quad s = \sqrt{\frac{1}{n-1}\sum_{i=1}^{n}(x_i - \bar{x})^2} \tag{1.37}$$

where s is a sample standard deviation and an estimator for σ and employs **Bessel's correction** (the use of $n-1$ instead of n) to correct for bias.
- **Variance**—The variance σ^2 of a sample space $\{x_1, \ldots, x_n\}$ is equivalent to the square of the standard deviation σ and also provides a measurement of dispersion.
- **Standard Error**—The standard error $SE_{\bar{x}}$ of a sample space $\{x_1, \ldots, x_n\}$ is the standard deviation of the error in the sample mean relative to the true mean and is defined as

$$SE_{\bar{x}} = \frac{s}{\sqrt{n}} \tag{1.38}$$

Student's *t*-Test

The Student's *t*-test is a statistical hypothesis test that is used to categorize the means of different samples. The test is most commonly used to determine if the means of two normally distributed samples are equal. This *t*-test, called the unpaired Student's *t*-test, assumes that the variances of the two samples are unknown and therefore must be calculated individually. By proposing a **null hypothesis** that the two means are equal, this test will generate a *t*-statistic that follows the Student's *t*-distribution. The **degree of freedom** (d.f.), a variable that describes the number of values that are free to vary, must also be calculated. Using d.f., a table of the distribution will correlate the *t*-statistic with a *p*-value, which gives the probability that the null hypothesis is true. Typically, the critical *p*-value is 0.05, which corresponds to 95% confidence, so when $p < 0.05$, the null hypothesis can be rejected with 95% confidence. If $p \geq 0.05$, then the null hypothesis cannot be refuted or proven, so further samples are needed. It is important to note that the null hypothesis can never be affirmed by the Student's *t*-test. It can only be refuted if the *p*-value is less than 0.05.

The mathematics behind the unpaired Student's *t*-test is quite long, but each step is well documented and simple to calculate in MATLAB. The variables needed for each sample are: the mean ($\bar{X}_1$ and $\bar{X}_2$), the number of points (n_1 and n_2), and the variances (s_1 and s_2).

$$\bar{X}_i = \frac{1}{n_i} \sum_{j=1}^{n_i} x_j \tag{1.39}$$

where $i = 1$ or 2, and x_j are the points in simple i

$$s_i = \frac{1}{n_i - 1} \sum_{j=1}^{n_i} \left(x_j - \bar{X}_i\right)^2 \tag{1.40}$$

where $i = 1$ or 2, and x_j are the points in simple i.

After calculating these variables, it is now possible to find the t-statistic and the degree of freedom.

$$t = \frac{\bar{X}_1 - \bar{X}_2}{\sqrt{\frac{s_1^2}{n_1} + \frac{s_2^2}{n_2}}} \quad \text{and} \quad d.f. = \frac{\left(\frac{s_1^2}{n_1} + \frac{s_2^2}{n_2}\right)^2}{\left(\frac{s_1^2}{n_1}\right)^2 \frac{1}{n_1 - 1} + \left(\frac{s_2^2}{n_2}\right)^2 \frac{1}{n_2 - 1}} \tag{1.41}$$

Using both the t-statistic and d.f., the p-value of the t-test can be determined. As a reminder, if $p < 0.05$, then the null hypothesis can be rejected. This means that the samples have different means. If $p \geq 0.05$, then no conclusion is reached. Thus, new samples with more data points are required to reach a conclusion.

Speckle Contrast

In optical imaging, an important concept similar to the t-statistic is called **speckle contrast**. While the t-statistic divides the mean by the standard deviation, speckle contrast is calculated by dividing the standard deviation by the mean.

$$C = \frac{\sigma}{I} = \frac{\sqrt{\overline{I^2} - \bar{I}^2}}{\bar{I}} \tag{1.42}$$

The speckle contrast is an important parameter that describes a speckle pattern, which is used in blood flow imaging. When a laser is incident on a living tissue, a random speckle pattern is generated. If the tissue is stationary, then the produced pattern will be static. Ideally, the standard deviation σ is equal to the mean intensity I, and the speckle contrast is 1. However, if the tissue is moving, then the produced pattern will be blurry. This blur will reduce the standard deviation of the pattern, resulting in a decrease in speckle contrast, so by tracking the changes in the speckle contrast over time, the speed of blood flow can be calculated. Figure 1.11 shows a static speckle pattern ($C = 1$) generated in MATLAB.

FIGURE 1.11
Static speckle pattern with speckle contrast of 1.

Statistical Distribution Functions

Often in statistics and engineering, characterization of random variables is achieved by distributions, which describe its probabilistic parameters. Two commonly used distribution functions are used to describe this variable: the **probability density function** (PDF) and the **cumulative distribution function** (CDF).

A variable's likelihood of taking on a particular value can be described by its probability density, typically portrayed by a distribution. There are many different types of distributions to characterize a variable's probability density. For example, the probability density function of the standard normal distribution is shown in Figure 1.11. The exact density function is given by

$$f(x) = \frac{1}{\sqrt{2\pi}} e^{-(x^2/2)} \tag{1.43}$$

The probability density function is also useful in finding the probability that the variable lies between two values. By taking the integral of the density function, the probability can be determined.

$$P(a \le x \le b) = \int_a^b f(x)dx \tag{1.44}$$

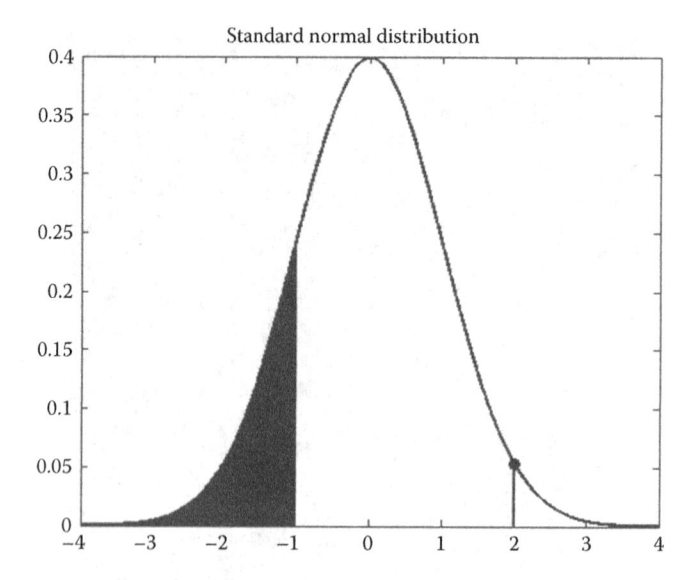

FIGURE 1.12
Probability density function of a standard normal distribution, with the cumulative distribution function of $F(x \leq -1)$ represented by the shaded area, and $f(x = 2)$ represented by the vertical line at 2.

The cumulative distribution function is a probability measure of a variable being less than or equal to a particular value. Thus, it can be calculated by taking the integral of the probability density function.

$$F(x) = \int_{-\infty}^{x} f(x)dx \qquad (1.45)$$

A visual representation of a standard normal distribution's CDF is seen in Figure 1.12, where the cumulative distribution up to a certain point is represented by the shaded area.

Part 2: Fundamental Biology

Vadim Backman

Basic Biological Concepts

For us to understand how light interacts with biological tissue, we first need an overview of the basic principles of biology underlying tissue organization and function. Essentially, what we are concerned with here is the science of functional anatomy. There are multiple levels of organization of living systems. At the largest-length scales, the organization of a living system is described by its gross anatomy. Although it is certainly important for a biomedical optics student to learn the gross anatomy of a human body, tissue optics typically interacts with smaller-length scales; thus, we will not focus on gross anatomy in this chapter.

At the tissue and cellular levels, we talk about microanatomy or *histology*. Understanding tissue structure and function at these length scales is key. This is because, as we will discuss in more detail in the following chapters, inelastic types of light–tissue interaction depend in part on the chemical (molecular) composition of tissue; light scattering in tissue is governed by tissue structure at length scales from tens of nanometers to microns, and light transport in tissue covers length scales from hundreds of microns to centimeters. Finally, if we were to zoom in even further, we would have to consider molecular structure and venture into the domain of *structural biology*. For all intents and purposes, this level of detail is beyond what is important for most applications of biophotonics. Therefore, for the purposes of this discussion, we will focus on histology.

This section is not intended to be a comprehensive overview of physiology or histology. For a more comprehensive review, the reader is referred to several excellent texts covering this complex subject. A detailed review of histology can be found in *Bloom and Fawcett: A Textbook of Histology* by D.W. Fawcett [2]. Robbins and Cotran's *Pathologic Basis of Disease* is widely considered a "Bible" of pathology and histopathology [3]; the text is regularly updated and is considered a must-have for any student, researcher, or practitioner working in health-related fields. However, our goal here is to review the basics of tissue composition and organization at a level that is critical to a biomedical optics student. So, let us begin!

Cell Structure

Generally speaking, all biological tissue consists of cells and extracellular matter. We start by examining the cell. The cell is the smallest unit capable of independent existence. There are hundreds of human cell types that can be identified by using light microscopy. Many of these cells are highly specialized, and, accordingly, their structures can be quite distinct from one another. Despite this diversity, many types of cells actually have a number of common features.

A cell is bounded by a cell membrane called the *plasmalemma*. The plasmalemma surrounding muscle cells or *myocytes* is referred to as the *sarcolemma* because of its specialized features. In a similar vein, the cell membranes of oocytes are called *oolemma*. The basic component of the plasmalemma is a *phospholipid bilayer* that separates the interior of a cell from the extracellular environment. The thickness of the bilayer is about 10 nm. In addition to phospholipids, the bilayer consists of glycolipids and cholesterols. (While we like to think of cholesterols as "bad" molecules that increase the risk of atherosclerosis, they are actually critically important for cell membrane function: for example, ensuring membrane stability.) The membrane forms what is called the liquid crystalline state with phospholipids held together via noncovalent interactions. The structure is dynamic. For example, individual lipid molecules undergo rapid motion in the lateral direction. The amount of each of the three basic components varies depending on the cell type. In most cell membranes (with the notable exception of red blood cells), phospholipids are the most abundant.

The picture is actually a little more complicated because the phospholipids come in a number of varieties: They may be saturated or unsaturated depending on the number of carbon bonds that are taken up by hydrogen atoms versus double or triple carbon–carbon bonds. While saturated lipids are fairly linear chains, unsaturated lipids have a "kink," which reduces their packing density. This has several important effects. In particular, a higher degree of unsaturation decreases the membrane's melting temperature (this is also referred to as increased membrane fluidity). As an example, differential membrane fluidity due to unsaturated-to-saturated lipid content plays a major role in the ability of algal cells that exist in symbiosis with animal cells forming a coral reef to withstand increases in water temperature: Algae with a higher unsaturated-to-saturated lipid content are at a higher risk of being damaged by temperature increase, which is one of the major factors behind the massive death of coral reefs as a result of global warming.

Although it appears that the main purpose of the cell membrane is to protect the cell and separate it from the external environment, this is only one of many functions that the plasmalemma performs. Indeed, cell membranes are involved in a variety of other cellular processes, including control of ion balance within versus outside of the cell through ion gating and conductivity, cell-to-cell connectivity, cell adhesion, intracellular and extracellular transport, and cell signaling. Furthermore, the cell membrane is a site of attachment of another key cellular structure, the *cytoskeleton*. This system plays a key role in signal transduction from the extracellular environment to the cytoskeleton and the cell nucleus, which in turn leads to genome regulation. In fact, mechanical forces acting on the cell surface affect cell shape and gene expression. It has been said that "a cell does what it touches"!

The multifunctionality of the cell membrane is in part ensured by the multitude of membrane-associated proteins that pierce the membrane. These are slightly larger structures with sizes on the order of 20–30 nm, thus bringing the overall thickness of the membrane to about 30 nm. There are three basic types of membrane-associated proteins: transmembrane, peripheral, and lipid-anchored. The transmembrane proteins span the entire thickness of the membrane. They may form ion channels, including passive channels, adenosine triphosphate (ATP)-driven pumps, and ion exchangers and G-protein-coupled receptors (GPCRs), and they play a critical role in membrane polarization (ion balance) and signaling. Lipid-anchored proteins are covalently bound to one or more lipid molecules but are not in direct contact with the membrane itself. Many of these proteins are G-proteins. Finally, peripheral proteins are associated with the peripheral parts of the lipid bilayer and typically have only temporary interactions with the membrane. Examples include enzymes and hormones.

From an optics perspective, the cell membrane has three primary modes of interaction with light. First, it is one of the denser cellular structures. As discussed in the next chapter, the refractive index of a biological structure is a leaner function of its molecular concentration (in g/cm^3). The plasmalemma and intracellular membranes (discussed later) are fairly dense structures whose concentration approaches 0.75 g/cm^3, which corresponds to over 50% of volume occupied by lipids. This translates into a high refractive index of a cell membrane approaching 1.41–1.44, which, relative to the refractive index of the cytosol that surrounds the membrane, is on the order of 1.05. This refractive index contrast would lead to elastic light scattering.

Second, lipids can be detected by a characteristic Raman scattering spectrum. Raman scattering is a specific type of inelastic scattering, and this source of contrast is discussed in Chapter 6. Finally, membrane proteins in general and membrane receptors in particular can be targeted by contrast agents (also referred to as molecular beacons) with an affinity for specific receptors. A variety of contrast agents have been developed, from fluorescent molecules to nanoparticles (nanospheres, nanorods, nanocages, etc.). This blends into *molecular imaging* with the goal of imaging the spatial location of specific molecular species. As receptors play a central role in cellular signaling and are frequently mutated or overexpressed in diseases, molecular imaging with sensitivity to specific receptors is of significant interest. Furthermore, imaging membrane receptors is actually easier compared to internalized receptors or intracellular components due to the relatively easier access of molecular beacons to cell membranes.

Next, we next consider the interior of the cell. The two major cellular compartments are the *nucleus* and the *cytoplasm*. The nucleus, the largest organelle in the cell, has a spheroid-like shape. In some cells (e.g., macrophages, polymorphonuclear cells, neutrophils), it has a lobulated shape so that on histological cross-sectional images it may appear as if a cell has several nuclei when, in fact, these are all interconnected parts of the same nucleus. Typically, the size of a nucleus is on the order of a few microns, with 5–10 μm being the most commonly observed size.

There are several intranuclear structures that can be identified, including *nuclear envelope*, *chromatin*, and *nucleolus*. Let us examine these three structures in more detail. The nuclear envelope is a membrane-like structure that bounds the nucleus. It is also known as the perinuclear envelope, nuclear membrane, nucleolemma, or karyotheca. There are a number of significant differences between the nuclear envelope and the cell membrane. First, as opposed to the plasmalemma, the nuclear membrane is not a single but a double lipid bilayer. The two bilayers are separated by a 30-nm gap, which is called the *perinuclear space* or *perinuclear cistern*.

Second, while the cell membrane seals the cell tightly and is impermeable to water and water-soluble molecules such as ions and larger molecules such as proteins (lipid-soluble molecules can diffuse through the lipid bilayer; the transport of other molecules that a cell needs to transport in or out of the cytoplasm is an active as opposed to a passive process), the nuclear envelope is filled with multiple nuclear pores.

The nuclear pores are large (about 100 nm in diameter) and serve to facilitate the exchange of materials between the nucleus and the cytoplasm. This exchange is a critically important process as a part of gene expression, as we will soon see. The exchanged molecules include various transcription factors going into the nucleus to regulate gene transcription and messenger RNA (mRNA) getting out of the nucleus to be read by the ribosomes in the cytoplasm as part of the translation process leading to protein synthesis. These processes are discussed later in more detail. The number of nuclear pores correlates with the metabolic activity of a cell. A metabolically silent cell may have as few as dozens of pores, whereas an active cell may have as many as thousands. Another function of nuclear pores is that they fuse the outer and the inner nuclear membranes; indeed, nuclear pores are not simple "holes" in the envelope but rather protein complexes called nucleoporins. Yet another function of the pores is active transport. Although small molecules freely diffuse through the pores, larger molecules are recognized by specific sequences, and their diffusion is assisted with the help of nucleoporins, which is known as the RAs-related nuclear protein cycle or RAN cycle.

The two-membrane organization of the nuclear envelope makes sense if we consider its function. The outer membrane of the nuclear envelope is continuous with the rough endoplasmic reticulum (RER), which is the network of membranes and ribosomes where protein synthesis occurs. On the other hand, nuclear membrane proteins reside on the inner nuclear membrane. The inner membrane serves as an attachment for the nuclear chromatin and participates in the regulation of chromatin structure and function. Specifically, the inner nuclear membrane is connected to the nuclear lamina, which is a network of intermediate filaments (sometimes referred to as the nuclear skeleton). In turn, chromosomes are attached to the lamina.

One of the biophotonics ramifications of the nuclear envelope is elastic light scattering due to its high refractive index contrast. Another consideration is that its large pores allow contrast agents to access nuclear chromatin as long as they are internalized into the cytosol.

Chromatin is a major part of the nucleus. Electron and light microscopy images frequently delineate two distinct types of chromatin: *heterochromatin* and *euchromatin*. On electron microscopy images, heterochromatin appears as a collection of irregularly shaped clumps that vary in size from 250 to 1000 nm. Euchromatin, on the other hand, appears fairly smooth. A divide between heterochromatin and euchromatin or other types of chromatin is by no means fixed in time. A gene may move from euchromatin to heterochromatin in a process that appears to be controlled by hypermethylation.

It used to be believed that heterochromatin is transcriptionally silent, while euchromatin is where transcription occurs. Understanding chromatin structure and function is currently a frontier science, and it is becoming increasingly clear how complex the functional anatomy of chromatin really is. Recent studies have shown that what was once a clear-cut distinction between heterochromatin and euchromatin in terms of their transcriptional activity is no longer unquestionable. For example, studies have demonstrated that transcription does occur in heterochromatin.

The supposition that heterochromatin is transcriptionally silent stems in part from its very dense appearance on electron microscopy images. When tissue specimens are stained with contrast dyes such as hematoxylin and eosin, heterochromatin appears dark. This led to a common understanding that heterochromatin is denser than euchromatin. It also appeared logical that a denser heterochromatin would not allow gene transcription to occur, in part due to limited diffusion of transcription factors. On the other hand, what we know about density distribution in chromatin has recently been questioned, and chromatin structure remains a bit of a mystery. It is possible that new research will uncover potentially counterintuitive facts. Based on confocal microscopy studies, it at least appears that euchromatin, which may occupy most of the nuclear space, is relatively homogeneous with little density variation at length scales above 500 nm.

Furthermore, recent studies have questioned the viewpoint that heterochromatin is dense, impermeable to transcription factors, and, as a consequence, transcriptionally inactive. The studies showed that transcription factors (typically a few nanometers in size) should be able to diffuse into heterochromatin. Furthermore, the denser appearance of heterochromatin on electron and light microscopy images might be at least in part due to the chemical affinity of contrast agents that are used to create these images instead of true density distribution. Some recent studies failed to observe a difference in density between heterochromatin and euchromatin. Future research will elucidate this subject.

In some recent studies, it was proposed that we should talk about not two but five different types of chromatin [4]. These types of chromatin were denoted by different colors: yellow, red, blue, black, and green. It is important to note that the colors were chosen for convenience only, and they have no functional or imaging significance; for example, red chromatin does not appear any redder than yellow chromatin on electron or light microscopy images. Yellow and red types of chromatin are considered transcriptionally active. They contain components of the transcription machinery, including necessary enzymes, that acetylate histones, thus opening deoxyribonucleic acid (DNA) for transcription (see discussion that follows). One of the main distinctions between yellow and red chromatin is that yellow chromatin primarily expresses the so-called housekeeping genes—the genes that control basic cell functions and are believed to be expressed in most cells regardless of specific cell function. Red chromatin, on the other hand, has a preference for genes that are specific to a cell or tissue type and enable the cell to perform its specialized function.

Blue chromatin is a less active type of chromatin. It contains polycomb-group proteins (PcG proteins), which repress transcription in particular genes involved in the regulation of developmental processes. While some genes are still transcribed in blue chromatin, black chromatin is almost completely repressive. Black chromatin actively represses transcription of many tissue-specific genes. It is not clear how black chromatin actually silences transcription, but some evidence indicates that it is achieved through specific histone proteins (see discussion that follows) that do not allow DNA to be exposed for transcription and lamins, which are associated with nuclear membrane and interfere with transcription. Indeed, black chromatin is frequently found at the nuclear periphery.

Finally, green chromatin is what used to be referred to as heterochromatin. It contains repetitive DNA sequences (as opposed to black chromatin, which rarely contains repetitive sequences). In addition, green chromatin is actually transcriptionally active and conductive to transcription to its natural genes. The classification of chromatin into red, yellow, blue, green, and black is based on proteins that are found in these types of chromatin. It is quite possible that further research will shed some new light on chromatin structure, and new means of classification will emerge. These questions remain to be answered.

Whether we classify chromatin into heterochromatin and euchromatin or into the five types discussed above, these constitute only high-level (large-scale) chromatin organization. There are at least three other levels of chromatin structure. It is easiest to visualize chromatin structure from the bottom up. At the molecular level, there is a double-strand molecule of DNA. The diameter of the helix in the transverse direction is close to 1 nm, and the period in the axial direction is ~2 nm. At the most basic level of chromatin organization, DNA wraps around *histone* proteins forming *nucleosomes*, which are approximately 10 nm in diameter. This is frequently referred to as the "beads-on-a-string" structure. Nucleosomes effectively help pack the large genomes (e.g., a mammalian genome is 2 m long) into a compact structure fitting within ~10-μm diameter cell nuclei. The beads-on-a-string configuration can be considered a flexible polymer, and the higher-level organization of the chromatin is governed by the laws of polymer physics.

Second, arrays of nucleosomes wrap into a 30-nm chromatin fiber, it is believed with the help of histone H1 (although the role of histone H1 has recently been brought into question), and these 30-nm fibers pack into the *higher-order chromatin structure*. In the latter process, scaffolding nuclear proteins are important. The higher-order chromatin structure has several levels of organization, which are an active topic of current research. The 30-nm chromatin fiber is believed to be a dynamic structure that readily unfolds into a 10-nm beads-on-a-string fiber ready for transcription. However, some recent studies question the existence of the 30-nm chromatin fiber, suggesting that it might be an artifact of chromatin preparation as opposed to a native structure that exists *in vivo* in live cells. As with other levels of chromatin organization, we have to wait until new research leads to a consensus.

Neutron-scattering studies have revealed that both nuclear proteins and DNA components have a mass fractal organization for the range of length scales from 15 nm to the size of entire nuclei with mass fractal dimensions between 2 and 3 [5]. Nuclear proteins appear to have the same fractal dimension for all length scales, whereas the DNA component is biphasic with a lower fractal dimension (close to 2) for short-length scales under 400 nm and a higher mass fractal dimension approaching 3 for larger-length scales. Larger mass fractal dimension is indicative of a more globular organization.

The third major component of the cell nucleus is the nucleolus. It is responsible for the transcription of ribosomal ribonucleic acid (RNA) and ribosomal assembly. A typical nucleolus ranges in size from 500 to 1000 nm and consists of the network of *pars granulosa*

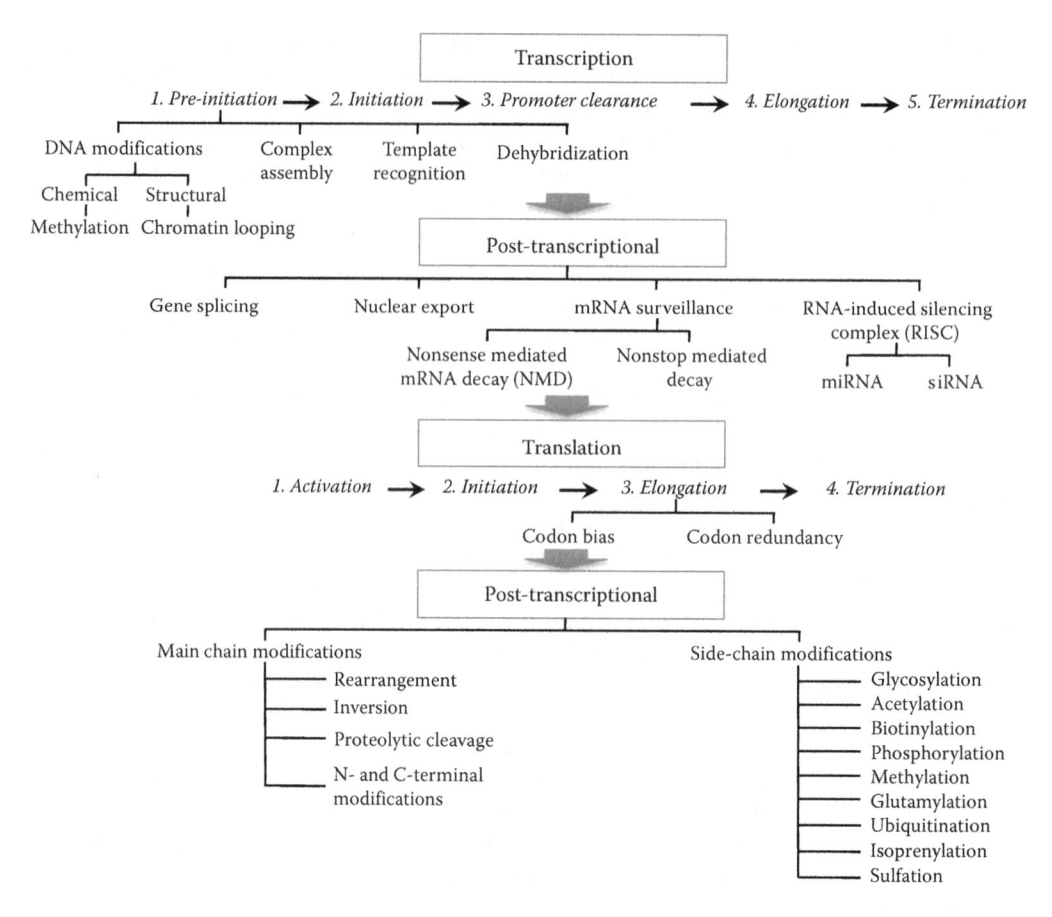

FIGURE 1.13
Major stages and steps of gene expression.

strands, which are made of 15-nm *ribonucleoprotein particles* and intranucleolar *fibrillar centers*, which are approximately 80 nm in size. According to some phase microscopy studies, the nucleolus is denser than the rest of the nucleus and may be the densest structure in a cell.

From a functional perspective, the central dogma of molecular biology is that as a result of gene transcription by an RNA polymerase, which reads out a DNA sequence, messenger RNA (mRNA) is produced. mRNA diffuses through the nucleus and into the cytoplasm through a nuclear pore. In the cytoplasm, mRNA is translated by a ribosome, which is essentially a protein assembly machine. In addition, ribosomes are being assembled in the nucleolus. This, of course, is a highly generalized and simplified chain of events. Multiple regulatory pathways control and modulate these processes.

For example, transcription itself consists of many steps (Figure 1.13). Transcription can occur only if a *core promoter sequence* in the DNA is present. Before transcription can occur, DNA of the promoter sequence has to be accessed by nuclear *transcription factors* in what is called the *transcription pre-initiation* process. Clearly, not all genes should be transcribed at any given time, and the accessibility of DNA to transcription factors is a tightly regulated process. Both nucleosomes and higher-order chromatin structure play key roles in this regulation.

In particular, nucleosomal organization protects DNA from being exposed to transcription factors and thus available for transcription. Accordingly, for transcription to occur, DNA must unwrap from a histone. This is accomplished by chromatin remodeling. In the process of chromatin remodeling, nucleosomes are moved along a DNA molecule assisted by enzymes. This is done by chromatin remodeling complexes such as SWI/SNF (SWItch/Sucrose NonFermentable). After a nucleosome is moved, transcription factors can get access to DNA.

How do nucleosome remodeling factors get to have access to DNA in the first place? It turns out that the process of DNA unwrapping is highly dynamic and, in fact, stochastic. Indeed, DNA does not have to be actively unwrapped from the nucleosome core. This is because the DNA–histone complex is a dynamic structure with two possible states: wrapped and unwrapped. Although DNA is in the wrapped state most of the time (~90%), in the remaining 10%, DNA is actually unwrapped and is accessible to transcription factors. Thus, the major impediment to entry of RNA polymerase into a nucleosome is not DNA unwrapping but rewrapping with nucleosome remodeling factors taking advantage of spontaneous nucleosome unwrapping and preventing the DNA from otherwise imminent and swift rewrapping.

Transcription factors are DNA-binding proteins. Some DNA-binding proteins are *activators*, while others are *repressors*. Their names tell it all: Activators increase the rate of gene transcription, while repressors do the opposite. These factors are critically important in transcription, as without their help, only a very low level of transcriptional activity would be possible. Other modulator molecules include *coactivators*—proteins that assist transcription factors to increase the rate of gene transcription—and *corepressors*—proteins that work with transcription factors to decrease the rate of transcription.

The best-studied promoter is a DNA sequence known as a TATA box. In this particular case, the TATA box becomes the binding site for a transcription factor called transcription factor II D (TFIID), which binds to the TATA box through TATA-binding protein. After TFIID binds to the TATA box, five more transcription factors and RNA polymerase form a *pre-initiation complex*.

In the pre-initiation of transcription, a pre-initiation complex contains the following: a core promoter sequence, transcription factors, DNA helicase, RNA polymerase, and various activators and repressors. Essentially, the promoter is the DNA site that allows transcription to start; transcription factors then significantly enhance the rate of transcription; the DNA helicase dehybridizes the DNA double strand, and the RNA polymerase reads out the gene sequence and helps produce mRNA, while activators, repressors, coactivators, and corepressors modulate the process.

Chromatin structure and the nanoscale environment (or nanoenvironment) in which these processes take place are important. Consider the following two examples. The access of chromatin to transcription factors depends on the surface area of the chromatin that is available for binding, which in turn depends on the local density. Thus, transcription factor accessibility may be modulated not only by molecular recognition mechanisms but through physical interactions provided by the nanoenvironment.

Another critical step is that a DNA double strand must be dehybridized ("unzipped") into two separate single strands of DNA ready for transcription. Dehybridization is accomplished by the work of a specialized transcription factor DNA helicase. The energy barrier that the DNA helicase has to overcome depends on the free energy of DNA hybridization. The latter depends, among other factors, on the local chromatin concentration. A higher affinity of the double strand would increase the work of the DNA helicase to dehybridize DNA, thus reducing the probability of transcription.

Contrary to old beliefs that cells transcribe their genes consistently through time, recent studies of transcription dynamics show that in many cells, transcription is irregular and stochastic. Long periods of transcriptional inactivity are followed by bursts of activity, a phenomenon called *transcriptional bursting* or *transcriptional pulsing*. This finding may indicate that open or closed chromatin structure may play an even more important role in gene expression than previously thought.

After transcription, mRNA diffuses through the nucleus into the cytoplasm. This is the time for *posttranscriptional regulation*. Interestingly, the concentration of mRNA reaching the cytoplasm correlates very poorly with their production as a result of transcription. In fact, only 5% of mRNA synthesized in the nucleus reaches the cytoplasm. Left unprotected, mRNA molecules are easily degraded by exonucleases. MicroRNA, short RNA sequences, are also known to recognize and degrade specific mRNA sequences. Thus, an RNA molecule is stable only if protected from degradation by means of an RNA-binding protein. As with transcription, these are not the only molecular-specific regulation processes that are important. Chromatin forms a complex fractal structure through which mRNA must diffuse, most likely through the process of anomalous diffusion. The residence time depends on the chromatin structure, which, therefore, may partially affect the probability of degradation of mRNA molecules.

Following translation on a ribosome, a protein must undergo posttranslational modifications and folding. Protein folding is an extremely complex process, as the three-dimensional structure of proteins is of crucial importance to their function. Chaperones are special protein complexes that help proteins fold correctly. High density of intracellular environment, called molecular *crowding*, helps increase the rate of folding. Without crowding, if left to their own devices, the time required for proteins to fold would be dramatically increased. Indeed, approximately 30% of the space inside a cell is physically occupied by macromolecules. Due to steric interactions and volume exclusion, most of the remaining space is excluded for macromolecular interactions. As a result, as little as 1% of cell volume remains for macromolecular processes such as mRNA diffusion and protein folding to take place. An increase or decrease in the available volume is expected to significantly change many cellular processes.

Until not long ago, the dominant viewpoint was a one-dimensional picture of gene expression regulation with genes running the entire show. We now know that this is far from the truth. Not everything is encoded in genes themselves, and extra-genetic factors play at least as, if not more, important a role in genome regulation. This is the domain of *epigenetics*. Critical not only from the molecular biology component, epigenetics has important implications and presents exciting opportunities at the interface of biology and biophotonics.

Epigenetics defines all acquired/heritable changes in gene expression that are not coded in the DNA sequence itself. Although the term was coined over 50 years ago, it is only in the last decade that the significance and potential of epigenetics were understood. It is now well accepted that epigenetic gene regulation plays a key role in multiple diseases, including cancer, with its significance extending from fundamental research to diagnosis and, more recently, to therapy. Initially, two epigenetic gene regulatory processes were identified: DNA methylation and histone modification. More recently, a third pathway, RNA-associated gene silencing (microRNAs), was described. Each of these processes may have multiple manifestations. Clearly, current understanding has only scratched the surface of these complex processes, and it is quite likely that other levels of epigenetic control will be identified in the future.

In the process of methylation, methyl groups are added to cytosine residues in gene promoters, thus silencing gene transcription. The best-studied type of histone modification is acetylation, which converts the positively charged amine group on a side chain of a histone into a neutral amide. This inhibits the binding free energy between the DNA and the histone and allows DNA to be more accessible by nuclear transcription factors that initiate transcription. Thus, histone acetylation is associated with more transcriptionally active chromatin.

RNA-directed epigenetic regulation works through several mechanisms. First, microRNA (miRNA) interference has been shown to participate in histone modification through establishing heterochromatic or, more accurately, transcriptionally inactive states. Second, miRNAs degrade mRNAs while they diffuse through the nucleus into the cytoplasm for translation, thus inactivating specific gene products despite their transcription. Third, miRNAs can bind to mRNAs, inhibiting translation. In addition to intracellular manifestations, RNA can spread from cell to cell across gap junctions.

It is becoming increasingly clear that the previously identified histone modifications (e.g., acetylation) are only one part of a wide array of regulatory processes. These are what can be called "one-dimensional" processes. More complex but potentially very powerful processes involve the entire three-dimensional structure of the chromatin, such as genome compartmentalization by the spatial colocalization of chromatin domains. Local concentration of macromolecules (including the density of chromatin itself), which is referred to as the nanoenvironment, may regulate DNA–histone interactions (thus affecting the accessibility of a gene for transcription) and, as we saw earlier, the access of chromatin to transcription factors and the free energy of DNA double-strand binding, thus modulating the work of the DNA helicase to separate opposing strands of DNA.

Recent studies have shown that chromatin is organized as a fractal globule. Thus, the contact probability for two genes separated by a given genomic distance may change depending on its fractal dimension. Therefore, the fractality of chromatin may change gene colocalization, while colocalized genes tend to be co-expressed. Altered chromatin structure may affect the residence time of mRNA during its intranuclear diffusion. Depending on mRNA lifetime, this may expose mRNA to degradation by miRNA and exosomal activity. Thus, 3-D chromatin organization may play a role in posttranscriptional regulation.

Finally, although not technically an aspect of epigenetics, the nanoscale structure of rough endoplasmic reticulum may play a critical role in translation of gene products. For example, increased crowding may lead to protein misfolding and amyloid formation, as in Alzheimer's disease. Taken together, these would have intragenic effects (e.g., modulation of expression of a particular gene depending on the local nanoenvironment) and intergenic effects (e.g., gene colocalization and co-expression), thus affecting the stability of the genetic circuit with potential impacts for the entire genome.

The significance of epigenetics goes beyond basic science. There are clear implications for disease diagnosis. Indeed, epigenetic involvement has been shown to be critical in neurodegenerative diseases, such as Alzheimer's and cancer. For instance, in colorectal cancer, gene methylation has been used for diagnostics and prognosis. Even more intriguingly, epigenetics may lead to therapy. For example, the absolute majority of genes mutated in cancer are not oncogenes but tumor suppressor genes. This severely limits the potential efficacy of gene or even pharmacological therapy—it is much more difficult to reintroduce tumor suppressor gene function rather than simply inhibiting a mutated proto-oncogene. Thus, attention is being focused on reexpressing silenced tumor suppressor genes via epigenetic interventions. Current approaches include demethylation or modulating miRNAs.

Understanding of other aspects of epigenetic mechanisms such as 3-D chromatin processes and the role of nanoenvironment may also potentially lead to new therapies.

Gene expression and epigenetic regulation in particular have a number of implications in biophotonics. Some examples include single-molecule techniques, imaging/measurement of nuclear nanostructure and nanoenvironment, measurement of dynamic transport processes, and techniques to identify the microarchitectural correlates of epigenetic silencing during carcinogenesis.

As we have already seen, nuclear structure in general and higher-order chromatin structure in particular play a critical role in gene expression and cell function. Because chromatin has a higher refractive index with the value dependent on its concentration, nuclear structure can be assessed by means of elastic light scattering. Several experimental techniques have been developed to do just that, including angle-resolved low-coherence interferometry (LCI) and partial wave spectroscopic (PWS) microscopy.

Single-molecule techniques can be used to measure DNA-histone, DNA double strand, and DNA-transcription factor interactions and how they are affected by biochemical (e.g., acetylation, methylation) and structural (e.g., nanoenvironment, higher-order chromatin structure) processes. Although there are molecular techniques that can measure at least some of these interactions (e.g., the flexibility of a particular DNA sequence can be measured using a cyclization assay, and accessibility of DNA to transcription factors can be ascertained through micrococcal nuclease digestion), optical techniques have the ability to provide complimentary and real-time information.

For example, Förster resonance energy transfer (FRET) can be used to quantify interactions between DNA and histones in the 1-D processes of histone modification, such as acetylation, as well as the 3-D processes such as chromatin compaction and macromolecular crowding. FRET can also be used to measure the dynamics of DNA melting in the pre-initiation phase of transcription and how it is affected by epigenetic regulation. Laser tweezers can be used to measure DNA mechanical properties. A caveat, of course, is that most of the single-molecule techniques have been developed for ideal experimental conditions that do not take into account realistic conditions such as crowding.

Another area where biophotonics has shown promise is quantification of nuclear/chromatin organization at the nanoscale. Conventionally, nanoscale imaging can be achieved using electron microscopy (EM) and, in particular, transmission EM (TEM) and its more advanced modality, scanning TEM (STEM). However, EM cannot be used to image live cells. On the other hand, the fact that the optical refractive index is a linear function of the local density presents an opportunity to measure the spatial distribution of chromatin density via optical approaches.

Examples include partial wave spectroscopic (PWS) microscopy (see Chapter 4), which measures the statistics of the spatial variation of local density with sensitivity to small-length scales. PWS has been used to accurately assess alterations in higher-order chromatin structures that are otherwise beyond the resolution of conventional light microscopy. In turn, alterations in higher-order chromatin structures measured by PWS are critical events in diseases such as cancer. Super-resolution techniques such as stimulated emission depletion (STED) microscopy and stochastic optical reconstruction microscopy (STORM) microscopy have the capability to image fluorescent contrast agent–stained targets.

The third direction includes optical methods that provide information about the transport of macromolecules within the nucleus and in the rough endoplasmic reticulum (RER). This can provide information regarding the diffusion of transcription factors (critical in understanding gene co-expression, which in turn governs many gene regulatory circuits), mRNA diffusion from a transcription site to the RER, and diffusion of miRNAs. This can be accomplished by use of fluorescence correlation spectroscopy (FCS).

The fourth application is to use optical means to identify the microarchitectural correlates of epigenetic silencing during pathophysiological processes, including cellular and extracellular matrix alterations. The list of techniques that have been implemented to address this need include low-coherence enhanced backscattering (see Chapter 6) to measure microscopic alterations in cells and extracellular matrix, and low-coherence interferometry to measure nuclear enlargement and alterations in its internal structure.

Finally, integration and co-registration of some of these optical techniques and other molecular assays will most likely be crucial. For example, structural nanoscale imaging (e.g., PWS) can be combined with molecular assays that measure gene co-localization (to understand how co-localization is affected by structure), fluorescence in situ hybridization or FISH (to localize specific sequences), DNase I sensitivity (to assess how DNA accessibility to TFs depends on the nuclear nanoenvironment), FCS (to measure TF, mRNA, and microRNA diffusion and mRNAs' lifetime), etc. While some optical techniques have been well established in genomics research (e.g., green fluorescent protein [GFP] fluorescence has been used to measure genetic noise), integration of these measures in the context of nanoenvironment (e.g., how genetic noise is affected by higher-order chromatin organization) is bound to provide crucial new information. Eventually, this comprehensive information may be interpreted by means of systems biology methods.

We now turn our attention to the organization and function of the cytoplasm. The first thing to know about the cytoplasm is that it is not a liquid suspension of organelles that flow freely in an empty cytosol. Quite the opposite: the cytoplasm is a highly organized and tightly regulated compartment. Its structural integrity is provided by the cytoskeleton and a network of intracellular membranes.

The functional role of the cytoskeleton is evident from its name: it is a network of filamentous proteins that provide mechanical stability and interconnectivity to the cell. The three basic elements of the cytoskeleton are *microtubules* (25 nm in diameter) 10-nm *intermediate filaments*, and 7-nm *microfilaments*. While microfilaments and microtubules can grow, intermediate filaments are fixed in size. The cytoskeleton performs many more functions than providing mechanical stability to the cell. Among them are intracellular transport, regulation of cell division, and signal transduction. Elements of the cytoskeleton enable cell motility and other functions such as cilia action.

Microtubules are formed by the polymers of alpha and beta tubulin. Microtubules originate ("radiate") from the *microtubule originating center* (MTOC), which lies near the nucleus in association with centrioles. The role of microtubules as structural elements is to resist compression. One can think of them as support beams. Microtubules are also found in cell flagella (thus enabling cell motility) and cilia. Cilia are structures protruding from the cell surface that are used to move extracellular material, such as mucus. This function, for example, is important in the respiratory epithelium and helps get rid of foreign microorganisms or inorganic microscopic objects that are trapped in the respiratory mucus in a process called mucociliary escalator. Microtubules also assist in mitosis. Their other function is intracellular transport. Associated with *motor proteins* dynein and kinesin, microtubules help transport organelles like mitochondria or vesicles across a cell. In this process, dynein and kinesin attach and move toward and from, respectively, a cell center. Furthermore, in cell division (mitosis), microtubules are part of the mitotic spindle, the structure that separates the chromosomes into daughter cells.

Intermediate filaments provide structural stability to cells and also connect desmosomes in epithelial cells, thus ensuring cell-to-cell and cell-to-matrix interconnectivity.

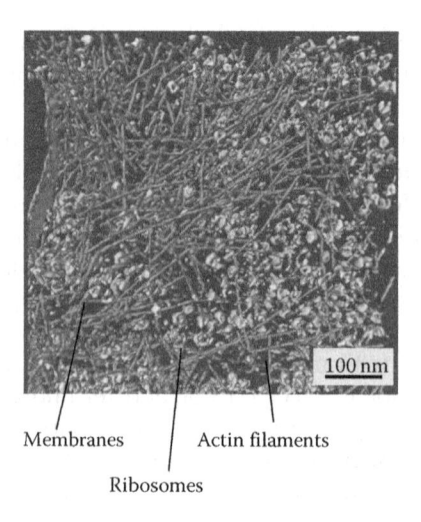

Membranes | Actin filaments

Ribosomes

FIGURE 1.14

Three-dimensional reconstruction of electron microscopy imaging of actin network, membranes, and cytoplasmic macromolecular complexes. A part of the nucleus is visible on the left and is cut out. (Adapted from O. Medalia et al., *Science* 2002;298(5596):1209–1213.)

Accordingly, intermediate filaments connect the cytoplasm from one cell–cell junction to another. Mechanically, they are tension elements (i.e., they function by bearing tension); thus, one can imagine them as cables that help provide structural integrity. Intermediate filaments provide anchoring points to organelles and are part of the nuclear lamina.

Microfilaments are made of actin (Figure 1.14). Most actin fibers are found in the cell periphery or cortex near the plasma membrane. They control cell shape, are used in muscle contraction (in association with myosin—actin—myosin movement is the central mechanism behind muscle contraction), and are present in microvilli. Although the cytoskeleton is frequently thought of as a purely cytoplasmic structure, actin is present in the nucleus. This nuclear actin is critical in gene transcription. Depending on the type of RNA polymerase involved, actin may act as a molecular motor (in association with myosin) or help formation of the pre-initiation complex (β-actin). Nuclear actin also serves as a component of chromatin-remodeling complexes, which we discussed above. In the posttranscription phase, actin plays a role in the nuclear export of mRNAs.

As we have seen, microfilaments and intermediate filaments are tension elements, while microtubules are compression elements. The structural principle that combines tension and compression elements is called *tensegrity*. The concept has been used in architecture: These are very rigid and stable structures given the typically light mass and small cross-section of their elements. Donald E. Ingber pioneered the idea that the cytoskeleton is based on the principles of tensegrity.

From a biomedical optics perspective, cytoskeleton filaments are just too small to give rise to substantial elastic scattering. However, multiple absorbing and fluorescent contrast agents and stains are available that have affinities for different elements of the cytoskeleton. Thus, the cytoskeleton can be readily visualized by bright-field fluorescence and confocal microscopy. Furthermore, as the cytoskeleton plays a key role in maintaining and regulating cell size, shape, surface profile, and internal structure, cytoskeletal events can be detected by observing changes in light scattering or by means of microscopy modalities that use the phase of light interacting with live cells as a source of contrast.

Let us now turn our attention to the cytoplasmic organelles. *Mitochondria* are the "lungs" of the cell and participate in the process called *internal respiration*. Structurally, mitochondria have spheroidal-like shapes. The length of the long axis of a mitochondrion typically ranges from 1 to 2 μm, although some mitochondria may be as big as 5 μm; this, however, is rarely observed. The small axis varies between 0.2 and 0.8 μm. Mitochondria are mechanically flexible and may easily change their shape. Each mitochondrion has a short life span of about 10 days.

The number of mitochondria in a given cell depends primarily on the metabolic activity of the cell. Relatively quiescent cells such as chondrocytes have few mitochondria. More active cells such as the membranous epithelial cells that line the gastrointestinal tract or the bronchial tree count mitochondria in double digits. Highly active cells, for comparison, such as hepatocytes, the main cell type in the liver, typically contain thousands of mitochondria, which are densely packed within the hepatocytes.

Inside each mitochondrion, there is a core matrix space that contains most of the mitochondrial enzymes participating in the glycolytic-made citric acid cycles. The matrix is surrounded by a folded *inner membrane*. The inner membrane is folded into *cristae* and contains the enzymes of the respiratory chain. These structures are surrounded by the *outer membrane*, with the intermembrane space separating the two membranes. Mitochondria are unusual in the sense that they contain their own DNA, which encodes about 20% of the proteins involved in oxidative phosphorylation. Not surprisingly, mitochondria are able to self-replicate.

From an optics perspective, a high concentration of membranes within mitochondria ensures their relatively high refractive index. Furthermore, mitochondria's size is on the order of the wavelength. Together, these two facts make mitochondria powerful light scatterers. Mitochondria can be detected by means of angle-resolved light scattering-based microscopy modalities such as optical scatter imaging (OSI). There are many fluorescence dyes that have affinity for mitochondria; thus, these organelles can be detected by means of fluorescence microscopy. Although most molecules found in nucleated cells neither absorb nor fluoresce in the visible part of the spectrum (i.e., with the notable exception of red blood cells—these are filled with hemoglobin molecules, which are strong light absorbers in the visible spectral range), a few products of cell metabolism do have some fluorescence activity. These include nicotinamide adenine dinucleotide (NAD), nicotinamide adenine dinucleotide reduced form (NADH), flavin adenine dinucleotide (FAD), and flavin adenine dinucleotide reduced form (FADH). However, the cross-section of endogenous fluorescence of these molecules is still very small compared to the much more strongly fluorescent molecules of extracellular tissue such as collagen and elastin, which are discussed later in this chapter. Finally, we point out that mitochondrial alterations have repercussions for the entire cell. Mitochondria are some of the first organelles to be affected in the course of programmed cell death or apoptosis. This changes not only cell behavior but cell structure as well.

Three organelles are the major players in the cell's biosynthetic machinery. *Endoplasmic reticulum* (ER) is a dense maze of interconnected tubules, flattened vesicles, and sheets of membranes with sizes ranging from 30 to 100 nm. The ER is classified as either the *smooth endoplasmic reticulum* (SER) or the *rough endoplasmic reticulum* (RER). The ER is the site where all transmembrane proteins and lipids for all a cell's organelles and membranes are synthesized. Although it appears random on electron microscopy images, the ER encloses a continuous space called *ER lumen*, also known as *ER cisternal space*.

The RER is located close—in fact, right next to—the nucleus; the membrane of the RER is continuous with the outer layer of the nuclear envelope. The Golgi apparatus is located closer to the cell membrane. The SER is a sparse network of interconnected tubules that is also located in proximity to the nucleus. The RER is the place where proteins are synthesized and folded. The SER helps with transporting proteins toward the Golgi apparatus, and the latter is charged with further processing of proteins and lipids and, like a mailman, sorts them to their final destination, whether this is another organelle or a cell membrane.

RER appears rough on electron microscopy images due to multiple *ribosomes* that dot its outer (cytosolic) surface. The average size of a ribosome is 25 nm. They are the sites where mRNAs are "translated" into proteins. After a protein is synthesized, it is extruded into the ER lumen for folding, oligomerization (polypeptide complex formation), and further transport. Protein folding and oligomerization is assisted by *chaperones*, molecules that surround a folding protein and ensure that it is fully folded. If a protein is not correctly folded, it is destined for degradation with the ER. As in other cellular processes, molecular crowding and the resulting nanoenvironment are important in protein folding. Crowding helps chaperones get access to the folding proteins and helps proteins fold quickly and correctly. On the other hand, too much crowding may lead to abnormal protein folding and amyloid formation, as in some deposition diseases such as Alzheimer's.

After proteins are freed from the chaperone, they leave the RER. Membrane-bound transport vesicles deliver the proteins to the *Golgi apparatus*. This structure has an overall size of about 100–400 nm and consists of stacked cisternae with one side (called the *cis* face) facing the ER and the other side (the *trans* face) facing the cell membrane. Proteins move from the cis to the trans face through the Golgi apparatus. As one can imagine, the Golgi apparatus is particularly prominent in cells active in secretions such as glandular epithelia or mucus-secreting epithelial cells. The Golgi apparatus is also responsible for posttranslational modifications of proteins such as glycosylation.

Similar to the RER, the SER is connected to the nuclear envelope. The SER, however, does not appear dense, at least not on electron microscopy images. It has multiple functions. One of its main purposes is as a transition zone from the RER to transport vesicles. The SER plays a key role in cholesterol and lipid synthesis. Its other functions include metabolism of carbohydrates and steroids and attachment of receptors on cell membrane proteins. Although inconspicuous in many cells, the SER is quite prominent in cells that synthesize steroid hormones (e.g., secreting cells of the adrenal glands) and lipid or lipoprotein-secreting cells such as those in the liver.

From an optics perspective, the ER is active in light scattering. This is due to its dense membranous network. Within the ER, the refractive index varies widely. However, these variations happen at a fairly small-length scale, on the order of tens of nanometers. Thus, sensing the ER requires special techniques and/or optical methods that are particularly sensitive to high special frequency refractive index variations. For example, backscattering of light is more sensitive to small-length scales.

The three main organelles responsible for waste disposal are *lysosomes*, *proteasomes*, and *peroxisomes*. Lysosomes are 250–800 nm in size and may have various shapes from smooth spheres and spheroids to randomly shaped structures resembling Gaussian random spheres. Peroxisomes are typically slightly larger, 200–1000 nm in size. They are spheroidal structures of lower inner density than lysosomes. Not surprisingly, lysosomes and peroxisomes are more abundant in metabolically active cells such as hepatocytes. These may contain hundreds of lysosomes and peroxisomes.

Lysosomes are the main centers of catabolism. Old organelles are delivered to the lysosomes for degradation. Lysosomes are filled with acid hydrolases (about 40 different enzymes) such as proteases (for protein degradation), nucleases, lipases (digestion of lipids), etc. Proteasomes are large protein complexes. Each proteasome is a cylinder-like structure composed of several different proteases. These are responsible for degradation of misfolded proteins. The latter are tagged by protein ubiquitin and fed to a proteasome. The role of peroxisomes is β-oxidation of fatty acids. Much like mitochondria, peroxisomes are self-replicating. The main difference is that peroxisomes lack their own DNA and ribosomes and depend entirely on the availability of proteins free-floating in the cytosol.

From an optics perspective, lysosomes and peroxisomes are denser than the surrounding cytosol and thus scatter light. Being sized on the order of the wavelength of light and having a sphere-like shape makes these organelles even more efficient light scatters. This makes them detectable by means of light-scattering techniques. It is not, however, well understood whether lysosomes and peroxisomes are denser or looser than mitochondria in terms of their macromolecular concentration and refractive index. Fluorescent dyes are available that target these organelles; thus, they can be imaged by means of fluorescence microscopy.

In addition to organelles, each cell contains numerous cytoplasmic *inclusions*, such as secretory granules, lipid granules, and pigment bodies. Many of these inclusions are nearly spherical, with surface roughness less than 40 nm, and have a huge variability in size from 20 to 500 nm.

Let us summarize the discussion of the relative sizes of organelles in a cell. The nucleus is by far the largest organelle. The size of most other organelles is smaller or close to 1 μm. Almost none of the organelles are internally homogenous. Instead, they have a complex internal organization where organelles are composed of smaller structures that in turn are conglomerates of even smaller building blocks, all the way down to macromolecules with sizes on the order of a few nanometers. Coincidently, the wavelength of light in the visible part of the spectrum is, on the order of magnitude, the borderline between the length scales associated with organelles and the length scales of macromolecular complexes with sizes below 100 nm. This is important because the diffraction limit of resolution of conventional light microscopy does not allow resolving structures smaller than approximately half the wavelength of light. This means that even the best state-of-the-art microscope would only be able to image the "organellar level" of a cell (at best) but not the "macromolecular level." As the understanding of cell function depends critically on our ability to probe the macromolecular level, specialized tools to sense these small structures have been and are being developed, which is one of the exciting thrusts of biophotonics. The tools include super-resolution fluorescence microscopy modalities, techniques whose contrast mechanism depends on intermolecular distances at the nanoscale (e.g., fluorescence resonant energy transfer or FRET) and light-scattering techniques, which attempt to detect but not visualize subdiffractional structures.

From a simplified perspective, living tissue can be viewed as a combination of cells and extracellular matrix (ECM). Having examined cell organization, it is now time to shift our attention to the ECM. The four main constituents of ECM are *fibers, adhesive glycoproteins, amorphous interfibrillar matrix*, and *tissue fluid*. There are two types of fibers: *collagen* and *elastin*. Some classifications identify a third type, *reticular fibers*, although the latter are a special type of collagen fibers.

Collagens are the most abundant fibers in the ECM. The elementary unit of a collagen fiber is tropocollagen or the collagen molecule. It is a strong triple-helix protein that is approximately 300 nm long and 1.5 nm in diameter. Collagen molecules form larger collagen aggregates such as fibrils. Collagen fibrils consist of staggered collagen molecules with a longitudinal period of about 67 nm. Because the length of each collagen molecule is 300 nm, each D-unit must contain four collagen molecules plus a portion of the fifth molecule. The transverse diameter of a typical fibril is 70 nm. The fibrils assemble into an even larger structure, collagen fiber. Most types of collagen fibers are unbranching (with the notable exception of collagen type III) and approximately 500 nm in diameter. Within a fiber, fibrils are separated from one another by about 100 nm. Collagen fibers provide tensile strength to ECM. There are several types of collagen fibers. Types I, II, III, and V are found in the interstitial matrix, while collagen type IV forms the basal lamina (see following discussion). Collagen types I and II are the two most abundant types of collagens in loose connective tissue.

While collagen provides strength to tissue, elastin fibers are flexible. It is elastin that provides elasticity to elastic arteries and is responsible for elastic recoil in lung tissue. These fibers are much smaller than collagen fibers: 10 nm in diameter. Another difference with these collagens is that they branch and form a loose three-dimensional network.

Although some classifications place reticular fibers in their own category, they are not altogether distinct fiber types. In fact, they are type III collagens. Unlike other collagen fibers, however, they are small, 10 nm in diameter, and branching (crosslinking). Reticular fibers form a fine network referred to as reticulin. The network supports soft tissues such as liver and bone marrow.

While fibers may be viewed as the main players in the ECM, providing strength and elasticity, they are helped by various "sidekicks." Most of these helpers are proteins, other individual macromolecules, or simple macromolecular complexes and thus are much smaller than the fibers. Adhesive glycoproteins link the ECM components to one another and also connect them to the cells. These are proteins such as fibronectin and laminin.

The amorphous interfibrillar matrix consists primarily of glycosaminoglycans (GAGs), which are linear polysaccharide chains. Their main function is to enable ECM to resist compressive forces. The two main types are proteoglycans, which help in collagen formation, and hyaluronic acid.

All these structures compose one of the two types of ECM: the *interstitial matrix* and the *basal lamina*. The former is secreted by and surrounds *mesenchymal* cells. Mesenchymal cells are the cells of connective tissue such as fibroblasts. These cells do not perform the main function of a particular organ but assist in the organ functioning (which is the role of the so-called *parenchymal* cells) through regulation of ECM. (To appreciate the difference between mesenchymal and parenchymal cells, as an illustration, consider the following example: In the liver, hepatocytes are the parenchyma and are tasked with performing the main secretory and other metabolic functions of the liver, while mesenchymal cells would help maintain the appropriate ECM supporting the function of hepatocytes.) The interstitial matrix consists of collagens type I, II, III, and V; elastin; fibronectin; proteoglycans; and hyaluronic acid.

The basal lamina, on the other hand, is synthesized by parenchymal cells. In particular, all epithelial cells (without exception) produce basal laminae that underlie all epithelia. In fact, some have suggested that it is the basal lamina–producing property of epithelial cells that is the principal, if not the only, common property of all epithelia. The basal lamina separates the epithelial lining from the underlying connective tissue. The main constituent of the basal lamina is collagen type IV. Other components include laminin and proteoglycans. While the interstitial matrix may have quite a few mesenchymal (e.g., fibroblasts) and inflammatory cells interspersed in it, basal lamina are completely acellular.

Functionally, ECM is not merely a "substance" that provides structural support for cells and tissues and cell attachment. The ECM plays a critically important role in tissue physiology. As a matter of fact, ECM governs, drives, and regulates many of the processes, with parenchymal cells frequently doing what ECM tells them to do. During the last decade, it has become increasingly clear that the microenvironment created in part by ECM is responsible for many physiological and pathophysiological processes. It is a two-way street, with cells affecting the matrix and the matrix affecting cells. The ECM determines cellular polarization, defines the apical and basal surfaces, controls cell growth through both molecular means (it stores growth factors) and mechanical stimuli (e.g., stiffness), and regulates cell differentiation.

For example, a proper microenvironment is required for oocytes to mature. Development of an abnormal microenvironment is one of the key events in carcinogenesis. The effects of microenvironment are due to several facets, including molecular species such as signaling molecules present in the ECM but also mechanical properties of the ECM such as its stiffness and elastic modules. The latter are regulated by both parenchymal and mesenchymal cells and external (environmental) factors through alteration of the matrix's structure such as the types of collagen, fiber orientation, crosslinking, etc. Many of these changes can be studied by means of optical tools.

Collagens and elastins are optically "active" and can be studied by many optical modalities. Starting with elastic-scattering processes, collagen fibers are quite optically dense due to a high concentration of proteins composing the fibers. Their overall size on the order of the wavelength of light makes them efficient scatterers. The ECM, in general, is one of the more optically dense media, and the ECM's structure can be assessed by measuring elastic light scattering. It is important to remember, however, that each collagen fiber is not a uniform structure and instead consists of multiple fibrils. Light interaction with each of the fibrils leads to scattering. Thus, light scattering carries information about not only the composition of ECM at the level of collagen fibers but also its subfiber structure. Light scattering coupled with an imaging technique such as optical coherence tomography (OCT) (e.g., inverse scattering OCT or ISOCT) is able to provide wide-field (on the order of millimeters) imaging of ECM while also preserving quantitative information about the network structure at submicron-length scales.

In addition to being an efficient scatterer, the collagen fiber network is randomly birefringent, meaning that the birefringence varies randomly from one point to the next. The degree of birefringence depends on collagen fiber orientation—the more linearly oriented the fibers are, the more birefringent the network is. Thus, birefringence becomes apparent only locally, within 100-μm-sized volumes. This, however, is sufficient to significantly alter light transport. Techniques that are able to measure light transport, especially at shorter-length scales where the information about the shape of the phase function (the angular scattering pattern; see the next chapter), can be sensitive to changes in collagen network orientation, crosslinking, and overall structure. An example of such a technique is low-coherence enhanced backscattering (LEBS; see Chapter 6).

In addition to being prominent scatterers, collagens have one of the highest cross-sections of fluorescence. In fact, if endogenous fluorescence is observed from an entire tissue, it is collagens (primarily type I and II) that would generate the highest signal that typically overwhelms the fluorescence generated by the rest of molecules. The fluorescence generated by other endogenous fluorophores such as NADH and porphyrins, for example, is all but negligible compared to that of collagens, in part due to the latter's fluorescence efficiency and high concentration in tissue. Specialized techniques are needed to differentiate other fluorophores from collagens. Most collagens have a peak fluorescence emission spectrum in the short wavelength part of the visible spectrum, below 500 nm. Elastin emits a similar fluorescence spectrum.

As if scattering and fluorescence were not enough, collagens are able to generate second-harmonic light scattering, which is a type of nonlinear optical phenomenon discussed in

Chapter 5. The second harmonic can be generated only when a laser light interacts with a material with a noncentrosymmetric molecular structure. Collagens are some of the few tissue molecules that have this property, which makes them easily detectable by this technique. Indeed, imaging the collagen network was the very first biological application of second harmonic generation (SHG) microscopy in 1986. Since then, SHG microscopy has become a powerful tool to image a collagen network in three dimensions with a high sensitivity and specificity to collagens. Furthermore, given the fact that collagen fibers have a locally preferred orientation, polarization properties of second harmonic–generated light can be used to quantify their orientation, as was demonstrated by Paul Campagnola and colleagues. We can say that collagen is not a shy molecule and certainly wants to be seen!

Tissue Structure

Having examined cell and ECM structure, we are now ready to take it to the next level. A combination of cells and ECMs compose tissue. Tissues, in turn, organize into organs and organs into organ systems (e.g., respiratory, gastrointestinal, circulatory, etc.). We start with defining four basic tissue types in our body. These are the *epithelium*, the *connective tissue*, the *nervous tissue*, and the *muscle*. Blood has a special standing in this classification scheme, as it is considered by some to be a specialized form of connective tissue, while others define it as a separate, fifth tissue type.

Epithelium is primarily composed of contiguous epithelial cells. There are two parallel classification schemes: macroscopic and microscopic. Macroscopic epithelia are divided into the membranous (or tissue layer) epithelia, which line all (with no exceptions) surfaces of the body, and the secretory epithelia. The latter are further subdivided into exocrine and endocrine. The exocrine glands are formed by contiguous layers of epithelial cells that form acini, which secrete in a lumen with secretions directed toward the cell's apical surface, which faces the lumen. Endocrine glands, on the other hand, secrete into blood or lymph, and the secretion is directed toward the cell's basolateral surface.

Microscopically, epithelia are classified based on three independent properties: the number of cell layers, the cell shape, and the free surface specializations. Based on the number of cell layers, epithelia are classified as simple (a single cell layer), stratified (multiple cell layers with cells getting progressively flatter as they approach the surface of the epithelial layer), pseudostratified (a single layer of cells that only appears to be stratified), or transitional (multiple cell layers with larger and flatter cells located close to the surface of the epithelial layer and taller, columnar-like cells on the bottom).

Based on the shape of the cells, epithelia are classified as squamous (flat and pancake-like with a thickness several times smaller than the diameter of a cell), cuboidal (cells of cuboidal shape resembling little cylinders), or columnar (also cylindrical cells but a few times taller than cuboidal cells). Free surface specialization implies specific features that might be present on the apical surface of cells. For example, epithelial cells that have cilia on their luminal surface are called ciliated. Squamous stratified epithelia such that cell form a keratin layer after their life cycle are called keratinized epithelia, while those that do not form keratin and are sloughed off are called nonkeratinized.

A particular type of epithelial tissue is identified through a combination of these two (or three if surface specialization is present) classifications. Simple squamous epithelium lines the inner surfaces of blood vessels (endothelium) and the surfaces of the peritoneal cavity (mesothelium). Simple cuboidal epithelium is found in many endocrine glands (salivary glands, pancreatic duct, mammary gland duct), the free surfaces of the ovary, and the tubes of the kidney. Simple columnar epithelium lines the luminal surface of the large and small intestines and endocervix (the part of the cervix that is close to the uterus). In the case of the intestinal epithelium, the lining folds into larger (100–500 μm) structures called crypts and villi. Crypts are invaginations into the tissue and are present in both the small and large intestines, while villi are invaginations into the lumen and are present in the small intestine only. The purpose of these structures is to increase the surface area to allow for more efficient absorption and secretion. Stratified squamous epithelia are some of the most omnipresent structures and are good for protection. This type of epithelium lines the surfaces of the skin (epidermis), oral cavity, esophagus, exocervix (the lower part of the cervix), the anus, and the upper respiratory tract. The skin epithelium is keratinized, while most other epithelia are considered nonkeratinized, although this classification is a bit confusing; thin layers of keratin are frequently found in the esophagus despite its conventional classification as a nonkeratinized type. Pseudostratified columnar epithelium lines the luminal surfaces of the large airways of the respiratory tract, in particular bronchi. Because of the cilia present on the apical surface of these cells, the full name of the epithelium is ciliated pseudostratified columnar epithelium. Finally, transitional epithelium forms the lining of the bladder. Its unique structure with larger cells on top allows it to withstand stretching.

Simple squamous epithelium is the thinnest type of epithelium and is only 1–2 μm thick. The surface cells of the squamous stratified type are just as thin but can be quite large in the transverse plane, approaching 50 μm or even greater in diameter. They do look like pancakes! The cells located at the bottom of a squamous stratified layer are more similar to cuboidal cells in terms of their shape. The overall thickness of the squamous stratified epithelium can be anywhere from 0.5 to 1 mm thick. Simple columnar cells are approximately 20 μm tall and 10 μm wide. However, one should remember that simple columnar epithelium is rarely a flat layer. Instead, it frequently forms three-dimensional folding and invaginating structures such as crypts in the intestines, with the thickness ranging from 200 to 600 μm. Finally, thickness of transitional and pseudostratified epithelia is on the order of 500 μm.

Epithelia are essentially layers of contiguous cells and do not have blood vessels or nervous fibers running through them. All blood supply is delivered through the underlying connective tissue, and epithelial cells derive nutrients and satisfy their oxygen demand from this microvasculature by means of diffusion from the blood vessels to the epithelial cells. This is one of the reasons epithelial layers are restricted in how thick they can grow; the bottom cells have access to the oxygen first, and the upper cells are destined to die. There are two outcomes of cell death: They are sloughed off into the lumen or form a dense layer of protein from the dead cells—this is the keratin that we have talked about.

The life cycle of epithelial cells almost always starts at the bottom. In the squamous stratified epithelia, the stem cells are located in the basal layer, close to the basement membrane, and in the columnar epithelia arranged in crypts; the stem cells are at the bottom of the crypts. In a normal-functioning epithelium, stem cells are the only dividing cells. They can undergo either a symmetric division as a result of which two other stem cells are formed or an asymmetric division with one of the daughter cells being a stem cell and the other being a non–stem cell destined to mature. The newly formed cells then migrate or are pushed up by junior cells to the upper layers (in case of squamous stratified epithelium) or up toward the lumen (in case of the columnar cells as part of crypts). On their way up, they differentiate and mature. On average, the higher the cells are located, the more mature they are. Finally, when cells reach the surface, they are ready to end their life cycle and undergo apoptosis, otherwise known as programmed cell death. Aberrations in the life cycle are signs of a disease state. For example, in carcinogenesis, some non–stem cells that are located in the parts of the epithelium where only mature cells are normally present are believed to retain some similarity to stem cells in terms of their genome expression profile as well as their structure. These cells are referred to as stem-like cells.

As we have seen, epithelium relies on the underlying connective tissue for life support. Connective tissues are found throughout the body and may play a supporting role to epithelia or form their own organs. At least eight types of connective tissue can be identified. These are loose (areolar), dense irregular, dense regular, adipose, reticular, cartilage, bone, and blood. Loose connective tissue is found in the mucosae and submucosae underneath epithelia. From the perspective of biomedical optics, these tissues are the most relevant. Loose connective tissue (just like any other connective tissue) has two components: ECM and cells. The cells are mostly parenchymal such as fibroblasts and blood cells or primarily inflammatory (white blood) cells such as macrophages.

An organ is formed by a combination of the four basic tissue types. The wall of the alimentary (or gastrointestinal) tract provides a good illustrative example. Its luminal surface is lined by epithelium. Underneath the epithelium, there is a basal lamina. In some organs, a layer of loose connective tissue is present that is called lamina propria. Below it, there is muscularis mucosa, which is a layer of smooth muscle cells. These first three layers (epithelium, lamina propria, and muscularis mucosa) form the mucosa. Underneath the mucosa, we find a layer appropriately named the submucosa—this is another layer of loose connective tissue. A distinction between the lamina propria and the submucosa is that the former's role is to support the function of the epithelium (which means microcirculation, e.g., pericryptal plexus of small arterioles and capillaries wrapping around crypts), while the latter contains larger blood vessels. Muscularis propria is typically a much thicker layer of muscle. In the case of the colon, it is this muscle that drives peristalsis. The outermost layer is either the serosa or adventitia. The serosa is a thin membrane composed of a layer of connective tissue lined, from the peritoneal cavity side, by mesothelial cells. The serosa is found primarily (with some exceptions) in the peritoneal organs (e.g., most of the gastrointestinal tract, with the exception of a part of the esophagus and part of the rectum). The adventitia, on the other hand, is a thicker layer and is a mixture of different types of connective tissue, including the loose connective tissue in which some adipose (fat) tissue might be present. The adventitia is found in the retroperitoneal organs (e.g., part of the esophagus and the rectum). Among the digestive organs, the thoracic esophagus, ascending colon, descending colon, and the rectum are bound by the adventitia, while the rest are bound by the serosa.

Other luminal organs have a similar organization. For example, blood vessels have a thin layer of simple squamous cells (endothelium) lining their inner surfaces. The combination of the endothelium and the underlying elastic lamina forms what is called the *intima*. In healthy blood vessels (in reality this only exists in children), the endothelium of large blood vessels is only one cell thick. Beneath it, there is a smooth muscle layer, the *media*. Depending on the size of the blood vessel, the media may be mostly muscular (in muscular arteries) or have a significant component of elastin fibers intermixed with smooth muscle cells. The latter is characteristic of elastic arteries. Elastic and muscular arteries serve different roles in our circulatory system. Elastic arteries are the largest arteries in our body, such as the aorta, carotid arteries, etc. Their job is to serve as pressure reservoirs during the heart relaxation phase (diastole) and to expand easily during heart contraction (systole). Accordingly, their walls are very elastic in order to be able to expand and generate elastic recoil. Muscular arteries, on the other hand, are charged with pressure regulation, and they contract and relax to rectify the blood pressure and flow to the end organs, thus their thick muscular walls. Below the media is located the layer of *adventitia*, which is a mixture of different types of connective tissue: loose connective tissue and some inclusion of the adipose tissue.

Blood has a special standing in tissue optics, as the hemoglobin packed in red blood cells (erythrocytes) is by far the most prominent absorber in the visible part of the spectrum. In many cases, it is the only absorber that needs to be considered. The absorption spectrum of hemoglobin is discussed in Chapter 4. Oxygenated, deoxygenated, and other forms of hemoglobin have slightly different absorption spectra, which lends itself to a powerful application of tissue spectroscopy that is able to measure not only total hemoglobin content but also oxygen saturation or tissue oxygenation. In the near-infrared part of the spectrum, hemoglobin loses its dominant position as the main absorber and competes with other substances such as fat and water. It is important to remember, however, that red blood cells are large (about 8 μm in diameter), dense, and fairly homogeneous cells that not only absorb but also scatter light. In addition to scattering and absorption spectroscopic contrast mechanisms, red blood cells can be detected by means of photoacoustic techniques such as photoacoustic tomography and microscopy, which takes advantage of light absorption by red blood cells.

References

1. O. Medalia, I. Weber, A. S. Frangakis, D. Nicastro, G. Gerisch, and W. Baumeister, Macromolecular architecture in eukaryotic cells visualized by cryoelectron tomography, *Science* 2002;298(5596):1209–1213.
2. D. W. Fawcett, *Bloom and Fawcett: A Textbook of Histology*, London, UK: Hodder Arnold Publishers, 1997.
3. V. Kumar, A.K. Abbas, and J. Aster, *Robbins & Cotran Pathologic Basis of Disease*. Philadelphia: Saunders, 2009.
4. B. van Steensel, Chromatin: Constructing the big picture, *EMBO Journal* 2011;30(10):1885–1895.
5. D. V. Lebedev, M. V. Filatov, A. I. Kuklin, A. K. Islamov, E. Kentzinger, R. Pantina, B. P. Toperverg, and V. V. Isaev-Ivanov, Fractal nature of chromatin organization in interphase chicken erythrocyte nuclei: DNA structure exhibits biphasic fractal properties, *FEBS Letters* 2005;579:1465–1468.

2

Optics Components and Electronic Equipment

The role optical instruments have played in the advent of the contemporary age is frequently underemphasized. While other modern marvels have taken the spotlight in recent years, the history of optics stakes a strong claim in the formation of the world we know. The telescope has been decisive in resolving the clash between geocentric and heliocentric conceptions, affirming the Copernican universe during Galileo's time. The microscope broadened the horizons for biology and medicine, opening windows on an entire new world that was previously unobservable. The camera ushered forth a perspective on the world more vivid and more real than any painter's brush could portray.

Day to day, we find ourselves in continuous contact with optical instruments and their products. These optical instruments, along with their modern electronics, are the cornerstones to shape and observe the world we live in. To fathom the world of optics is to understand the fundamentals behind their operation. To explore the design and function of various sophisticated optical instruments is to comprehend the individual lenses, mirrors, and other optical devices that compose them.

This chapter begins with an introduction to various commonplace optical devices at their simplest, namely lenses, mirrors, and objectives. While this section primarily relies on a diagrammatic description to suffice for illustrating fundamental physics concepts, several accompanying photographs of optics equipment are included for reference. In In addition to to manipulating the geometry and spatial organization of light, properties such as polarization and phase are tantamount to intensity and direction when describing optics. Introductions to waveplates and linear polarizers will offer descriptions of how such properties of light are controlled and adjusted.

An important concept in optics deals with the transmission and efficient delivery of light. This topic primarily concerns optical fibers, their composition and fabrication, and their functions and importance in modern optics. The various mechanical equipment that supports the operation and calibration of sensitive optical arrays and apparatuses are termed "optomechanics" and is another topic covered in this chapter. These optomechanical devices come in various forms, including stages, mounts, and tables, each carefully manufactured to perform a specific function.

The generation and detection of light itself involves sophisticated equipment that will be the focus of the remainder of this chapter. The principles of lasers and their action will be briefly described in the context of solid-state mediums. While these mechanisms of coherent light generation represent only a portion of the many types of laser technologies that are currently available, they prove sufficient material for further discussion of detection and analysis. Various electro-optical equipment such as oscilloscopes, power meters, and spectrometers will be introduced, and the fundamentals of their operation will be explained. Included in the sections covering electronic equipment will be principles of data acquisition and signal conditioning and analysis. There are many well-organized textbooks for comprehensive explanations of signal processing; this chapter provides only a brief introduction sufficient in the context of optics. Finally, the chapter concludes with a discussion of issues and precautions in laser safety.

Lenses

Lenses are optical components designed to focus or diverge light, and they are the most basic components in any optical experimental system. A lens at its simplest consists of a single optical element made of glass or transparent plastic. As light passes through a lens, it is affected by the lens's profile or substrate and is deflected appropriately. While there are many categories of lenses, this textbook will focus on **spherical lenses**, where the two lens surfaces are parts of the surfaces of spheres. Each surface can be **convex** (protruding outward from the lens center), **concave** (depressed into the lens), or **planar** (flat). Simple lenses are classified by the curvature of the two optical surfaces. The line joining the centers of the spheres that make up the lens surfaces is referred to as the **axis** of the lens, and for all spherical lenses, it will pass through the lens's physical center.

Convex Lenses

A lens is *biconvex* (or just *convex*) if both surfaces are convex and *equiconvex* if both surfaces exhibit equal radii of curvature R_1 and R_2 (Figure 2.1). If one of the surfaces is convex while the other surface is flat, the lens is referred to as *plano-convex*.

A major difference between plano-convex and biconvex lenses involves an optical distortion phenomenon known as *spherical aberration*. Specifically, it refers to an increased refraction of collimated light rays when they strike a lens surface near its edge relative to those that strike near its center (Figure 2.2a). A consequence of spherical aberration is an imperfect focal point, which results in a distortion of the produced image. Provided proper orientation, a plano-convex lens can oftentimes be subjected to less spherical aberration than a standard biconvex lens (Figure 2.2b).

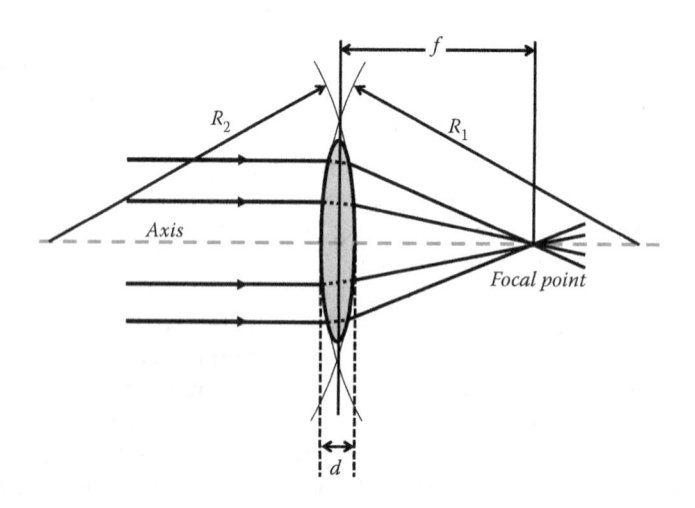

FIGURE 2.1

Diagram of light striking a biconvex lens. R_1 and R_2 correspond to the respective radii of curvature. The lens thickness is referred to as d, and the lens focal length is denoted by f and represents the distance from the lens center to its focal point. Such a lens is capable of converging a collimated beam of light traveling parallel to the lens axis and is referred to as a positive or converging lens.

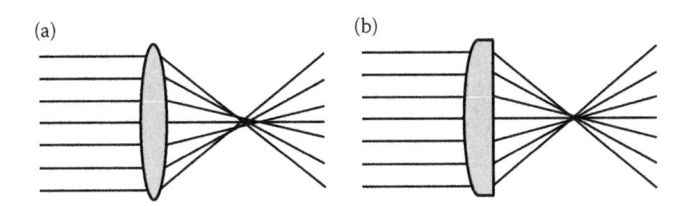

FIGURE 2.2
The phenomenon of spherical aberration demonstrated with collimated light striking a biconvex and a plano-convex lens. (a) Light refracted near the biconvex lens's edge deflects at a sharper angle than light near the center, resulting in positive spherical aberration and a weaker focus. (b) This effect is reduced with a plano-convex lens, where the peripheral rays are deflected to a lesser degree, resulting in a stronger focal point.

Concave Lenses

Concave lenses are typically thinner at the center than at the edges and are capable of diverging collimated light. The diverging rays appear to emanate from a point in front of the lens called the **principal focus**. Similar to convex lenses, if both surfaces are depressed inward, the lens is referred to as a *biconcave* lens (Figure 2.3). If one of the surfaces is flat, it is referred to as *plano-concave*.

Compound Lenses and Arrays

Compound lenses consist of multiple simple lenses sequentially arranged one after another and sharing a common axis. Oftentimes, the collection of simple lenses will consist of a variety of materials and shapes, each with its own refractive index. The act of combining simple lenses in arrays leverages their complementary aberrations to correct or compensate for imperfections when the lenses are standalone. Of course, combining different

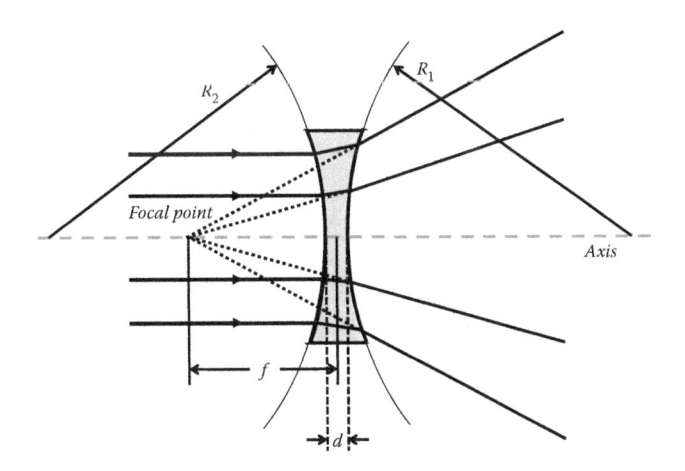

FIGURE 2.3
Diagram of light rays striking a biconcave lens. R_1 and R_2 correspond to the respective radii of curvature of both surfaces. As collimated light passes through such a lens, the beams appear to originate from a point on the axis in front of the lens. The distance from this point to the lens center is the focal length and is denoted by f but is negative with respect to the focal length of a convex lens. The ability of this lens to diverge incoming light allows it to be called a negative or diverging lens.

lenses introduces changes into the overall optical system's parameters, particularly that of *focal length*.

If two thin lenses of focal lengths f_1 and f_2 are placed in contact, the combined focal length f of the lens array is given by

$$\frac{1}{f} = \frac{1}{f_1} + \frac{1}{f_2}. \tag{2.1}$$

If two thin lenses are separated in air by some distance d, the combined focal length f is given by

$$\frac{1}{f} = \frac{1}{f_1} + \frac{1}{f_2} - \frac{d}{f_1 f_2}. \tag{2.2}$$

Equation 2.2 applies only if the distance d is smaller than f_1 (the focal length of the first lens).

A typical compound lens is the **objective**, which usually is composed of a combination of several optical elements. Commonly used in microscopes (Figure 2.4a), it acts as a high-powered magnifying glass with a very short focal length. The many lenses in a microscope objective are designed to be **parfocal**, where the lens maintains focus while magnification or focal length is changed. The various parameters of an objective are usually directly inscribed (Figure 2.4b).

There are several ways to characterize objectives and optical arrays. Each optical system features an *entrance pupil*, or the aperture where light can enter the system, and an *exit pupil* or *back aperture*, where light leaves. *Magnification* of an objective or optical system is a dimensionless number and reflects the ratio between the apparent size of an object seen through the system and its true size. The optical system's *numerical aperture* is another dimensionless number that describes the range of angles over which the system can accept or emit light. In most areas of optics, the numerical aperture is denoted *NA* and is defined by

FIGURE 2.4
Photograph of typical compound (objective) lenses for optical microscope. (a) Compound microscope objectives and stage. (b) Standard oil immersion objective with 40× magnification and an NA of 0.65. PL: positive low-phase contrast.

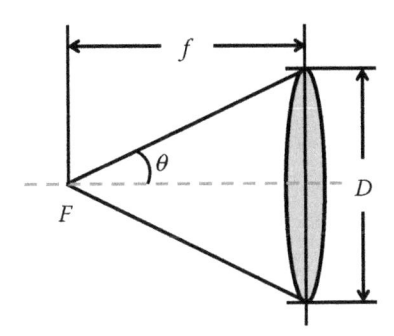

FIGURE 2.5
The half angle of the maximum cone of light that can enter or exit the lens. f: focal length, D: diameter of lens, F: focal point.

$$NA = n\sin\theta \qquad (2.3)$$

where n represents the index of refraction of the medium in which the lens is functioning, and θ denotes the half angle of the maximum cone of light that can enter or exit the lens (Figure 2.5).

A lens or optical system with a larger numerical aperture is capable of visualizing finer details than a lens with a smaller NA. Having a larger NA also means the lens or optical system is capable of collecting more light and therefore producing a brighter image, but the *depth of field*, or the distance between the nearest and farthest objects that appear acceptably sharp, is shallower.

An optical system's *f-number* represents the ratio between the lens's focal length f and the diameter of the entrance pupil D, so that it is defined by f/D. As such, it is a dimensionless number and is frequently denoted by $f/\#$, where the f-number replaces the "#." For example, a 100-mm focal length lens with a pupil diameter of 25 mm will have an f-number of 4, denoted $f/4$. Lenses featuring low f-numbers have high light-collecting ability, while lenses with higher f-numbers exhibit lower light-collecting ability.

An objective or optical system's *working distance* defines the maximum distance at which the objective or system can focus. This distance is physically measured from the front surface of an objective or lens and allows for a quick approximation of its focusing capacity. In general, the working distance decreases as the NA of the objective or system increases. Increasing the magnification power of the objective also serves to decrease the working distance. Typically, when selecting microscope objectives, working distances should be kept as large as possible, provided that NA requirements are satisfied.

Maintenance and Care of Optical Lenses

The delicate nature of optical components requires that special procedures be followed to maximize their performance and lifetime. Through everyday use, optics can come in contact with contaminants such as dust, water, and skin oils. These contaminants increase scatter of the optical surface and absorb incident radiation, which can create hot spots

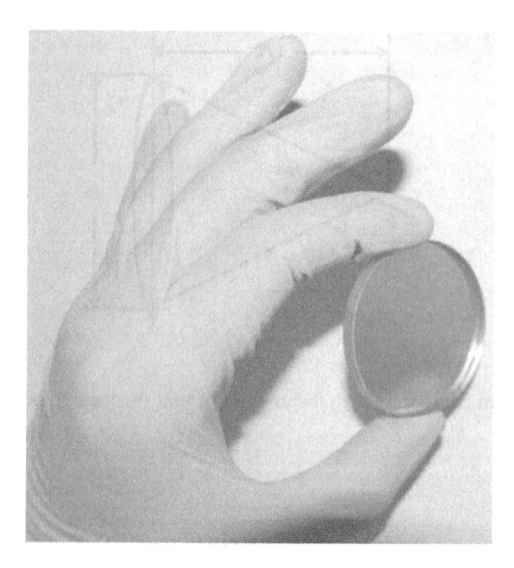

FIGURE 2.6
Grip optical components along ground edges.

along the optical profile, resulting in permanent damage. Optical components with coatings are particularly susceptible to this sort of damage.

Handling

By practicing proper handling techniques, you will decrease the need to clean your optics and thus maximize their lifetime. Always unpack or open optics in a clean, temperature-controlled environment. Never handle optics with bare hands, as skin oils can permanently damage the optical surface quality. Instead, wear gloves or finger cots; alternatively, for smaller optical components, it may be helpful to use optical or vacuum tweezers. During normal everyday use, hold the optic only along nonoptical surfaces, such as the ground edges of a lens or mirror (Figure 2.6).

Storage

Never place optics on hard surfaces because any contaminant on the optic or the surface will be ground in. Instead, most optics should be wrapped in lens tissue and then stored inside an optic storage box designed for the optic (Figure 2.7). Typically, the box should be kept in a low-humidity, low-contaminant, and temperature-controlled environment. Optics are easily scratched or contaminated, and some optical coatings are hygroscopic, so proper storage is important for preserving the optical component.

Cleaning and Maintenance

Always read the manufacturer's recommended cleaning and handling procedures if available. Before cleaning an optic, take time to inspect the optic to determine the type and severity of the contaminants. Optics can be permanently damaged if cleaned or handled incorrectly.

FIGURE 2.7
Dedicated storage box with foam padding and lens tissue paper.

Dust and other loose contaminants usually should be blown off before any other cleaning technique is employed. A canister of inert dusting gas or a blower bulb is needed for this method. Do not use your mouth to blow on the surface because it is likely that droplets of saliva will be deposited on the optical surface.

If blowing off the surface of the optic is not sufficient, using clean wipes and optical-grade solvents are viable alternatives. Wipes should always be moist with an acceptable solvent and never used dry. Acceptable wipes (in order of softness) are pure cotton (such as Webril Wipes or cotton balls), lens tissue, and cotton-tipped applicators. Typical solvents employed during cleaning are acetone, methanol, isopropyl alcohol, and TravelSAFE Precision Optical Cleaner. Use all solvents with caution since most are poisonous, flammable, or both. Read product data sheets and material safety data sheets (MSDSs) carefully before using any solvents.

Waveplate

Optical waveplates (also called *retarder plates*) are transparent plates with a carefully chosen amount of birefringence. A substance is birefringent if its refractive index varies based on the polarization and propagation direction of entering light and is quantifiable by the maximum difference in refractive indices achievable within the material. Waveplates are mainly used for manipulating the polarization state of a light beam, where the phase between two perpendicular polarization components of an incident light wave can be shifted.

A waveplate has a *slow* and a *fast* axis, both perpendicular to the surface of the waveplate as well as the propagation direction of the incoming light beam. The phase velocity of light is slightly higher for polarization along the fast axis, while the polarization component along the slow axis is phase delayed. The end result is a propagation profile that is polarized differently due to one of its components undergoing a phase shift. The desired value of optical retardance must be achieved within a limited wavelength range due to the finite surface area of the waveplate.

The amount of relative phase Γ the waveplate introduces between the two different polarization components can be calculated by

$$\Gamma = \frac{2\pi \Delta n L}{\lambda_0} \tag{2.4}$$

where Δn is the birefringence of the waveplate crystal, L represents the crystal thickness, and λ_0 is the vacuum wavelength of the incident light.

The two most common types of waveplates are *quarter-waveplates* and *half-waveplates*. Quarter-waveplates introduce a difference of phase delay between the two linear polarization directions equal to $\pi/2$. This has the effect of changing linearly polarized light to circular and vice versa. The half-waveplate introduces a difference of phase delay equal to π and effectively retards one polarization component by half a wavelength. This has the effect of changing the polarization direction of linearly polarized light by 90°.

Linear Polarizers

A polarizer is a special type of optical filter that passes light of a specific polarization while attenuating or reflecting waves of other polarizations. Specifically, a linear polarizer is a device that selectively allows specific orientations of plane-polarized light. By definition, such a device can convert a beam of light of mixed polarizations into a beam of well-defined polarization. In imaging applications, polarizers can reduce glare from reflections and eliminate hot spots, and simultaneously enhance the final image contrast.

The simplest linear polarizer consists of a regular array of fine parallel metallic wires, known as a *wire grid polarizer*. This wire grid is typically placed perpendicular to the incident beam. Incoming electromagnetic radiation with electric field components that are parallel to the wires on the grid excite electrons that subsequently move along the length of the wire. Since the electrons have a large range of movement along the wire direction, the grid behaves like the surface of a metal and reflects these wave components backward along the incident beam.

However, electrons excited by waves with electric field components perpendicular to the wire can only move very little along the width of the wire. As a result, little energy is reflected, and the incident wave passes through the grid with little loss in intensity. These waves that successfully pass through the grid contain only electric field components that lie perpendicular to the wires and are thus linearly polarized (Figure 2.8).

Mirrors

Mirrors are commonplace optical components used to manipulate the path of light by means of reflection. They are also used to filter out unwanted wavelengths while

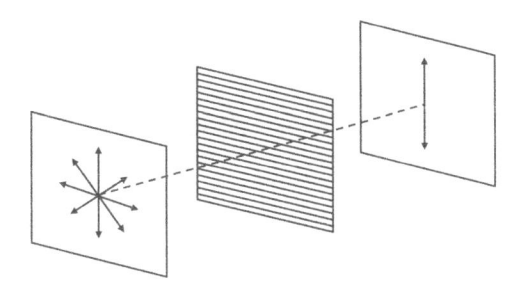

FIGURE 2.8
A wire grid polarizer permits only plane polarized light with electric field components perpendicular to the wires.

preserving other wavelengths of interest in the reflection. The manufacturing process for a mirror typically begins with a sheet of glass or plastic that is formed into a desired shape. One side of the pane is covered in a reflective substance (typically aluminum). Special mirrors with specific light-reflection properties may be coated with any number of other chemicals, each with its own individual properties.

The simplest and most familiar type of mirror is the *plane mirror*, which features a flat surface. Other types of mirrors involve curved surfaces, which can produce magnified or diminished images, as well as focusing or distorting the reflection. Such mirrors are referred to as *spherical mirrors* (or *curved mirrors*) and, similar to their lens counterpart, they can be *convex* or *concave*.

Plane Mirror

Light rays striking the mirror's flat surface reflect at an angle that is similar to its incident angle (Figure 2.9). A plane mirror forms an image that appears to be behind the plane in which the mirror lies. The formed image is always a *virtual image* in that light rays do not actually originate from the image upright and of the same shape and size but are laterally inverted.

Convex Mirror

Convex mirrors (or *diverging mirrors*) are curved mirrors where the reflective surface reflects lights outward and thus cannot focus light. In a manner similar to a convex lens,

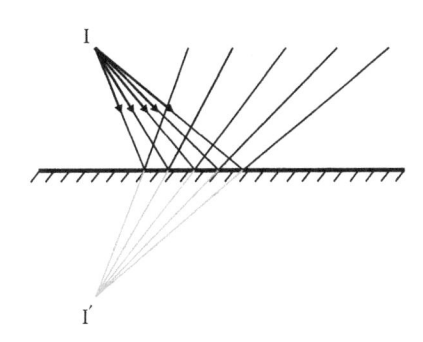

FIGURE 2.9
Light rays reflecting off a plane mirror. I denotes the location of the object, and I′ denotes the location of the image formed. Note that the angles of incidence are equal to the angles of reflection.

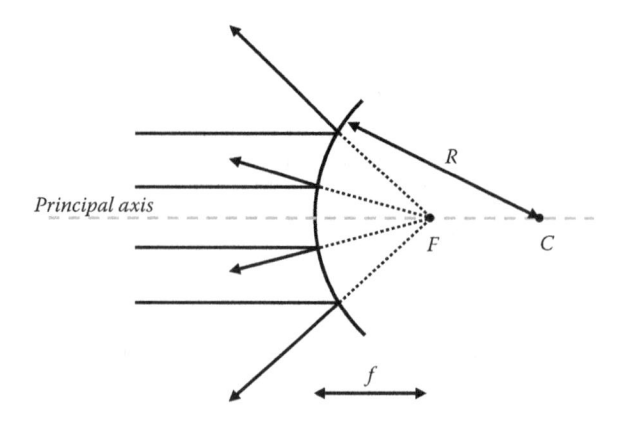

FIGURE 2.10
Parallel light rays reflecting off a convex mirror. *R*: radius of curvature, *F*: focal point, *C*: center of sphere defining mirror surface, *f*: focal length.

the reflective surface of the mirror protrudes toward the light source. The image formed by a convex mirror is always virtual in that the light rays do not physically pass through the image. The image will also always be diminished in size, regardless of the distance between the object and the mirror.

Even though a convex mirror cannot physically focus an image, the focal point *f* of its surface represents the apparent origin of light rays (Figure 2.10). Furthermore, its radius of curvature *R* is always twice that of the focal distance *f* so that

$$R = 2f. \tag{2.5}$$

Concave Mirror

Concave mirrors (or *converging mirrors*) are curved mirrors where the reflective surface depresses inward and away from the light source. Unlike convex mirrors, concave mirrors are used to reflect light inward and thus are capable of focusing incoming light rays (Figure 2.11).

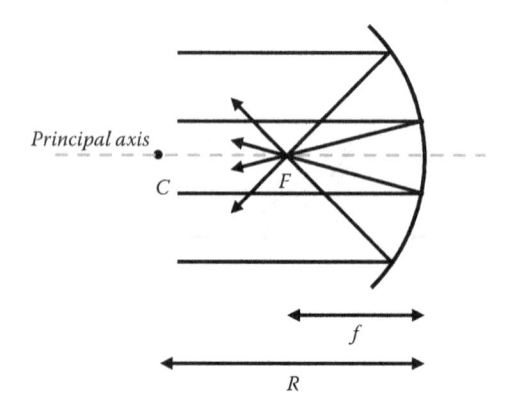

FIGURE 2.11
Parallel light rays reflecting off a concave mirror. *C*: center of sphere defining mirror surface, *F*: focal point, *f*: focal length, *R*: radius of curvature.

Depending on the object distance from the concave reflecting surface, different image types are formed with varying orientation, magnification, and entity (real or virtual).

The focal point F marks the position where all reflected parallel beams converge. Equation 2.5 also holds true for concave mirrors, where the spherical center C is located twice as far from the reflecting surface as the focal point.

Optical Coating

An optical coating is composed of one or more thin film layers deposited on the surface of an optical component (such as a lens or mirror) to enhance its transmission or reflection properties. The performance of the coating depends on the number of layers and their respective thickness, as well as their composition. The refractive index differences between multiple layer interfaces provide for their unique transmission or reflective behavior.

There are many different types of coatings available for precision optics. The simplest coatings consist of a thin layer of metal and exhibit varying reflective properties across the visible and near-infrared range. *Antireflection* coating is designed to reduce unwanted reflections from surfaces and is commonly used on refractive optics to increase throughput. *High-reflective* coatings accomplish the opposite of antireflection coatings and can maximize reflectance at either a single wavelength or a broad range of wavelengths. Finally, *filters* can be used to transmit light at specific wavelengths but absorb or attenuate light at others, conferring properties of dichroism.

Metal Coatings

Thin layers of metals deposited onto glass substrates can produce mirror surfaces and represent the simplest category of optical coatings. This process is known as *silvering* and can alter the reflective characteristics of the mirror, depending on the type of metal used. Aluminum is the cheapest and most commonly used coating, yielding a reflectivity of around 90% over the visible spectrum. Silver is more expensive than aluminum, but it provides a reflectivity of up to 98%, even into the far infrared (Figure 2.12). However, silver coatings suffer from decreased reflectivity (<90%) in the blue and violet spectral regions.

Even more expensive is gold, which gives superior reflectivity throughout the infrared spectrum, but it suffers limited reflectivity at wavelengths shorter than 550 nm. Not surprisingly, this attribute of gold mirrors results in their typical gold color.

FIGURE 2.12
A backside-polished, protected silver mirror with a diameter of 1 inch.

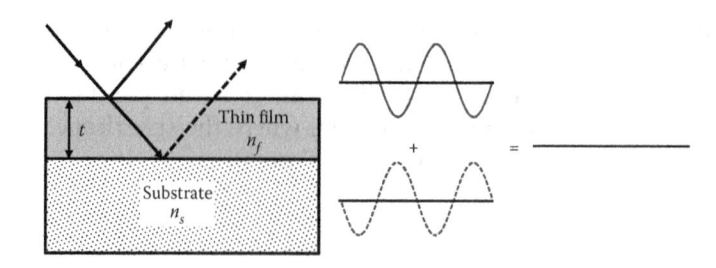

FIGURE 2.13
Light ray striking an antireflective thin film coating with refractive index n_f and substrate with refractive index n_s. The reflected beams from both interfaces destructively interfere, yielding minimal reflective intensity. The "optical thickness" of the thin film can be calculated by multiplying the physical thickness t with its refractive index n_f.

Antireflection Coatings

Antireflection (AR) coatings applied to the surface of optical devices serve dual purposes. Hazards caused by reflections traveling backward through the optical system can be reduced by AR coatings, and the efficiency of the system can be improved due to decreasing loss from reflection.

The basic principle behind AR coatings is the selection of thin film structures with alternating layers of contrasting refractive indices. The layer compositions and thicknesses are chosen to allow for destructive interference of the reflected beams from the various interfaces (Figure 2.13) and constructive interference in the transmitted beams.

High-Reflection Coatings

High-reflection (HR) coatings function the opposite way from antireflection coatings. Rather than destructive interference between the reflected beams, thin films are chosen to yield constructive interference. The most common type of HR mirror is known as a *dielectric mirror*, where multiple thin layers of alternating dielectric material are deposited on a substrate such as glass or plastic.

In general, the multiple layers of dielectric material are collectively known as a *dielectric stack*, consisting of thin layers with a high refractive index interwoven with thick layers of a lower refractive index. The thicknesses of the layers are chosen so that the path length differences for reflected light beams between interfaces are integer multiples of the beam's wavelength, resulting in constructive interference. By this principle alone, dielectric mirrors can confer ultra-high reflectivity: Values of 99.999% or better over a narrow range of wavelengths can be achieved.

Filters

Optical filters are used to selectively transmit certain wavelengths of incident light while blocking the rest. These filters can be divided into two groups: colored filters, which transmit light of the wavelength corresponding to the filter's color; and neutral density filters, which nonselectively block a portion of the incident light. Absorption filters also exist,

FIGURE 2.14
Photo of a neutral-density filter.

which absorb selective wavelengths of light while transmitting the rest. The colored filters used in optics, however, are bandpass filters; they allow only a certain range of wavelength through while stopping the rest of the light; that is, a red-colored filter will allow only a wavelength of 610 nm through. The range of the bandpass could be as general as ± 8 nm or as precise as ± 0.2 nm. Note that the central wavelength of bandpass filters does not need to fall within the visible spectrum. The term *colored filter* simply refers to a specific subset of bandpass filter that selectively allows wavelength in the visible spectrum; there exist bandpass filters for wavelengths of 340 nm (ultraviolet) to 4.75 μm (infrared).

Neutral density filters (ND filters) are used to reduce the intensity of light for all wavelengths. Typically, the filter is grey, as can be seen in Figure 2.14.

The attenuation factor of the ND filter is described by its optical density (d). This is calculated by taking the negative log of the ratio of the exit intensity (I) to the incident intensity (I_0)

$$d = -\log_{10}\left(\frac{I}{I_0}\right) \tag{2.6}$$

For example, the ND filter shown in Figure 2.14 has an optical density of 2.0. This corresponds to an intensity ratio of 0.01. Thus, the output intensity is only 1% of the incident intensity.

In photography, ND filters are used to lower the intensity of light entering through the lens. This allows the photographer the ability to use larger apertures and create effects such as motion blur. In optical microscopy, however, ND filters are used to reduce the intensity of light and to prevent overexposure. Typically, a laser's power cannot be changed without affecting other properties, such as collimation. Thus, ND filters can attenuate the laser without changing any other parameters.

Prisms

In optics, a prism is a general term that describes any transparent object that refracts light. The shape and surface angles of the prism determine its optical properties and its applications. Typically made of glass, prisms are used to disperse, deflect, or reflect the incoming light beam. The most recognizable use is to disperse light, which means to break up light into its spectral components. A common image of this is a white beam of light entering a triangular prism and exiting as a rainbow of light beams. This is possible because the refractive properties of light depend on the wavelength. Longer wavelengths, such as red light, will be bent less than shorter wavelengths, such as blue light.

In addition, prisms are commonly used to bend the direction of light through deflection or reflection. These bends can change the direction or the parity of the incoming light. These prisms are grouped into three main types: *reflection prisms*, *rotational prisms*, and *displacement prisms*. They are prominently used in optical imaging. *Reflection prisms* are used to bend the direction of travel for light rays, typically in angles of 45°, 60°, 90°, or 180°. While these changes can also be achieved through a system of mirrors, the use of a prism often consolidates and simplifies the optical setup. When using reflection prisms, however, the parity of the images must be taken into consideration. Parity describes the orientation of an image in reference to the readability of texts. For example, even parity is defined as the orientation of the texts that can be read. On the other hand, if you view the texts through a mirror, the mirror image will be in odd parity. For prisms, in general, if the light is reflected an even number of times, then the outgoing image will be of the same parity as the incoming image. In contrast, an odd number of reflections mean the outgoing image will have the opposite parity. This concept is illustrated in Figure 2.15. Rotational prisms are used to rotate images by factors of 90°. They are typically used after the image has been inverted or reflected by other prisms. Finally, displacement prisms are used to shift the axis of the light rays but still maintain the direction of the rays. One example of a displacement prism is a rhomboid prism, shown in part b of Figure 2.15.

Since the only function of dispersion prisms is to separate the spectral components of the incident light, they are often used in spectrometers. Other types of prisms, however, are widely used in microscopic and telescopic setups.

Combination Prisms

Combination prisms, as the name implies, involve two or two prisms that are combined to form a prism system. The most widely used combination prism is the cube beamsplitter, made of two right-angle prisms. These two prisms are adhered together to create the beam splitter, and a dielectric coating is applied to the junction. This coating causes the prism to reflect a part of the incident light while transmitting the rest through the entire

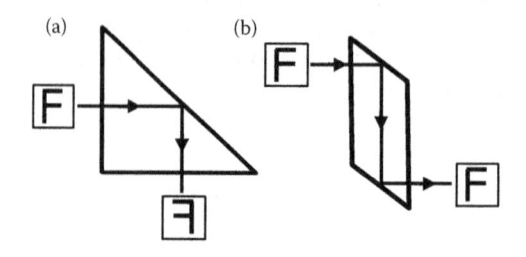

FIGURE 2.15
Effect of prisms on parity changes: (a) A right-angle prism with one reflection causes a change in parity; (b) A rhomboid prism with two reflections does not change the parity.

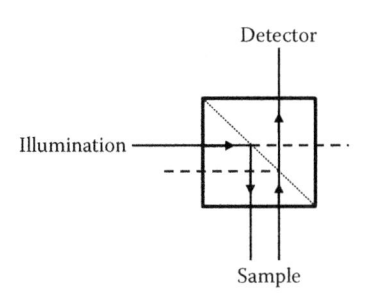

FIGURE 2.16
Simple schematic of a beamsplitter setup in a microscope.

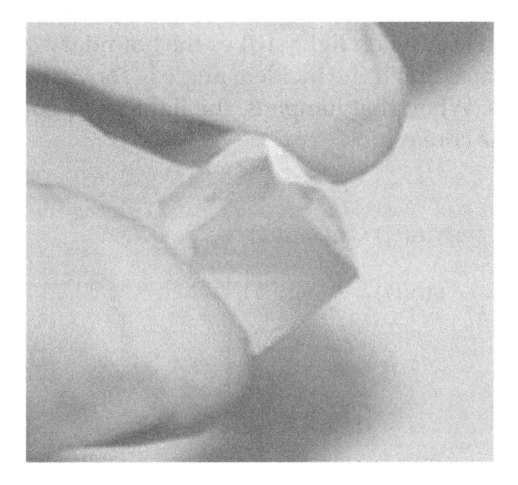

FIGURE 2.17
Photo of a cube beam splitter.

prism. The percentage of light reflected versus the percentage of light transmitted is determined by the type of coating. Typically, the percentages are 30/70, 50/50, or 70/30. The coating also changes what wavelength or polarization of light is reflected and transmitted through the prism. Cube beam splitters are commonly used in interferometers and optical microscopes. They are essential when the illumination path and the detection path share a common route, as shown in Figure 2.16. Figure 2.17 shows a photo of a cube beam splitter.

Optical Fibers

Optical fibers are transparent fibers made from silica that are extremely important in optical microscopy and telecommunication. They are used to transmit light from one end to the other, serving as flexible and portable waveguides. The fibers transport light effectively due to their unique structure.

Structure

At the center of an optical fiber is the core, a cylindrical region made of glass. The core is characterized by its radius (a) and its index of refraction (n_1). It is important to note that

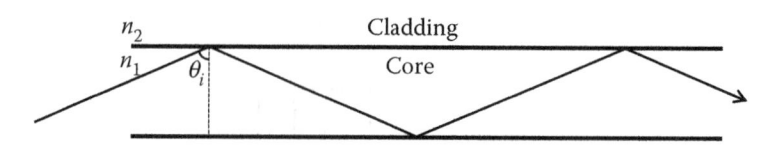

FIGURE 2.18
Light propagating in the core region of the optical fiber.

light primarily travels inside the core region. Surrounding the core is a layer of material called the cladding. The cladding is typically made of glass or plastic, and it is characterized by its index of refraction (n_2). The cladding serves as a barrier, effectively trapping the light inside the core region through a phenomenon called total internal reflection. For this to occur, the index of refraction of the cladding must be less than the index of refraction of the core ($n_2 < n_1$). Thus, when light strikes the boundary between the core and the cladding, Snell's law dictates that if the incident angle is above the critical angle, then total internal reflection occurs. When that happens, the light ray is completely reflected and is then contained within the core region.

$$n_1 \sin(\theta_i) = n_2 \sin(\theta_t)$$

$$\sin(\theta_i) = \frac{n_2}{n_1} \sin(\theta_t) \quad \text{when } \theta_t = 90°.$$

$$\theta_c = \arcsin\left(\frac{n_2}{n_1}\right)$$

(2.7)

Thus, given the indices of refraction of the core and the cladding, a critical angle could be calculated. This specifies at what angles light can be effectively coupled into the optical fiber. A diagram of the light propagation inside a fiber is shown in Figure 2.18.

Outside the cladding is a layer of buffer coating, which protects the inner fiber from physical damage. In a typical optical fiber, the bare fiber is then surrounded by a cable jacket filled with Kevlar threads. The Kevlar threads increase the durability and the flexibility of the fiber, while the jacket is color coded to denote the type of optical fiber. At either end of the jacket are the fiber connectors used to join the optical fiber to the laser setup. Photos of optical fibers can be seen in Figures 2.19 and 2.20.

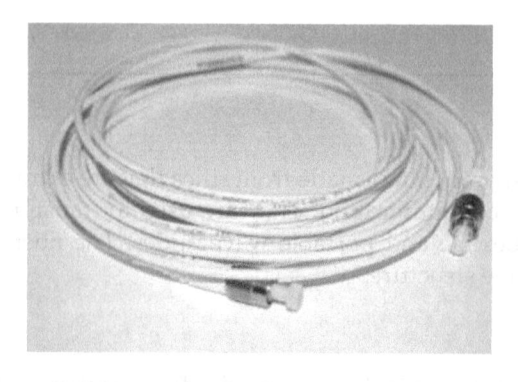

FIGURE 2.19
Photo of an optical fiber.

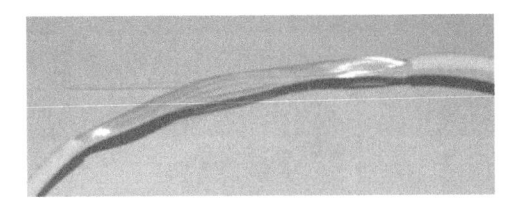

FIGURE 2.20
Photo of an optical fiber showing the Kevlar threads and the bare fiber.

Single-Mode and Multimode Fibers

There are two types of optical fibers based on the number of propagation paths or transverse modes: single-mode fibers (SMF) and multimode fibers (MMF).

For single-mode fibers, there is only one propagation path that light can travel. There is only one mode because the core diameter is usually much smaller than the cladding layer. Typically, the core diameter is around 8 microns, while the cladding layer is 125 microns thick. Because light can only propagate in one way, there is very little loss into the cladding layer. Thus, SMF has very low attenuation and is ideal for long-range signal transmission or high-precision optical microscopy. For easy identification, the outer jacket of SMF is colored yellow.

Unlike single-mode fibers, multimode fibers have many propagation paths. This allows multiple light rays to travel inside the core region. To accommodate for propagation paths, the core diameter is around 50–100 microns. Because of this feature, multimode fibers are great for transmitting large amounts of information. However, allowing multiple modes also increases the dispersion loss. Thus, MMF is used in communication systems for buildings and infrastructures. The jacket of MMF is colored blue or orange to differentiate it from SMF.

Fiber Connectors

To connect optical fibers to any telecommunications or optical system, the right type of connector must be used. There are more than 10 different types of fiber connectors, each designed for different equipment and applications. Typically, an optical fiber cable will already have two connectors at each end. These fiber cables are called patch cables. There are two fiber connectors that are important for single-mode fibers in optical setups: ferrule connector/angled physical contact (FC/APC) and ferrule connector/physical contact (FC/PC). For multimode fibers, the typical connector used is the SMA905. The first part of the name denotes what type of connector it is, and the second part describes the type of polish at the end of the connector.

Ferrule connectors (FC) are the most common connector used in optical setups. This connector type contains a white ceramic ferrule at the end of the connector where the bare fiber can exit. FC connectors offer precise positioning of the fiber cable with their locatable notch and threaded receptacle. They are typically used for single-mode fibers. A picture of an FC connector is shown in Figure 2.21. The latter half of the connector, whether it's FC/PC or FC/APC, describes the polish or shape of the ferrule. Different polishes result in different back reflections of the fiber. Back reflection is a measure of the light reflected off the ferrule tip in units of negative decibels. Physical contact, PC, means the tip is slightly rounded and typically results in a back reflection of around −40 dB. The fiber connector shown in Figure 2.21 is an FC/PC connector. Angled physical contact, or APC, has the tip

FIGURE 2.21
Photograph of a FC/PC connector.

slanted at an 8-degree angle. This further reduces back reflection, resulting in a value of around −60 dB.

SMA905 connectors look very similar to FC connectors, which are mainly used for multimode fibers. SMA905 also has a ferrule at the tip of the connector, but it is usually made of stainless steel instead of ceramic. The ferrule design also makes it ideal for bare fibers with large core diameters ranging from 125 microns to 1250 microns.

Optomechanical Equipment

Optomechanics represents a broad category of optical devices and parts, all of which play a specific role in the construction and function of an assembly or setup. They typically allow the experimenter to make accurate and minute adjustments to an existing system, especially to provide motions along multiple axes. Optomechanical equipment encompasses everything from large optical tables to small translational stages.

Optical Table

Optical systems usually require flat and stable surfaces, especially due to their extreme sensitivity to the smallest of vibrations. An optical table is a specialized workbench that provides many advantages for supporting such systems. The surface is typically stainless steel with a rectangular grid of tapped holes, providing a basis for the accurate alignment of various optical components (Figure 2.22). Many tables often use pneumatic bearings to not only keep the work surface flat under nonuniform loads, but also to act as a low-pass filter for external vibrations.

Translational Stages

A translational stage (or *linear stage*) is a common optical component used to restrict an object to round-trip motion along a single axis. All translational stages consist of a platform and a base, joined by some form of a guide or actuator so that the platform moves relative to the base, to which it is often fixed (Figure 2.23). Such stages found in optics usually employ a fine-pitch screw or a micrometer that presses on the stage platform. Rotating the screw or micrometer pushes the platform forward, while a spring provides a restoring force to maintain contact between the platform and screw tip. Both the base and the

FIGURE 2.22
Optical table with pneumatic bearings.

FIGURE 2.23
Linear dovetail translational stage made of anodized aluminum.

platform possess tapped holes for the purposes of fixing the stage and mounting components onto the moving platform, respectively.

Motorized linear stages may replace the manual micrometer knob with a stepper motor, which moves in fixed increments, or a servo motor, which translates the stage based on sent instructions. Both options provide fine automated control over stage translation and are commonly found in many optical setups. Usually, motorized linear stages with servo motors have better precision and less backlash due to the feedback control in motors.

Rotation Stages

Rotation stages (or *rotary stages*) are an optical component that restricts an object to a single axis of rotation. Similar to translational stages, rotation stages are composed of a platform that moves and a base that is fixed. The platform and the base are usually

FIGURE 2.24
A Newport Corporation 481-A rotation stage.

joined by a guide so that the platform rotates about a single axis with respect to the base (Figure 2.24).

Several mechanisms of operation exist for rotation stages, ranging from manual turning of the platform by hand to a linear actuator for precise positioning. Many rotation stages employ a *worm drive* for improved position control, where a worm wheel fixed to the rotating platform meshes with a worm in the base. Rotation of the worm is accomplished by a manual or motorized control knob.

Mirror Mounts

A mirror mount is a device frequently encountered in many optical setups. It is specifically designed to hold a mirror in a precise position while providing room for tipping and tilting the mirror in a controlled manner.

The most common type of mirror mount is the *kinematic* mount, which consists of a fixed frame and a moveable frame (Figure 2.25). The moveable frame holds the mirror and typically pivots on a ball bearing set into a hole on the fixed frame. The two frames are also kept in close proximity by a series of screws designed to maintain contact between the moveable frame and the two micrometers on the fixed frame. By adjusting the two micrometers, the profile of the moving frame can be changed to accommodate the desired angle and position of the mirror.

Lasers

A laser is a device that generates a monochromatic, coherent beam of electromagnetic radiation by means of optical amplification based on the stimulated emission of photons. All lasers consist of a *gain medium*, which is a material with properties allowing for the amplification of light with specific wavelengths. The gain medium is usually supplied with power from an electric current or light at different wavelengths (provided by a flash lamp or diode). A series of mirrors confine an optical resonator cavity where light may travel through the gain medium hundreds of times before exiting through a partially reflective mirror as a coherent beam.

One of the simplest lasers used in daily life is the *laser pointer*. Although laser pointers have relatively low power output, they can be very useful and handy in many simple optical experiments, especially those dealing with basic optical properties and phenomena such as refraction, polarization, and diffraction.

FIGURE 2.25
Kinematic mirror mount with two adjustable micrometers.

The laser beam in a laser pointer is often generated by a *laser diode*, which uses a semi-conductor as the gain medium. Specifically, the semiconductor consists of a P-N junction, made from doping two thin layers on the surface of a wafer. The top layer is doped to be a P-type region, while the bottom layer is an N-type region. When the diode is powered by electric current, the holes in the P-type region and the electrons in the N-type region move toward the gap between the layers. There, the two charge carriers annihilate each other, releasing energy and emitting photons. This spontaneous transmission then causes further annihilation events in the junction. This creates the stimulated emission necessary to create a laser beam. The wavelength of the laser pointer depends on the semiconductor material used in the two doped layers. Early laser pointers could emit only red light with a wavelength of 633 nm. Today, laser pointers are made as diode-pumped solid-state lasers. This new and improved technology offers higher-quality laser beams and narrower spectral bandwidths. It also allows for the production of different colored laser pointers, including green, blue, and yellow.

Solid-state lasers are lasers that use solid gain mediums as opposed to liquids used in dye lasers or gases used in gas lasers. One of the most common types of solid-state lasers is the neodymium-doped yttrium aluminum garnet (Nd:YAG) laser (Figure 2.26).

The yttrium aluminum garnet crystal ($Y_3Al_5O_{12}$) is doped with triply ionized neodymium, Nd(III), which replaces a small fraction of the yttrium ions. The amount of neodymium dopant in the material is usually around 1% by atomic percent. Nd:YAG lasers typically emit light with a wavelength of 1064 nm; in many situations, Nd:YAG lasers work in a frequency-doubled manner to produce an output wavelength of 532 nm.

If an experiment requires changing the wavelength of the optical illumination, then a *tunable laser* is often required. As the name suggests, a tunable laser can shift its output wavelength over a certain range. Depending on the tuning mechanism, the tuning range varies from a few nanometers to hundreds of nanometers. Wavelength tuning needs an optical dispersion assembly, such as a prism or a diffraction grating, and then selects the

FIGURE 2.26
1,064-nm diode-pumped Nd:YAG infrared laser.

desired wavelength to be further amplified and output. The tuning capability is extremely useful if an experiment involves a frequency spectrum of a parameter, where only a single-laser system is needed instead of using a multitude of lasers with different wavelengths. The two most commonly used tunable lasers, categorized by their gain medium, are solid-state tunable lasers and liquid dye lasers.

A *dye laser* uses organic dye solution as the gain medium (Figure 2.27) in both oscillator and amplifier to achieve wavelength tuning. Dye lasers are considered "passive lasers" and work with *pump lasers* that provide energy. A commonly used pump laser is often a high-power frequency doubled Nd:YAG laser, as shown in Figure 2.28. Like a gas laser, dye lasers use mirrors to amplify the spontaneous emission of photons within the gain medium. The mirror at the output will usually be 80% reflective, while all other mirrors are coated to be 99% reflective. This system creates the amplification needed for stimulated emission, and the 20% transmittance of the output mirror allows the laser beam to exit.

FIGURE 2.27
Tunable dye laser operating under a pump wavelength of 532 nm and Styryl 11 dye (758–826-nm tunable range).

FIGURE 2.28
532-nm pump laser for downstream tunable dye lasers.

Another advantage of the dye laser, besides its tuning capability, is the ability to swap the dyes. By switching the dyes in the laser, different wavelengths of light can be produced. Thus, a dye laser with a set of different dyes can output an extremely wide range of wavelengths, eliminating the need for additional lasers.

Aside from the gain medium, lasers are also characterized by the pulse width of the beam. Some lasers emit **continuous waves** (CW), which means the laser produces a constant output. These lasers are simply controlled by an on-off switch. Examples of these lasers include laser pointers, as the laser beam can only be turned on or off. If the laser beam is pulsed, then its main parameter is the pulse width of the laser. There are a total of three different methods to produce pulse trains for a laser beam, each for generating pulses of different width. A method called **Q-switching** is used to create long pulses with high energy per pulse. While this means the pulse repetition rate is low, Q-switched mode can generate a beam with a much higher power than a continuous-wave mode. This is the main advantage of a pulsed laser, as many experiments and optical phenomenon require lasers with high-power density. The other two methods of creating a pulsed laser beam are called **gain-switching** and **mode-locking**. These are used to generate pulse trains with extremely short pulse widths, on the order of picoseconds or femtoseconds (Figure 2.29).

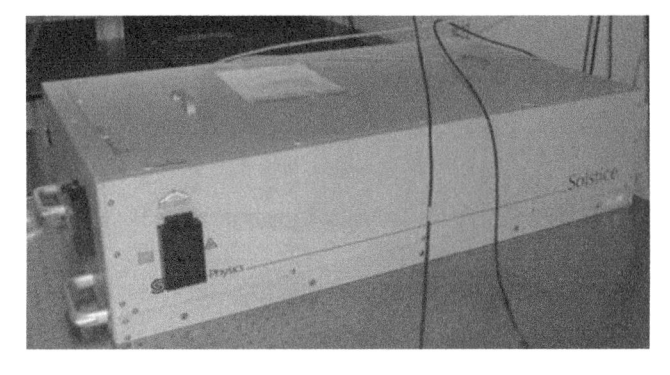

FIGURE 2.29
Spectra Physics® Solstice® one-box ultrafast femtosecond laser featuring a pulse width of 100 fs.

The lasers also have high repetition rates and are primarily used for nonlinear optics and nanotechnology.

Photodetectors

Photodetectors (or *photosensors*) are used primarily as optical receivers to convert light energy into a voltage or current. The underlying principle behind photodetectors is the *photoelectric effect*, which explains the emission of electrons from matter upon absorbing energy from incident electromagnetic radiation. There are many devices classified as photodetectors, ranging from *photodiodes* to *charge-coupled devices (CCDs)*.

Photodiodes

The photodiode is capable of converting light energy into electrical energy through the use of P-N junctions. When a photon from incident electromagnetic radiation strikes the photodiode, an electron is excited to a higher energy state, creating a negatively charged *free electron* and a positively charged *electron hole*. In the simplest sense, the two charged carriers are capable of driving electric current given the positioning of a cathode and an anode on either end of the junction.

The electric current driven by the migration of charge carriers to the cathode and anode is referred to as the *photocurrent*. Photodiodes can operate under different modes depending on how this photocurrent is manipulated. In *photovoltaic mode* (*zero bias*), the flow of photocurrent out of the junction is restricted, and a voltage builds up. In *photoconductive mode*, the diode is reverse biased so that the voltage in the cathode is greater than the anode. The junction's capacitance is reduced, which results in fast response times, but also exhibits more noise due to the thermal agitation of the charge carriers (*Johnson–Nyquist noise*).

The material used to fabricate a photodiode greatly affects its properties. Different compounds exhibit different bandgaps between their conduction and valence bands, which require specific photon energies to cross and create photocurrents. Common materials for photodiode active elements include silicon (Figure 2.30), germanium, indium gallium arsenide, and lead (II) sulfide.

Phototransistors

Phototransistors are very similar to photodiodes in that the observed photosensitive effects are much the same. Rather than a P-N junction, a phototransistor consists of a base and a

FIGURE 2.30
Thorlabs amplified photodiode with silicon active element and switch-selectable gain.

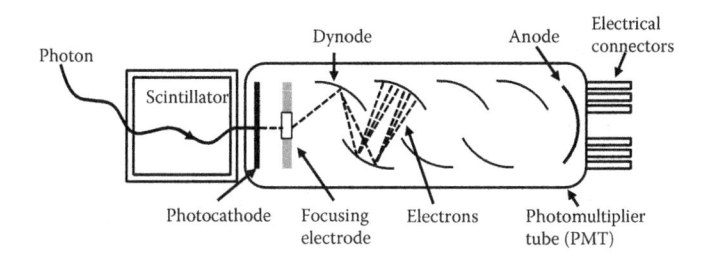

FIGURE 2.31
Schematic of a photomultiplier tube (PMT) coupled to a scintillator.

collector and functions very similarly to an N-P-N bipolar junction transistor. Rather than a voltage being delivered to the base, hole-electron mobility due to incident light striking the base provides the voltage. In essence, a phototransistor amplifies variations in the light striking it.

Photomultiplier Tubes

A photomultiplier tube (PMT) is a sophisticated vacuum tube designed to detect light in the ultraviolet, visible, and near-infrared ranges. Apart from detection, PMTs can amplify the current produced by the incident light via multiple dynode stages by secondary emission. The resulting output from a PMT is a high-current electrical signal (amplified by as much as 100 million times) that corresponds to the detection of a single photon.

Oftentimes, PMTs are coupled to scintillators (Figure 2.31), which are materials that exhibit luminescence when excited by ionizing radiation. The photocathode (negatively charged electrode) in the PMT reabsorbs the light emitted from the scintillator and produces electrons as a consequence of the photoelectric effect. A focusing electrode directs the electrons toward a series of dynodes, which are a series of electrodes in which each has a more positive electric potential than its predecessor. As electrons migrate toward each dynode, they are accelerated by the electric field and strike the dynode surface with increased energy, releasing large numbers of low-energy electrons, which continue to produce even more electrons. The geometry of the dynode series is such that an electron cascade occurs with increasing numbers being produced at each stage, all eventually striking the anode and generating a sharp current pulse.

Charge-Coupled Devices

A charge-coupled device (CCD) is a light-sensitive integrated circuit that stores and displays data belonging to an image where each pixel corresponds to an electric charge the intensity of which is proportional to a particular color. Implementation of CCDs is commonplace in cameras used for image acquisition (Figure 2.32). When an image is projected onto the lens of a CCD camera, a capacitor array (or *photoactive region*) inside the CCD accumulates electric charge where each individual capacitor's charge is proportional to the light intensity at a corresponding location. Following charge accumulation, a control circuit prompts each capacitor to transfer its charge to its neighbor until the last capacitor

FIGURE 2.32
A CCD camera connected to an optical assembly. Data is output via a CAT5 cable.

FIGURE 2.33
The sensor head of a commercial power meter.

in the array deposits its charge into a charge amplifier. The charge amplifier converts the charge into a voltage, and this process is repeated until the image information is encoded as a sequence of voltages. These voltages can then be sampled, digitized, and stored for data acquisition purposes or output to a display.

Power Meter

A power meter is an electrical device that measures the power of a light beam. A typical commercial power meter consists of two parts: a display that shows the power output and a sensor head that is placed in the optical setup. The sensor head, shown in Figure 2.33, is mounted on a stand and should be positioned in the line of the light beam to be measured. The display, which is connected to the sensor, will then show the power of the input beam. Note that there is a zeroing function for the equipment, which will account for any background light. Figure 2.34 shows the entire system.

Spectrometers

Spectrometers are instruments used to identify materials by spectroscopic analysis, which concerns the interaction between matter and radiated energy. A spectrometer examines light over a specific portion of the electromagnetic spectrum, where the intensity of each optical wavelength is recorded within the spectral range of the incoming light. Figure 2.35 shows a miniature optical fiber–coupled portable spectrometer with universal serial bus (USB) communication.

Typically, a sample is heated to the point of incandescence, which causes it to emit light characteristic of its atomic makeup. Other times, a broad-spectrum light is shined through

FIGURE 2.34
A complete power meter system from a commercial vendor.

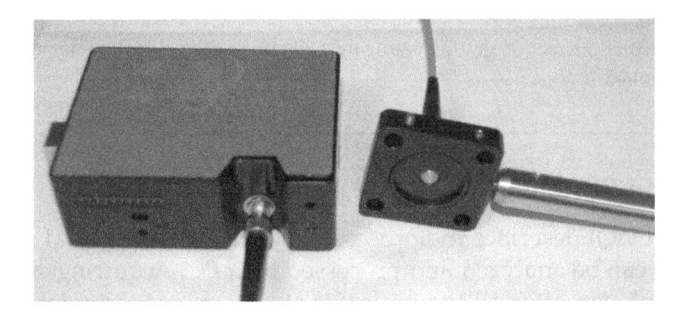

FIGURE 2.35
A commercial USB-driven miniature fiber optic spectrometer with fiber port.

the sample, and the sample absorbs specific wavelengths while transmitting others. The resultant signature from the sample is then passed through a dispersion optics (grating or prism), which splits the light according to optical wavelengths that travel different directions. Beam separation and direction heavily depend on its wavelength, as well as the dispersion optics itself. Photographic film, photosensor arrays, and photomultiplier tubes are often used to record the dispersed light.

The dispersion of the light rays from the sample into separate beams is necessary for individual analysis by spectrometers. This is often accomplished by a diffraction grating, which is an optical component with a periodic structure. The form of the light diffracted by a grating depends on the structure of the elements and the number of elements present. Many diffraction gratings used in optical systems feature finely spaced patterns (Figure 2.36).

Electrical Equipment

Electrical equipment for optics involves a broad collection of instruments that perform functions such as acquiring data, generating synchronization signals, and supplying power. Instruments range from the digital oscilloscope, which helps visualize the digital output, to function generators, which create output in a predefined manner.

FIGURE 2.36
A commercial high-density diffraction grating featuring 1,800 line pairs per millimeter at 840 nm, mounted on an aluminum rotation stage.

Data Acquisition System

Data acquisition (DAQ) describes the process of converting a physical phenomenon to a digital signal that can be analyzed and processed by a PC. Acquiring signals correctly is the most important step after all the optical systems are properly designed and implemented. Data acquisition typically involves a sensor that first converts the physical phenomenon into an electrical signal, then an amplifier or multiple amplifiers that condition the signal, and a final instrument that samples the electrical signal and converts it to a digital signal. In optical experiments, the most commonly used sensor in data acquisition is a photodiode, which converts light into electrical signals such as voltage or current as previously described. A flowchart of this process is shown in Figure 2.37.

Amplifiers

After a photodiode has transduced light into an electrical signal, the resulting electrical signals may be too weak and need to be amplified before further conditioning or digitization. In optics, most conditionings are usually accomplished by three types of amplifiers: DC, AC, and differential. All three of these types of amplifier can be acquired commercially, and the most important parameters that need to be clearly understood when selecting an amplifier are its frequency bandwidth and its gain across the frequency spectrum. Make sure that the input signal falls within the correct frequency range of the desired amplifier.

Most instrument amplifiers are differential amplifiers. A differential amplifier takes two input signals and amplifies the difference between the two inputs. The capability to amplify only the difference can be very beneficial when there is a common value between these two inputs. A differential amplifier can easily be built using a commercial operational amplifier (op-amp). Figure 2.38a shows a schematic of a differential amplifier based on a single op-amp, and more complicated designs of differential amplifier are also available. In this simple schematic, the output voltage is calculated by Equation 2.8, and the gain is controlled by the ratio of the resistors used in the circuit. One thing to keep in mind is that the output voltage cannot go beyond the value of the op-amp power supply. In most

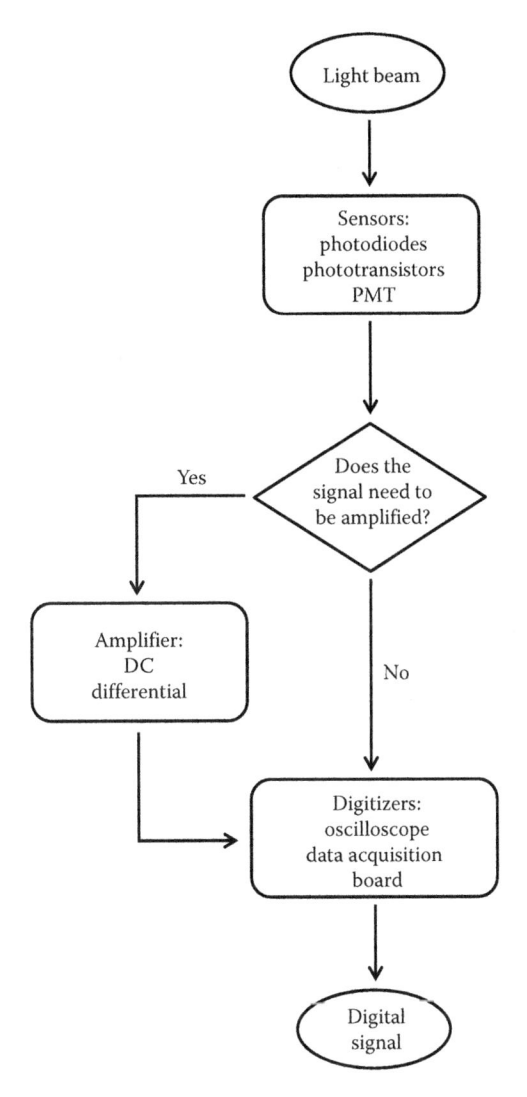

FIGURE 2.37
Flowchart of the data acquisition process.

situations, however, a low-cost commercial amplifier can easily be obtained. Figure 2.38b is a commonly used wideband low-noise amplifier with a gain of ~12 dB. The user can simply connect power to the unit, and it is ready to use.

$$V_{out} = \frac{(R_1 + R_2)R_4}{(R_3 + R_4)R_1} V_{in}^+ - \frac{R_2}{R_1} V_{in}^-; \quad \text{when } \frac{R_2}{R_1} = \frac{R_4}{R_3}, \quad V_{out} = \frac{R_2}{R_1}\left(V_{in}^+ - V_{in}^-\right) \quad (2.8)$$

Digitizer

The final part of a data acquisition process is the digitizer. Its function is to accurately sample the electrical signal and produce the corresponding digital signal for further processing. There are two types of commonly used digitizers: an oscilloscope and a data

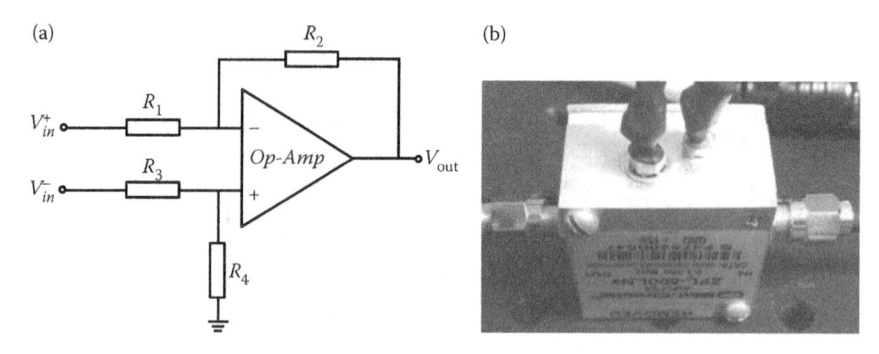

FIGURE 2.38
(a) Schematic of a simple differential amplifier with balanced resistors using a single pp-amp.

acquisition board. An oscilloscope graphs the output signal on a screen for visualization during an experiment. The data acquisition board is a module that must be plugged into a computer, allowing direct data transfer after digitizing.

The oscilloscope can measure varying voltages from multiple signals and display them simultaneously on the screen. The input signals are connected to the oscilloscope with probes, and both signals can be displayed on the same graph for comparison. Several controls can be changed to better visualize the signal. The vertical and horizontal knobs can be used to shift the graph to focus on a particular signal segment. There are also voltage/div and time/div functions to alter the time and the voltage scale of the display. Essentially, using these knobs allows the user to zoom in or zoom out of the signal. A picture of an oscilloscope is shown in Figure 2.39. For more information on how to select and operate modern digital oscilloscopes, readers can visit http://www1.tek.com/Measurement/programs/301913X312631/ for an excellent introduction.

A data acquisition board is a PC board that must be directly plugged into a computer to be used. The sampled data would then be directly saved onto the computer for postprocessing. Figure 2.40 shows a picture of a typical data acquisition board. When deciding on whether to choose a digital oscilloscope or an acquisition board, here are some criteria: First, if fast acquisition and data storage are required, users should choose an acquisition board because data transfer between a digital oscilloscope and PC is usually very slow. If

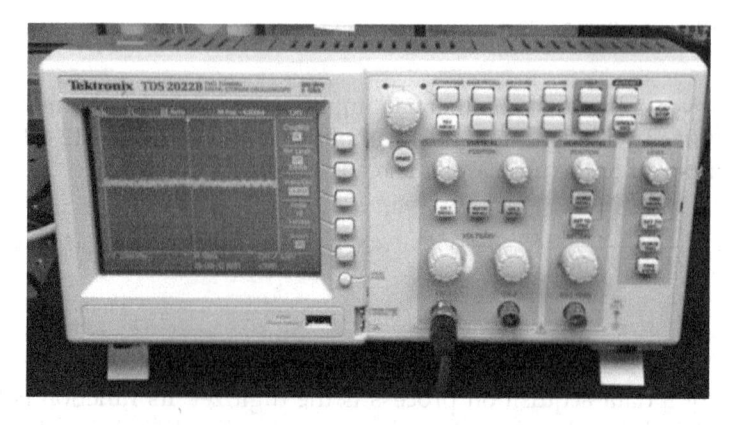

FIGURE 2.39
Photograph of a Tektronix 200 MHz digital oscilloscope.

FIGURE 2.40
Picture of a PCI bus 14-bit data acquisition board.

simultaneous signal display with large data length and extensive data manipulation are needed, a digital oscilloscope becomes essential.

When selecting a digitizer, specific parameters are needed for both types to correctly sample the target signals. These parameters are bandwidth, sampling rate, dynamic range, and equivalent noise level. The bandwidth describes the frequency range of the input signal that can be digitized or displayed. To properly capture the target signal, the digitizer must have a bandwidth that is larger than the maximum frequency of the signal. Oftentimes, the price of the digitizer is higher when the bandwidth is large. Maximum sampling rate is another key parameter to be considered. The Nyquist theorem requires that in order to digitally sample an analog signal, the sampling rate must be at least twice the maximum frequency component. In practice, however, it is better to sample at four or five times the maximum frequency to confidently recover the signal. For example, if a light modulation has a frequency of 40 MHz, the digitizer is expected to have a sampling frequency of 150–200 MS/s.

The dynamic range describes the range of amplitude that the digitizer can measure. Dynamic range is also related to the variable gain of the digitizer channel and the resolution of the A/D convertor, which is the number of bits encoded in each digital sample. For example, if an 8-bit digitizer measures a 20-Vpp input range, the smallest detectable voltage is 78 mV. However, using a 16-bit digitizer will allow for a smallest detectable voltage of 0.305 mV.

The final parameter to consider is the equivalent noise level of a digitizer. This describes the amount of internal noise the machine generates, measured in decibels. Thus, if a digitizer has 10 dB of equivalent noise level, that means the equipment itself introduces 10 dB worth of noise. Of course, always try to minimize the equivalent noise level.

Signal Generation

While many electrical instruments have individual adapters that allow them to use the traditional wall power outlet, other electronics such as circuit boards and amplifiers require specific and often unique input voltages to function properly. In these cases, instruments such as a power supply or a function generator can solve the problem.

Power Supply

A power supply is a device that outputs a specific DC voltage and can be connected to a board through banana connectors. Usually, a power supply can provide a range of voltages, such as ±25 V, which can be continuously adjusted. The instrument also provides a ground voltage that is essential for many circuits. Figure 2.41 shows a photo of a simple DC power supply. The ground is supplied by the black banana plug, while the voltage is

FIGURE 2.41
Photo of a typical DC power supply with banana output plug.

supplied by the red banana plug. When selecting an appropriate power supply, the maximum output current should be taken into consideration. For example, a power supply with 25 V with 2 A can provide power up to 50 W, but if an electrical device such as a servo motor needs 4 A, this power supply will not function properly.

Function Generator

Many electrical instruments need specific inputs to operate, and a function generator serves this purpose well. A function generator can provide predefined waveforms at a given frequency and amplitude defined by users. Example waveforms include sinusoidal waves, square waves, sawtooth waves, and pulse outputs. Some function generators also provide a DC offset voltage and the capability to define arbitrary waveforms. A photo of a function generator is shown in Figure 2.42. Recently, PC-based function generators have become increasingly popular because they can provide multiple output signals, for example, 16, simultaneously, and each output channel can be individually programmed. Such PC-based function generators are often referred to as analog output boards and are commercially available from several manufacturers.

FIGURE 2.42
Photograph of a typical function generator.

Electrical and Light Safety

Electrical and optical hazards are commonly found in photonics laboratories and represent major safety concerns. Existing rules that regulate behavior in your photonics laboratory should be followed in conjunction with these additional precautions that apply to relevant hazards.

Electrical Safety

Electrical current exposes workers to a serious, widespread occupational hazard. The Occupational Safety & Health Administration (OSHA) outlines a regulatory agenda that lists standards designed to protect individuals exposed to dangers such as electric shock, electrocution, fires, and explosions (http://www.osha.gov/SLTC/electrical/index.html).

Electricity requires a complete circuit to continuously flow. As an electric current is conducted through any material, heat is generated proportional to the voltage present and the resistance presented by the material. Tissues in the human body exhibit resistance levels that are dramatically high, providing the possibility of significant and permanent damage to internal organs and other sensitive areas.

A safe workplace is vital for working with electronics and can minimize life-threatening hazards. Workbenches should not feature exposed metal and should be sturdy enough to prevent movement if accidentally bumped. In a typical photonics lab, the various electrical measurement equipment presents minimal risks for shock and electrocution. However, its placement on optical workbenches and connection to nearby power sources should be considered to minimize cluttered workspaces and possible accidents.

Light Safety

Light safety is a major concern in a photonics laboratory. Light sources can come in a variety of designs ranging from incoherent sources, such as light bulbs, to coherent sources, such as various lasers. An incoherent source is light and relatively safe to use, but in some cases, much like looking directly into the sun, can still be dangerous despite the fact that it is incoherent. Lasers emit coherent light and therefore have a higher intensity, resulting in more damage per area. This is a safety concern because the retina in our eye is made of photosensitive tissue that when damaged does not repair itself. The cornea and the lens of the eye work as an optical lens and focus the light onto the retina. When a strong coherent beam passes through the lens and the cornea, it will induce permanent damage to the photoreceptors and the supporting pigmented cells.

Lasers may also emit radiation out of the visible range, resulting in hazards that we cannot directly sense. Laser wavelengths shorter than the lower limit of the visible range (<400 mm) fall within the ultraviolet (UV) radiation range. Exposure to UV radiation causes skin inflammation and possibly even skin cancer and permanent eye damage. Radiation wavelengths longer than the upper limit of the visible range (>700 nm) are referred to as infrared radiation and are mostly perceptible as heat. While infrared radiation contains less energy than visible or UV radiation, prolonged exposure to the eyes may cause an irreversible opacity of the lens. When the light intensity is high, permanent damage to the retina can still occur.

When working with lasers, never, ever look directly into the operating end or aperture of a laser or optical fiber. Protective eyewear, typically made of filter glass, can attenuate radiation in the dangerous wavelengths. Always wear appropriate laser goggles when operating a laser. When selecting laser goggles, users should read the optical density (OD) number for each optical spectral range marked on the goggle. To be safe, the OD number should be higher than 6 for the operating optical wavelength. Always ensure that laser beams are terminated by a suitable nonreflective surface. Be careful to prevent accidental reflections and refractions that may send the laser beam in unintended directions. Keep metallic objects (such as watches, rings, and pens) out of a laser beam. Whenever possible, arrange the experiment so that you are not working with lasers at eye height. Unattended lasers should always be switched off. For more information on laser safety, readers are encouraged to read more on the ANSI laser safety standard from the Laser Institute of America (http://www.lia.org/store/ANSI%20Z136%20Standards).

3

Fundamental Light–Tissue Interactions: Light Scattering and Absorption

Principles of Light Scattering and Absorption

In Chapter 2, we discussed the organization of biological tissue. All the intracellular and extracellular structures interact in different ways with light due to their different chemical compositions and molecular structures. Indeed, there is a myriad of possible interactions. Almost any light–matter interaction described in a physical optics textbook can be observed in tissue. Some of these interactions, however, are key players in tissue and can be easily observed even without specialized equipment, while others are extremely difficult to detect.

First, light may be *elastically scattered* by a tissue structure. This happens if a structure has a different refractive index from its surroundings. The term "elastic scattering" refers to the type of interaction in which the wavelength of light is not changed in the process of scattering. We are all familiar with this type of scattering and observe it in everyday life: It explains the blue color of the sky, the white color of clouds, the yellow sun in the middle of the day, and the red sun at sunset. Generally speaking, in tissue, elastic scattering is a phenomenon that is related to the structure of the tissue and to a lesser degree its molecular composition.

Light *absorption*, which leads to a "loss" of a photon and the transfer of its energy into other forms such as fluorescence or thermal fluctuations, is another key light–tissue interaction process. We also see it in everyday life; for example, it explains why the color of water changes with depth. Contrary to elastic scattering, absorption is a function of chemical composition of tissue and not so much of its structure (although we will see in Chapter 6 that the spatial organization of absorbers does play a part in how much light gets absorbed). Although most tissue molecules have light-absorbing properties, most of the molecules have a negligible absorption cross-section; thus, their light-absorbing properties do play a significant part in light–tissue interactions. A handful of molecules, however, absorb light efficiently—typically within a specific bandwidth. The single most prominent absorber in the visible part of the spectrum is hemoglobin contained in red blood cells or erythrocytes. Hemoglobin absorbs prominently in the blue and green parts of the spectrum, thus giving perfused tissue its pinkish color. (What is your skin color when you blush?) This property of hemoglobin makes it easy to detect by means of visible light spectroscopy. Other absorbers in the visible spectrum include bilirubin and beta-keratin.

In the near-infrared part of the spectrum, however, hemoglobin no longer has the dominant position in terms of light absorption, and other molecules may have comparable absorption, such as lipids. This is the principle behind near-infrared (NIR) tissue spectroscopy and NIR optical tomography that can measure or image concentrations of substances such as hemoglobin and lipids in tissue.

Other interactions include *fluorescence*, a variety of *inelastic scattering* phenomena (i.e., a scattering event in which the wavelength of light changes as a result of the scattering) such as Raman scattering, and dynamic inelastic scattering (e.g., Brillouin scattering). As a general rule, these processes are also chemically specific, and some of them—particularly Raman scattering and its modalities—are much more chemically specific than absorption.

Elastic scattering and absorption are the two most dominant types of light–tissue interaction and play the most important role in light propagation in tissue. In this chapter, we provide an overview of the theory of light scattering and absorption. These key processes are the underlying physical mechanism behind light transport, which is discussed in Chapter 6. Throughout this chapter, we refer to elastic light scattering simply as either "scattering" or "light scattering."

Refractive Index of Biological Tissue

Light scattering affects light propagation in tissue, including the spectroscopic, polarization, or angular features of scattered light emerging from tissue. When light interacts with tissue (or any other turbid medium), scattering is inevitable. What causes light to be scattered? Any spatial heterogeneity of the optical refractive index would lead to light scattering. Refractive index heterogeneities with sizes on the order of magnitude of the wavelength of light—a broad range of sizes from a few tens of nanometers to microns—are the most efficient scatterers, but this notion will be further qualified and quantified later in this chapter. In turn, the refractive index of tissue at a particular location depends on the concentration and, as we will see later, to a lesser extent, the chemical composition of molecules at the particular location. Tissue structures such as subcellular organelles and collagen fiber in the extracellular matrix may have a refractive index that is different from that of their surroundings, thus giving rise to a spatially heterogeneous distribution of the refractive index. However, it is not only the easily recognizable structures such as organelles that have a different refractive index. As we saw in Chapter 2, almost any organelles are not homogeneous themselves and instead have a complex internal structure. For example, although we can assign an average refractive index to a cell nucleus, it is generally a heterogeneous structure with chromatin density varying through the nucleus. It is tempting to think of mitochondria as micron-sized football-shaped structures with a higher refractive index than the cytosol, but this, again, would be an approximation, as any mitochondrion has a complex structure with inner and outer membranes, cristae, etc. Therefore, our first and foremost step is to understand what determines a local refractive index.

Tissue structure and refractive index are interlinked. Consider several liquid media with different refractive indices. The Gladstone–Dale equation allows one to calculate the refractive index of the mixture, n

$$n = 1 + \rho \sum_i \alpha_i m_i,$$

where ρ is the density (typically in g/cm^3) of the mixed medium, the liquid components are mixed in mass fractions m_i, and α_i are refractive index increments of the pure components (sometimes referred to as molar refractivity) and, in our chosen units, in cm^3/g. This relationship stems from a physical principle that it is the electron density of molecules within a medium that determines its refractive index. The Gladstone–Dale equation was formulated in 1864 and is used for many practical applications today. For example, the Brix refractometer determines the alcohol content of a water-alcohol mixture by means of measurement of the index of refraction of the mixture.

Now consider an elementary (and negligibly small) volume in a live biological cell or, more generally, biological tissue. The media that are "mixed" are water and several chemically distinct solid components, such as different types of proteins, lipids, nucleic acids, etc. Following the above equation, we arrive at a simple relationship expressing a local optical refractive index n through the local molecular density of cell solids

$$n = n_0 + \sum_i \alpha_i \rho_i \tag{3.1}$$

where n_0 is a refractive index of a solvent medium (i.e., water, $n_0 = 1.333$ for wavelength 589 nm), ρ_i denotes the weight/volume concentration of the ith solid component, and a proportionality coefficient α_i is the refractive index increment for this particular component. ρ_i essentially quantifies the portion of the local mass that is due to tissue solids such as proteins, DNA, RNA, lipids, etc. Naturally, n_0 and α_i are the functions of the wavelength of light (see discussion that follows).

If ρ_i is measured in the convenient units of g/cm^3, for most biological molecules, $\alpha \sim 0.17$–0.2 cm^3/g. Unfortunately, different studies have reported slightly different values for the refractive index increment even for the same types of molecules—this might be in part because it is experimentally difficult to measure α with high precision given three-dimensional conformation of proteins and the fact that experimental conditions might have varied among the studies. Recent studies suggest that a refractive index increment is a unique characteristic of a protein with changes in protein structure such as conformational changes producing a small but measurable change in its refractive index increment not associated with a change in volume. However, these changes are quite small and below significance for most practical applications.

What is actually remarkable is not the uncertainty associated with the range of α values but the fact that this range is fairly narrow for quite a wide variety of biological molecules! Indeed, while protein species certainly have different three-dimensional structures, the differences are much greater if we compare proteins with nucleic acids and lipids. Still, it is generally accepted that it is reasonable to assume the same value of $\alpha \sim 0.182$ cm^3/g for essentially any biological macromolecular species. Thus, Equation 3.1 reduces to a convenient and simple expression

$$n = n_0 + \alpha \rho, \tag{3.2}$$

where ρ is the weight/volume concentration of the total local solid component (macromolecules). Perhaps the best explanation for the similarity among different refractive increments is that, at the molecular level, tissue is composed of a limited number of basic molecular species: mostly proteins, lipids, DNA, and RNA. Despite their dramatic chemical and biological differences, all these macromolecules are composed of optically similar carbon-based chemical units. Ensemble average over any relevant volume, even as small as a few tens of nanometers (refractive index variations within even smaller volumes are not likely to change tissue optical properties) further homogenizes the optical behavior of biological structures.

Note that ρ is a weight/volume concentration and not a volume portion of solid components. The difference is that while 1 g of water fills $\sim$1 cm^3, 1 g of protein corresponds to $\sim$0.75 cm^3. If expressed in terms of partial volume occupied by solids, ρ_V, Equation 3.2 becomes $n = n_0 + \alpha_V\rho_V$ with $\alpha_V\sim$0.24.

For a volume completely filled with protein, this equation implies that the pure protein would have refractive index $\sim$1.577. Although this value may appear too high—it is higher than the refractive index of certain types of glasses—it is actually lower than what was observed in experiments with protein crystals, which reported a refractive index close to 1.61. The reason for the discrepancy is not well understood, but it is most likely related to the fact that pure-protein concentration of 1.33 g/cm^3 is an underestimate. Indeed, even pure protein in its natural three-dimensional confirmation does not fill the space completely, which means that some space is filled with a surrounding medium such as water or air. Proteins have been shown to have mass-fractal organization. A mass fractal is an object such that the portion of its mass contained within a sphere of radius R increases with R as $M(R)\propto R^D$, with D referred to as the mass fractal or simply the fractal dimension. D depends on the space-filling properties of the object. Typically, the higher the portion of space filled by the object's mass, the higher D is. For an object that completely occupies a three-dimensional volume, $M(R)\propto R^3$ and thus $D = 3$. Fractal dimension below 3 would imply that at least some part of the space is not filled. Fractal dimensions of proteins range from approximately 2.3 to 2.7, with the average $D\sim$2.5 and proteins composed of a greater number of amino acids having higher D values. This implies that about 20%–30% of protein is actually empty. (Of course, these "voids" are tiny: Sizes of typical proteins are between 1 and 10 nm, and the voids are on the order of few angstroms). Thus, if the protein were completely crystallized with no voids at all, its concentration would be $\sim$1.78 g/cm^3, which would drive the refractive index of pure protein up to $\sim$1.64. It is, however, rather impossible to synthesize proteins that would fill the space completely, and we thus conclude that the range of refractive index values that is defined by Equation 3.2 is quite reasonable, and experimentally observed values do fall within this predicted range.

Experiments have shown that the linearity and the values of the increment hold until $\rho\sim$0.4–0.6 g/cm^3. For higher ρ, α is a decreasing function of ρ, consistent with similar behavior for many solutions of higher-refractive index molecules.

What does Equation 3.1 tell us about the refractive index of biological tissue? Denser tissue structures such as collagen fibers and cell membranes are expected to have a higher refractive index. Lipid membranes (e.g., external and internal cell membranes), have $\rho_V \approx$ 30%–35% and thus $n\sim$1.42. Cytosol (without organelles) has $\rho_V\sim$10%, which translates into $n\sim$1.36. In a cell organelle, on the other hand, ρ_V might be as high as $\sim$30% and $n\sim$1.41. Some of the denser small organelles are believed to be mitochondria and lysosomes, which may reach this value of refractive index. Some cell compartments such as the nucleolus and heterochromatin probably have even higher ρ_V, with as much as 50% of their volume occupied by macromolecules, with n approaching $\sim$1.45. Of course, all these values are approximate because we still do not know with certainty the mass density of most intracellular and extracellular structures. However, these estimates are close to the experimental values of the refractive index that have been obtained by use of techniques such as phase microscopy, low-coherence interference, and differential interference contrast microscopy.

Until now, we treated refractive index as a real number. Generally, it may have both real and imaginary parts

$$n = n' + in''.$$

While the real part of the refractive index of a tissue structure defines its scattering properties, the imaginary part is responsible for light absorption. Refractive indices of most epithelial cells and connection tissues have a negligible imaginary part in the visible part of the spectrum. On the other hand, proteins, DNA, and RNA absorb very strongly in the ultraviolet range. Hemoglobin molecules have significant absorption in the visible spectrum.

Lastly, we turn our attention to the wavelength dependence of the refractive index. Due to the effect of dispersion, the refractive index of water as well as most nonabsorbing substances decreases with wavelength—another reason is light absorption by most biological molecules in the ultraviolet and infrared parts of the spectrum, which leads to a decrease of the refractive index as a function wavelength between these spectral regions due to the *Kramers–Kronig relation*. The refractive index of water decreases from 1.345 at wavelength 400 nm to 1.331 at 700 nm.

The refractive index increments of biological tissue solids decrease with wavelength in a similar way. The normal dispersion is described by the Sellmeier equation

$$n = P + \frac{Q}{\lambda^2} + \frac{R}{\lambda^4},$$

where P, Q, and R are constants. For most biological molecules, the fourth-order term can be neglected. The values of constants P and Q for pure biologically relevant substances are provided in Table 3.1. These values should be considered approximate because accurate determination of the refractive index for pure tissue constituents is difficult for the reasons discussed above.

As detailed in Chapter 4, the real and imaginary parts of the refractive index are not independent and are related by the Kramers–Kronig relationship. Typically, around the absorption band, n' decreases from its "baseline" value (i.e., that is far from the absorption band) for wavelengths shorter than that of the absorption band and then increases for longer wavelengths, eventually reaching a maximum and then approaching its baseline value far from the band.

TABLE 3.1

Sellmeier Constants for Pure Biological Constituents

Constituent	P	Q (nm²)
Protein	1.578	7530
Carbohydrate	1.517	4410
Fat	1.457	4530
Water	1.324	3170

Basics of Theory of Light Scattering and Absorption

In this section, we review some of the main principles of light scattering as relevant to light transport in biological tissue. This review is by no means extensive. For a more complete description of the subject, the reader is referred to other excellent texts for the theory of light scattering. Written in the 1960s, *Light Scattering by Small Particles* by H. C. van de Hulst is still a relevant text focused on the theory of light scattering by isolated particles. C. F. Bohren and D. R. Huffman's *Absorption and Scattering of Light by Small Particles* is another comprehensive text on the same topic. Although focused on light transport, another classical text by A. Ishimaru, *Wave Propagation and Scattering in Random Media*, has several excellent chapters on the theory of light scattering and its approximation, including the treatment of isolated particle scattering as well as scattering in media with continuous refractive index fluctuations (as we will see, the latter case is more relevant to what happens in biological tissue). *Principles of Optics* by Born and Wolfe provides an excellent and comprehensive overview of physical optics with a chapter on light scattering with an emphasis on the integral representation of scattering problems and the Born approximation. Finally, mathematically and physically inclined students will enjoy *Scattering Theory of Waves and Particles* by R. G. Newton. It is one of the most comprehensive monographs treating scattering problems in both classical and quantum physics contexts.

Consider a light wave with unit amplitude propagating in direction $\mathbf{s}_0$ that encounters a region of spatial variations of the refractive index—a region where the refractive index differs from that of the surrounding space. For simplicity and for the time being, we will refer to this region as a "scattering particle." Note that this definition of scattering particle does not imply that the particle is internally homogeneous or even has well-defined boundaries. Later in the chapter, we will see how this definition of a scattering particle is actually associated with what in scattering theory called the scattering potential. In the far field, the process of scattering generates a spherical wave $\mathbf{E}_s(\mathbf{r})$ that, for any far-field point $\mathbf{r}$, propagates in direction $\mathbf{s} = \mathbf{r}/r$, where $r = |\mathbf{r}|$. Note that $\mathbf{E}_s$ is a complex vector. Its magnitude describes the strength of the field (and is directly related to light intensity, as discussed below), its direction describes the polarization of the scattered wave, and its phase (thus a complex quantity) is due to a shift in phase (e.g., delay in time) that a scattered field may undergo. This can be written as

$$\mathbf{E}_s(\mathbf{r}) = \mathbf{f}(\mathbf{s},\mathbf{s}_0)\frac{e^{ikr}}{r}, \tag{3.3}$$

where $\mathbf{f}(\mathbf{s},\mathbf{s}_0)$ is referred to as the *scattering amplitude*. Note that despite its name, $\mathbf{f}(\mathbf{s},\mathbf{s}_0)$ is not exactly an *amplitude*: It is, after all, generally a complex vector. However, this name is commonly accepted. Although in this section we focus on elastic processes, $\mathbf{f}(\mathbf{s},\mathbf{s}_0)$ describes not only elastic scattering but also absorption.

Scattering amplitude is one of the most fundamental characteristics and the basis of light-scattering theory. $\mathbf{f}(\mathbf{s},\mathbf{s}_0)$ depends on the spatial arrangement of the refractive index in the region of space that scatters light or, phrased differently, the scattering particle. In other words, $\mathbf{f}(\mathbf{s},\mathbf{s}_0)$ provides the connection between the underlying structure (e.g., local tissue structure) and light scattering. Indeed, consider a particular (microscopically small) part of tissue. This tissue would be composed of various macromolecular components, which gives rise to having locally spatial heterogeneity of density $\rho(\mathbf{r})$. (From a physicist's perspective, $\rho(\mathbf{r})$ is *the* tissue structure.) Since, according to the Gladstone–Dale equation and its corollary Equation 4.2, the tissue refractive index at any point $\mathbf{r}$ is a linear function of the density at this particular point, the density heterogeneity translates into the spatial variations of the refractive index $n(\mathbf{r}) = n_0 + \alpha\rho(\mathbf{r})$. Deviations of $n(\mathbf{r})$ from the mean refractive index $\delta n(\mathbf{r}) = \alpha(\rho(\mathbf{r}) - \rho_0)$, where ρ_0 is the average tissue density, are responsible for light scattering, which, in turn, can be characterized by the scattering amplitude $\mathbf{f}(\mathbf{s},\mathbf{s}_0)$.

There are two consequences. First, if tissue structure (i.e., $\rho(\mathbf{r})$) is known, it is, in principle, possible to calculate $\mathbf{f}(\mathbf{s},\mathbf{s}_0)$. This is a *direct scattering problem*. When $\mathbf{f}(\mathbf{s},\mathbf{s}_0)$ is known for the entire tissue, all aspects of light transport in tissue can be predicted. Although this is possible in principle, solving the direct problem in most realistic applications is practically impossible—the structure of typical tissue is just too complex, with components spanning length scales from nanometers to tens of microns (relevant to scattering) with heterogeneous distribution of its optical properties continuing at longer length scales (relevant to light transport). Thus, to address the problem, one of two approaches (or their combination) is typically taken. One of these strategies is based on implementation of physical approximations that relate $\mathbf{f}(\mathbf{s},\mathbf{s}_0)$ to $\rho(\mathbf{r})$ for each scattering event, followed by another set of approximations that describe light transport as governed by individual scattering events. Some of these approximations are discussed in this section and in Chapter 4. Another approach is to use comprehensive and accurate computational simulations that model light scattering and transport. For example, the finite-difference time-domain (FDTD) method can accurately model light scattering and propagation in a tissue with essentially an arbitrarily complex structure from the first principles, that is, Maxwell equations. Other computational methods vary in their complexity and accuracy, allowing users to choose a compromise between the level of detail captured by the simulations and computational complexity. Relevant computational tools are discussed in detail in Chapter 6.

While solving the direct scattering problem is difficult, the *inverse scattering problem* is even more complex. It deals with learning about tissue structure (e.g., measurement of $\rho(\mathbf{r})$) based on an experimentally measured $\mathbf{f}(\mathbf{s},\mathbf{s}_0)$. In practice, the complete function $\mathbf{f}(\mathbf{s},\mathbf{s}_0)$ is never known, which further complicates the problem. Interest in these inverse scattering problems stems from the fact that it is comparatively easy to make light-scattering measurements either on excised or living tissue: The method is nondistractive and does not require cumbersome preparation as would be required for, as an example, electron microscopy. In tissue optics, inverse problems are typically solved by the use of physical approximations that relate $\rho(\mathbf{r})$ and particular scattering characteristics that are observed in an experiment. In this section, some of the relevant approximations are discussed.

As one can see from Equation 3.3, the scattering amplitude has a dimension of inverse length—this makes sense, as the intensity of the scattered light I_s is proportional to the square of the field and decreases with distance as $1/r^2$

$$I_s(\mathbf{r}) = \frac{\left|\mathbf{f}(\mathbf{s},\mathbf{s}_0)\right|^2}{r^2}.$$

As a reminder, for simplicity, we assumed that the incident field has unit amplitude. If the intensity of the incident light is I_0, this equation becomes $I_s(\mathbf{r}) = (|\mathbf{f}(\mathbf{s},\mathbf{s}_0)|^2/r^2)I_0$.

An alternative but physically equivalent description of the scattered field is based on the concept of the *scattering matrix*. Before we introduce the scattering matrix, we must introduce a few additional notations. A plane defined by vectors $(\mathbf{s},\mathbf{s}_0)$ is referred to as the *scattering plane*. Both incident and scattered fields can be projected onto the scattering plane, yielding projection components: the components of the incident wave E_{i2} and E_{i1} that are parallel and perpendicular to the plane of scattering, respectively, and the components of the scattered wave E_{s2} and E_{s1} that are parallel and perpendicular to the plane of scattering. For historical reasons, the parallel components are denoted with a subscript 2. A scattering matrix S relates these four components

$$\begin{pmatrix} E_{s2} \\ E_{s1} \end{pmatrix} = \frac{e^{-i(kr-kz)}}{ikr} \begin{pmatrix} S_2 & S_3 \\ S_4 & S_1 \end{pmatrix} \begin{pmatrix} E_{i2} \\ E_{i1} \end{pmatrix}, \tag{3.4}$$

where $\mathbf{s}_0$ is chosen along the $\mathbf{z}$-direction, $r = r(\theta, \varphi)$, θ, and ϕ are polar angles in the spherical system of reference associated with the scattering particle (or, more generally, the region of the refractive index variations); and $S_j(\theta, \phi)$, $j = 1, \ldots, 4$ are complex functions. The angle $\theta = \cos^{-1}(\mathbf{s} \cdot \mathbf{s}_0)$ is called the *scattering angle*. If a particle is cylindrically symmetrical in respect to the direction of propagation of the incident light (i.e., the $\mathbf{z}$-direction in our system of reference), components $S_3 = S_4 = 0$ and S_1 and S_2 are all functions of the scattering angle θ only and do not depend on the azimuthal angle ϕ. If an observer is to pass the scattered light through a set of polarizers to measure the intensities of scattered light that are polarized along and orthogonal to the scattering plane, here defined as $I_{\|s}$ and $I_{\perp s}$, one would find that they are proportional to the respective components of the incident light $I_{\|i}$ and $I_{\perp i}$

$$I_{\|s} = \frac{|S_2(\theta)|^2}{k^2 r^2} I_{\|i}, \quad I_{\perp s} = \frac{|S_1(\theta)|^2}{k^2 r^2} I_{\perp i}. \tag{3.5}$$

The *differential (scattering) cross-section* is defined as

$$\sigma_s(\mathbf{s}, \mathbf{s}_0) = |f(\mathbf{s}, \mathbf{s}_0)|^2.$$

As opposed to the scattering amplitude, the differential cross-section lacks information about a phase shift that the scattered light has undergone in the process of scattering. It is, however, one of the fundamental characteristics of light scattering, and it is the differential cross-section that is measured in many experiments—the intensity of light scattered in a given angle is proportional to the differential cross-section.

The *scattering cross-section* is a very convenient and perhaps the most widely used property characterizing a scattering event. It is the geometrical cross-section of a particle that would produce the amount of scattering equal to the observed scattered power in all directions. It can be expressed through the scattering amplitude or the scattering matrix elements

$$\sigma_s = \int_{(4\pi)} |f(\mathbf{s}, \mathbf{s}_0)|^2 \, d\Omega = \frac{1}{2k^2} \int_0^{2\pi} \int_\pi^0 (|S_1 + S_4|^2 + |S_2 + S_3|^2) \, d\cos\theta \, d\phi. \tag{3.6}$$

Here, $d\Omega = d\cos\theta \, d\phi$ is an element of a solid angle over which the integration is performed.

A related quantity, the *total cross-section*, is the sum of the scattering and absorption cross-sections

$$\sigma_t = \sigma_s + \sigma_a.$$

In the literature on light scattering, the following definitions are frequently used: The ratio of the scattering to total cross-sections (σ_s/σ_t) is called the *albedo* (or the *scattering albedo*). *Scattering efficiency* is defined as $Q = (\sigma_s/G)$, where G is the geometrical cross-section of a scattering particle.

While the scattering cross-section quantifies the total scattering power of a scattering particle, it does not tell us about the angular distribution of the scattered field, which might be drastically different even for the same scattering cross-section. For example, two particles may scatter the same amount of light relative to the incident light (thus having the same scattering cross-section); the first particle may scatter light isotropically, while the scattering pattern of the other particle might be so highly forward-directed (which is typical of scatters that are large compared to the wavelength of light, as discussed below) that experimentally, it would be difficult to conclude that the particle scattered any light at all.

The angular distribution of scattered light can be described by the *differential cross-section* $\sigma_s(\mathbf{s},\mathbf{s_0})$, as discussed above, and the *phase function* $p(\mathbf{s},\mathbf{s_0})$

$$p(\mathbf{s},\mathbf{s_0}) \propto \frac{|\mathbf{f}(\mathbf{s},\mathbf{s_0})|^2}{\sigma_t} \tag{3.7}$$

There are two alternative means to normalize the phase function. According to one notation, the phase function is normalized such that the integral over all scattering angles equals the albedo

$$\int_{(4\pi)} p(\mathbf{s},\mathbf{s_0})d\Omega = \frac{\sigma_s}{\sigma_t} \tag{3.8}$$

In this case, the phase function has the meaning of the probability of scattering into a given direction with the posterior probability that the scattering event has occurred taken as 1. Alternatively, the phase function can be normalized such that the integral equals the total solid angle 4π

$$\int_{(4\pi)} p(\mathbf{s},\mathbf{s_0})d\Omega = 4\pi\frac{\sigma_s}{\sigma_t}.$$

Both notations are common, and when reading a text or paper dealing with light scattering, it is important to pay attention to which definition is used.

For scattering structures found in tissue, a typical phase function is peaked in the forward direction. The width of the peak is inversely related to its size. The phase function decreases for larger (side) angles and may (and typically does in tissue) slightly increase for angles approaching the backward direction (i.e., ~180° scattering). Thus, as a general rule, larger scattering particles (relative to the wavelength of light) have more forward-peaked phase function, while smaller scatterers scatter more isotropically.

In the absence of absorption, the total cross-section σ_t can be calculated by integrating the differential cross-section over a solid angle. Alternatively, the *forward scattering theorem* can be used

$$\sigma_t = \frac{4\pi}{k} \text{Im}\{\mathbf{f}(\mathbf{s}_0,\mathbf{s}_0)\}\mathbf{e}^{(i)}, \tag{3.9}$$

where unit vector $\mathbf{e}^{(i)}$ is the direction of polarization of the incident field $\mathbf{E}^{(i)}$ and the symbol $\text{Im}\{\,\}$ stands for an imaginary part. If the incident wave is unpolarized (e.g., a scalar wave), $\sigma_t = (4\pi/k)\text{Im}\{f(\mathbf{s}_0, \mathbf{s}_0)\}$, where f is the scalar scattering amplitude. This is a remarkable and *a priori* not intuitive result. The forward scattering theorem will be used later in the section to derive the total scattering cross-section of large particles.

Other useful theorems of scattering are the *unitary relation* for nonabsorbing particles

$$f(\mathbf{s}_2,\mathbf{s}_1) - f^*(\mathbf{s}_1,\mathbf{s}_2) = i\frac{k}{2\pi} \int_{(4\pi)} f(\mathbf{s},\mathbf{s}_1)f^*(\mathbf{s},\mathbf{s}_2)d\Omega,$$

and the *reciprocity relation*

$$f(\mathbf{s}_2,\mathbf{s}_1) = f(-\mathbf{s}_1,-\mathbf{s}_2).$$

In tissue optics, light propagation is conventionally parameterized by four key parameters: the *scattering coefficient* μ_s, the *absorption coefficient* μ_a, the *anisotropy coefficient* (also referred to as *anisotropy factor*) g, and the *reduced scattering coefficient* μ_s'. If one were to count how frequently different scattering characteristics are used in tissue optics literature, these four would certainly make the top of the list. The reduced scattering coefficient and the scattering coefficient are used to define how strongly scattering a tissue is. The absorption coefficient does the same in terms of the absorption properties of tissue, and the anisotropy coefficient parameterizes, via a single parameter, the shape of the phase function.

We start with defining the scattering coefficient. If a scattering medium is particulate (i.e., composed of isolated particles) with the scattering particles located in each other's far field

$$\mu_s = \sigma_s \rho_N, \tag{3.10}$$

where $\rho_N = (N/V)$ is the number density of scattering particles (N particles per volume V), and σ_s is the scattering cross-section of each particle. When a scattering medium is not composed of isolated scattering particles but instead has a continuously spatially fluctuating refractive index (this description is more appropriate for biological tissue), the following definition of μ_s can be used:

$$\mu_s = \frac{\sigma_{s,\delta V}}{\delta V}, \tag{3.11}$$

where $\sigma_{s,\delta V}$ is the scattering cross-section of a volume δV. Thus, the scattering coefficient is the scattering cross-section per unit volume. It is easy to see that these two definitions converge to exactly the same result in the case of a particulate medium.

The scattering mean free path length is defined as the inverse of the scattering coefficient

$$l_s = \frac{1}{\mu_s}.$$

Conceptually, l_s is the mean distance between two scattering events. In tissue, l_s may range from a few tens of microns (in highly scattering tissues such as connective tissue) to a few hundred microns (in weaker-scattering tissues such as certain types of epithelia).

The absorption coefficient μ_a is defined through the absorption cross-section of individual absorbing particles σ_a for a particulate medium, or as the absorption cross-section per unit volume when absorbing molecules are distributed throughout the medium and it is not convenient to talk about isolated absorbing particles

$$\mu_a = \sigma_a \rho_N = \frac{\sigma_{a,\delta V}}{\delta V}.$$

Accordingly, absorption mean free path length is defined as

$$l_a = \frac{1}{\mu_a}.$$

It is interesting to note that while the spatial variations of the real part of the refractive index, which leads to elastic scattering, are typically not concentrated within well-defined particles, and it is more appropriate to talk about a continuously fluctuating refractive index, in the case of absorption, the situation is typically the opposite. Most of the tissue molecules are relatively weak absorbers in the visible part of the spectrum, and the imaginary part of their refractive index can be neglected. Examples include proteins, lipids, DNA, and RNA. There are certainly indigenous molecules, particularly small molecules such as NAD and NADH, that have a nontrivial absorption cross-section, but they are not present in large concentrations. (These molecules are indigenous fluorophores.) Weak absorption applies to most epithelial cells and connective tissues. On the other hand, light absorption in tissue is primarily concentrated in particular cells, with red blood cells being the most prominent absorbers. These biconcave disk-like cells are about 8 microns in size and filled with hemoglobin, which is responsible for absorption. Thus, in most practical circumstances, it is appropriate to consider absorption as being concentrated in isolated particles.

As a function of wavelength, hemoglobin absorption has a complicated shape with multiple absorption peaks (bands) throughout the spectrum. As shown in Figure 3.1, absorption profiles of oxygenated and deoxygenated hemoglobin molecules are slightly different. The epignomonic absorption spectra of hemoglobin are widely used in tissue optics to measure total hemoglobin concentration and oxygenation, which is difficult to accomplish by the use of nonoptics techniques. The y-axis of Figure 3.1 shows values for the molar extinction coefficient E (in units of $cm^{-1}/(mole/liter)$; the original data were compiled by Scott Prahl and are publicly available at the Oregon Medical Laser Center website (http://omlc.ogi.edu/)). To convert E to μ_a

$$\mu_a = 2.303 \frac{C_{Hb}}{W_{Hb}} E,$$

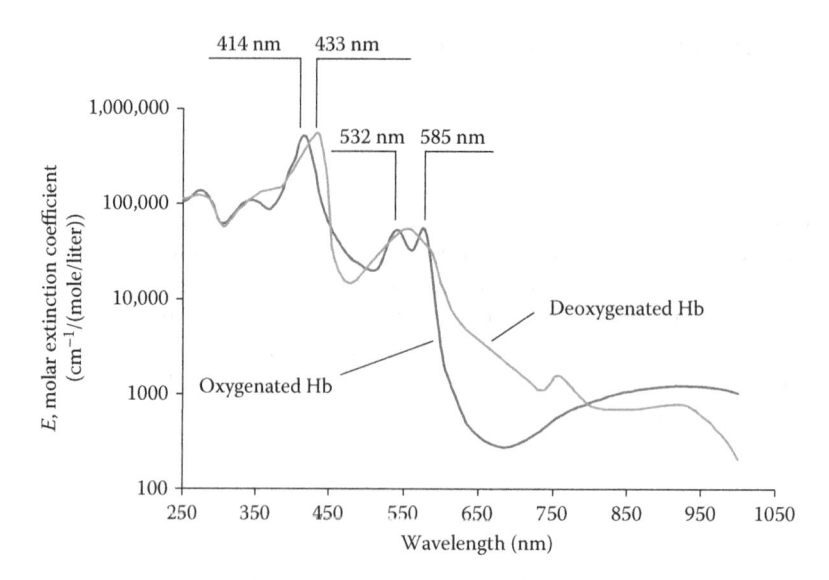

FIGURE 3.1
Molar extinction of hemoglobin as a function of wavelength.

where C_{Hb} is the number of grams of hemoglobin per liter (in whole blood, a typical value is 150 g Hb/liter) and W_{Hb} is the molecular weight of hemoglobin (64,500 g Hb/mole).

Coefficients μ_s and μ_a have units of inverse length. If a collimated beam of light (or a plane wave) of initial intensity I_0 travels through a slab of tissue of thickness L, the intensity of the transmitted collimated light is exponentially attenuated

$$\frac{I}{I_0} = e^{-(\mu_s + \mu_a)L}. \tag{3.12}$$

What this equation tells us is that not only can light (collimated transmission) be attenuated due to absorption but also due to scattering. This phenomenon is referred to as *attenuation by scattering*.

Equation 3.12 holds for distances L such that the tissue is optically thin; that is $(\mu_s + \mu_a)L < 1$. For longer distances, multiple scattering becomes significant, and attenuation as a function of tissue thickness becomes muted. Note that exponential attenuation is relevant only for the transmitted *collimated* light. As a result of multiple scattering, some of the light first scattered out of the collimated beam would be redirected and may end up emerging from the tissue propagating in the direction of the incident light. This multiple scattered component, however, is not accounted for in the above equation.

The anisotropy coefficient is the average cosine of the phase function

$$g = \frac{\displaystyle\int_{(4\pi)} p(\mathbf{s}, \mathbf{s}_0)(\mathbf{s} \cdot \mathbf{s}_0)d\Omega}{\displaystyle\int_{(4\pi)} p(\mathbf{s}, \mathbf{s}_0)d\Omega} \tag{3.13}$$

g approaches 1 for highly forward scattering patterns and is exactly zero if the scattering is completely isotropic. It most tissues, g ranges from 0.8 to 0.95 but occasionally may approach even higher values, up to 0.99. High anisotropy, for example, exists in scattering by red blood cells due to their large size—most of the light is scattered almost without deviation from its incident propagation.

The parameters l_s and μ_s tell only a part of the whole story: In a tissue with $g\sim 1$, a large value of μ_s may not necessarily be indicative of how strongly light is deviated from its original direction through scattering. Given that tissues with the same μ_s may have dramatically different scattering-phase functions and thus affect light transport in quite different ways, a useful measure of how scattering actually affects light transport is the reduced scattering coefficient

$$\mu_s' = \mu_s(1-g). \tag{3.14}$$

The corresponding *transport mean free path length* is

$$l_s' = \frac{1}{\mu_s'}.$$

If μ_s is a property of a single scattering event, μ_s' is critical in describing light transport in a multiple-scattering regime. Conceptually, l_s' can be understood as the distance over which light is being randomized in direction due to multiple scattering, and $1/(1-g)$ is approximately the number of scattering events required to randomize the light. In tissue, l_s' may range from a few hundreds of microns to millimeters, with a typical value being around 1 mm.

As a function of wavelength, μ_s' is typically smooth and decreases monotonically with wavelength. Many experiments have reported that μ_s' follows an inverse power-law dependence as a function of wavelength. As discussed in Chapter 4, the power-law dependence is due to a large g in tissue with the exponent of the power law being dependent on the type of the autocorrelation function of the refractive index in a given tissue.

Scattering and Absorption of Light in Tissue by Small Particles

In this section, we discuss how the scattering parameters introduced above are related to refractive index variations in tissue. We start with the definition of the scattering amplitude. The scattering amplitude can be found by solving Maxwell equations, and any scattering problem can be expressed as an integral equation. Accordingly, the scattering amplitude can be expressed through the field that exists inside a scattering region of volume V [1]

$$\mathbf{f}(\mathbf{s},\mathbf{s_0}) = -\frac{k^2}{4\pi}\int_V \mathbf{s}\times(\mathbf{s}\times\mathbf{E}(\mathbf{r}'))\left(n_1^2(\mathbf{r}')-1\right)e^{-ik\mathbf{sr}'}d\mathbf{r}' \tag{3.15}$$

$$\mathbf{E}_s(\mathbf{r}) = \mathbf{f}(\mathbf{s},\mathbf{s_0})\frac{e^{ikr}}{r},$$

where k is the wave number in the surrounding medium ($k = k_{vacuum} n_0$), $n_1 = n/n_0$ is the relative refractive index within the scattering volume V, and the integration is taken over the volume of the particle. (Without the loss of generality, we can assume that n_0 is the refractive index of water.)

The function

$$F(\mathbf{r}) = k^2 \frac{n_1^2(\mathbf{r}) - 1}{4\pi}$$

is called the scattering potential.

As always, the scattering and total cross-sections can be calculated from the scattering amplitude

$$\sigma_s = \int\limits_{(4\pi)} |f(\mathbf{s}, \mathbf{s}_0)|^2 \, d\Omega$$

$$\sigma_t = \frac{4\pi}{k} \text{Im}\{\mathbf{f}(\mathbf{s}_0, \mathbf{s}_0)\} \mathbf{e}^{(i)}$$

with $\mathbf{e}^{(i)}$ being the unit vector in the direction of polarization of the incident field $\mathbf{E}^{(i)}$.

If the refractive index has both real and imaginary parts, both add to scattering in Equation 3.15

$$n_1^2 = n_1'^2 - n_1''^2 + i2n'n''$$

If the imaginary part of the refractive index n_1'' is not zero, the particle not only scatters but absorbs the incident light, and the absorption cross-section is given by the following relation

$$\sigma_a = 2k \int\limits_V n_1 n_1'' |\mathbf{E}(\mathbf{r}')|^2 \, d\mathbf{r}'.$$

The integral equation has a clear physical interpretation. The function $G \equiv (e^{ikr} e^{-iksr'}/r)$ is a spherical wave that is originated at a point $\mathbf{r}'$ inside the scattering particle and recorded in the far field at a distance r from the particle. This is a wave propagating with no sources and no scattering. The other terms of the equation can be grouped as

$$n^2 = \varepsilon, \quad \frac{n^2 - 1}{4\pi} = \frac{\varepsilon - 1}{4\pi} \equiv \chi, \quad \frac{n^2 - 1}{4\pi} \mathbf{E} = \chi \mathbf{E} \equiv \mathbf{P},$$

where ε is the electrical permittivity, χ is the electrical susceptibility, and $\mathbf{P}$ is the electric polarization. Thus, Equation 3.15 can be represented as

$$\mathbf{E}_s \propto k^2 \int\limits_V \mathbf{P}G d\mathbf{r}'$$

The field that exists inside the particle as a result of the incident field as well as all scattered waves creates oscillating electric polarization vectors **P** at each point in the particle space. In turn, oscillating electric polarization implies oscillating electrical dipoles in the scattering medium, which results in a scattered spherical wave G being emitted from each point inside the particle. The spherical waves from all points in the scattering object interfere in the far field at point **r**, which is accounted for by taking an integral over the volume of the particle.

Equation 3.15, of course, is just that, an integral equation, because the integration requires knowledge of the field inside the scattering volume, and to calculate this field, one would need to solve Maxwell's equations. This is the scattering version of the "catch-22." A solution to this predicament is to approximate the field inside the scattering particle by making certain assumptions based on *a priori* available information. If one is lucky and the approximation is a good one, the integration can then be carried out and the scattering amplitude calculated.

We first consider the case in which scattering particles can be approximated as independent scatters. This approximation is frequently used in tissue optics. Depending on the shape and the internal structure of a scattering particle, Equation 3.15 can be greatly simplified, and, in some cases, its closed-form solution exists.

The two most widely used approximations consider a particle as either a weak perturbation (see the definition that follows) or large relative to the wavelength of light. In the former case, the scattering problem can be solved by applying the first-order *Born approximation*. The Born approximation has been widely used to solve a great variety of scattering problems, from quantum mechanics to atmospheric science and tissue scattering. Under the Born approximation, the field inside a scattering particle is approximated as being the same as the incident field. In this case, Equation 3.15 can be rewritten as

$$\mathbf{f}(\mathbf{s},\mathbf{s}_0) = -\frac{k^2}{4\pi}\int_V \mathbf{s}\times(\mathbf{s}\times\mathbf{E}^{(i)}(\mathbf{r}'))(n_1^2(\mathbf{r}')-1)e^{-ik(\mathbf{s}-\mathbf{s}_0)\mathbf{r}'}d\mathbf{r}' = -\frac{k^2}{4\pi}(\mathbf{s}\times(\mathbf{s}\times\mathbf{e}^{(i)}(\mathbf{r})))VR(\mathbf{s},\mathbf{s}_0), \quad (3.16)$$

where

$$R(\mathbf{s},\mathbf{s}_0) = \frac{1}{V}\int_V \left(n_1^2(\mathbf{r}')-1\right)e^{-ik(\mathbf{s}-\mathbf{s}_0)\mathbf{r}'}d\mathbf{r}'$$

and $\mathbf{e}^{(i)}$ is the unit vector in the direction of polarization of the incident field $\mathbf{E}^{(i)}$ and the vectorial product

$$-\mathbf{s}\times(\mathbf{s}\times\mathbf{e}^{(i)}(\mathbf{r})) = \sin\chi$$

where χ is the angle between the direction of polarization of the incident wave and the direction of propagation of the scattered wave

$$|\sin\chi| = 1 - \sin^2\theta\cos^2\varphi.$$

Thus, the scattering amplitude is a Fourier transform of the *scattering potential,* and since

$$n'^2(\mathbf{r}) - 1 \approx \frac{2\alpha}{n_0}\rho(\mathbf{r}),$$

it is a Fourier transform of the spatial distribution of tissue mass density. As we have discussed earlier, this provides connection between scattering and tissue architecture.

Importantly, the Fourier transform is taken at frequency

$$k_s = k|\mathbf{s} - \mathbf{s}_0| = 2k \sin \frac{\theta}{2},$$

where θ is the scattering angle, $\mathbf{k}_s$ is a *momentum transfer* vector, and the Fourier transform is evaluated at the frequency that is equal to the magnitude of the momentum transfer.

Thus, in principle, the scattering potential, the refractive index profile, and the density profile can be recovered by taking an inverse Fourier transform of scattering amplitude as a function of $\mathbf{k}_s$. This requires knowing function $\mathbf{f}(\mathbf{k}_s) = \mathbf{f}(\mathbf{s},\mathbf{s}_0)$ for a range of $\mathbf{k}_s$. This can be achieved by keeping the wavelength of illumination and collection constant and recording the scattered field for a range of scattering angles. In this case, k_s can, in principle, vary from 0 (forward scattering) to $2k$ (backscattering). Alternatively, one can fix an angle of scattering (e.g., make all observations in the backscattering direction) and do spectroscopic analysis of the scattered field, that is, measure the scattering amplitude as a function of wavelength. In this example, the range of k_s would be restricted by the spectral bandwidth from $2k_1$ to $2k_2$.

In any case, we can say that $\mathbf{f}(\mathbf{k}_s)$ is known within a particular frequency space $\Pi(\mathbf{k}_s)$, and the recovered scattering potential $F'(\mathbf{r})$ would be different from the true scattering potential $F(\mathbf{r})$

$$F'(\mathbf{r}) = F(\mathbf{r}) * P(\mathbf{r}),$$

$$P(\mathbf{r}) = \frac{1}{(2\pi)^3} \int \Pi(\mathbf{k}_s) e^{i\mathbf{k}_s \mathbf{r}} d\mathbf{r}$$

where the asterisk denotes convolution and P is the Fourier transform of the bandwidth function $\Pi(\mathbf{k}_s)$. If the scattered field were known throughout the entire frequency space (i.e., $\Pi(\mathbf{k}_s) = 1$ for all $\mathbf{k}_s$), then $P(\mathbf{r}) = \delta(\mathbf{r})$, and the result of the convolution is $F'(\mathbf{r}) = F(\mathbf{r})$. In any practical application, however, $\Pi(\mathbf{k}_s)$ is bounded, and, as a result of the convolution, $F(\mathbf{r})$ is filtered (moving window averaged) with high spatial frequency details of refractive index fluctuations being averaged out. The narrower $\Pi(\mathbf{k}_s)$ is, the wider the filtering function $P(\mathbf{r})$. If we consider a fixed wavelength and measure $\mathbf{f}$ for all scattering angles, from forward to backscattering, and then repeat the measurements for all directions of illumination, $\Pi(\mathbf{k}_s)$ would be sphere of radius $2k$ and is referred to as the *Ewald sphere*. In this case, the filter P is a Fourier transform of a uniform sphere.

The conclusion, then, is that knowing the scattered field within a finite range of frequencies is equivalent to performing a low-pass filter on spatial variations of $n(\mathbf{r})$. This does not, however, mean that small length scale features in $n(\mathbf{r})$ are completely lost. Instead, the sensitivity to these small length scale features is diminished. This is because the Fourier spectrum of an arbitrarily small length scale feature in $n(\mathbf{r})$ extends to low frequencies. However, the portion of the total power of its Fourier spectrum that is bound within a particular lower-frequency band (e.g., $\Pi(\mathbf{k}_s)$) is diminished.

As an illustration, we consider a one-dimensional example of a perturbation $n_1(x)$ that has a shape of a random superposition of top-hat functions: $n_1(x) = n_1$, if $(x - x_i) < a$ and 1 elsewhere with random positions x_i. The Fourier transform of this perturbation is a sum of sinc functions. Most of the power is concentrated within a band of frequencies around $k_0 \sim 1/a$. If $\Pi(k)$ covers only lower frequencies up to $k_{max} < k_0$, the sensitivity of scattering to this perturbation will be greatly diminished. However, some of the power is spread over even the lower frequencies, and this would affect the scattered field. Thus, there would be a finite amount of scattering proportional to $\sim(n_1 - 1)k_{max}a$ for the field and $\sim(n_1 - 1)^2(k_{max}a)^2$ for the intensity, but recovering the shape of the scattering potential would be impossible. In other words, small length scale perturbations $n(\mathbf{r})$ are *detectable but not resolvable*.

In fact, the Fourier transform of any bounded perturbation would have an infinitely broad frequency spectrum with at least some spectral power content in any $\mathbf{k}$-bandwidth. Consider an example. If a scattering medium is unbound and the scattering potential has a profile of a harmonic wave (e.g., $n_1 - 1 \propto \sin(k_0 x)$ in a one-dimensional case), the Fourier transform is a delta function $\delta(k - k_0)$, and if frequency k_0 is outside the bandwidth $\Pi(k)$, any sensitivity to this perturbation is completely lost. Interestingly, this applies only to a fairly unrealistic case of unbounded scattering "particles." In a more practically important case, a scattering object is bounded within a volume $\Pi_n(\mathbf{r})$ (mathematically, this means that F is a function of finite support) and the Fourier transform of this bounded harmonic wave perturbation equals $P_n(\mathbf{k} - \mathbf{k}_0)$, where $P_n(\mathbf{k}) = \int \Pi_n(\mathbf{r})e^{i\mathbf{k}\mathbf{r}}d\mathbf{r}$. Thus, the spatial frequencies get "spread" over a larger range of spatial frequencies, including a lower $\mathbf{k}$ range, and these frequencies can pass through the low-pass filter $\Pi(\mathbf{k}_s)$; the information about the high frequencies that compose the scattering potential can now be probed even if the observation bandwidth does not include $\mathbf{k}_0$.

While the scattering amplitude is a Fourier transform of the scattering potential, the scattered intensity is a Fourier transform of the autocorrelation function of the refractive index $B_{n_1}(\mathbf{r})$

$$I \propto \int_V B_{n_1}(\mathbf{r}')e^{-ik(\mathbf{s}-\mathbf{s}_0)\mathbf{r}'}d\mathbf{r}',$$

where

$$B_{n_1}(\mathbf{r}) = \frac{1}{V}\int_V n_1(\mathbf{r}')n_1(\mathbf{r}'+\mathbf{r})d\mathbf{r}'.$$

The same discussion that we had earlier in regard to the feasibility of recovering the scattering potential (or the refractive index profile) based on an experimentally measured scattering field applies to the problem of measuring the refractive index autocorrelation function based on the intensity measurements within a particular bandwidth $\Pi(\mathbf{k}_s)$.

The absorption cross-section under the first-order Born approximation is found as

$$\sigma_a = 2k\int_V n_1(\mathbf{r}')n_1''(\mathbf{r}')d\mathbf{r}'.$$

The same solution can also be presented by use of the scattering matrix formalism

$$\begin{pmatrix} S_2 & S_3 \\ S_4 & S_1 \end{pmatrix} = \begin{pmatrix} \cos\theta & 0 \\ 0 & 1 \end{pmatrix}\frac{ik^3}{4\pi}VR(\theta,\varphi), \tag{3.17}$$

where $\mathbf{z}$ is chosen to be parallel to $\mathbf{s}_0$ and $\mathbf{s} = \mathbf{s}(\theta,\varphi)$.

The validity criterion for the Born approximation is that the phase shift gained by the field propagating inside the particle must be small

$$k\left|\int (n_1(z)-1)dz\right| \ll 1,$$

where the integration is performed over the size of the particle. The same criterion can be written as

$$2ka\,|n_1 - 1| \ll 1,$$

where a is a characteristic size of a particle (e.g., its radius), and n is its average refractive index. In other words, the approximation is valid when the perturbation to the field propagating through the particle due to the refractive index variations is negligible. In some cases, however, the Born approximation is valid even when this strict criterion is not satisfied. The error due to the approximation increases as $|k\int(n(z) - 1)dz|^2$; thus, it is the square of the phase shift rather than its absolute value that is more pertinent as an estimate of the accuracy and validity of the approximation. This makes the validity requirement somewhat less strict.

Another consideration is that the first-order Born approximation works better for side scattering and backscattering, while its error is more significant for forward scattering. This is because the approximation neglects the fact that neighboring dipoles generate waves that are in phase to each other and the incident field propagating within a particle. This leads to an incorrect estimation of the phase of the scattered amplitude in the forward direction. This is also apparent from the following example. In Equation 3.16, the magnitude of the scattering amplitude is proportional to the integrated scattering potential: $F(s_0,s_0) \propto 1/V \int_V (n_1^2(\mathbf{r}')-1)\mathrm{d}\mathbf{r}'$. If n_1 is always a real number, the scattering amplitude and the total scattering cross-section $\sigma_t = (4\pi/k)\mathrm{Im}\{f(\mathbf{s}_0,\mathbf{s}_0)\} = 0$. This, of course, does not make sense. On the other hand, σ_t can still be fairly accurately estimated by integrating $\mathbf{f}(\mathbf{s},\mathbf{s}_0)$ over all scattering directions. This limitation of the Born approximation can be corrected either by considering higher-order corrections or by modifying it by considering a traveling wave inside a particle, which is the basis for the *interior wave number approximation,* which is also discussed later in this section.

The physical interpretation of the Born approximation should be clear from that of the original integral, Equation 3.15. The incident field, while propagating inside the particle, creates oscillating electrical dipoles in each and every point inside the particle. Each dipole emits a spherical wave that travels unperturbed to the point of observation. In this approximation, the impact of the scattered field on the dipoles is neglected, which is why the first-order Born approximation is the *single scattering approximation.*

With this physical picture in mind, the first-order Born approximation can be modified to solve a scattering problem when either the illumination is not a plane wave or the collection does not allow for all scattered waves to be recorded

$$R(\mathbf{s},\mathbf{s}_0) = \frac{1}{V}\int_V \left(n_1^2(\mathbf{r}')-1\right)E_{eff}(\mathbf{r}')e^{-ik\mathbf{s}\mathbf{r}'}d\mathbf{r}'.$$

In the case of nonplane wave (e.g., focused) illumination, E_{eff} is the incident field. In the case of a collection geometry where the scattered field is recorded only from a portion of a scattering particle (e.g., only light scattered from a microscopic focal spot is being recorded), E_{eff} is the field that would be created inside the particle if we propagated the collected light back into the particle. An alternative but mathematically equivalent formulation would involve applying a first-Born approximation for a plane-wave illumination and then integrating all relevant illumination and collection angles.

In a special case of a plane-wave illumination and a nonplane wave collection, R becomes

$$R(\mathbf{s}, \mathbf{s}_0) = \frac{1}{V} \int_V \left(n_1^2(\mathbf{r}') - 1 \right) C(\mathbf{r}') e^{-ik(\mathbf{s}-\mathbf{s}_0)\mathbf{r}'} d\mathbf{r}',$$

where the function $C(\mathbf{r}')$ accounts for the strength of the dipoles that are excited by the incident field at point $\mathbf{r}'$ and emit scattered waves that are recorded. The intensity, in this case, is

$$I \propto \int_V B_{n_1}(\mathbf{r}') B_c(\mathbf{r}') e^{-ik(\mathbf{s}-\mathbf{s}_0)\mathbf{r}'} d\mathbf{r}',$$

where B_c is the autocorrelation function of C: $B_c(\mathbf{r}) = (1/V) \int_V C(\mathbf{r}') C(\mathbf{r}' + \mathbf{r}) d\mathbf{r}'$.

For example, in the case of a plane wave illumination and a collection by means of an objective lens (e.g., a microscope objective), $C(\mathbf{r}')$ accounts for a focal volume—although all dipoles in the scattering particle generate scattering waves, not all these waves contribute to the recorded field, and it is only waves that are generated by the dipoles within the focal volume that matter. This approach allows modeling of the scattering signal that is recorded in bright-field, confocal, and other microscopy modalities.

Consider an illustrative example of how this modified-Born approximation can be applied to model a signal detected in a focused-beam collection, as in microscopy. If the depth of focus is longer than the thickness of a cell, and we neglect a phase shift due to an angular distribution of light (e.g., small numerical aperture approximation), it is convenient to go to a cylindrical coordinate system (z, ρ, φ), where $C(\mathbf{r})$ depends only on the coordinates in the transverse plane, that is, the transverse radial coordinate ρ. Thus, the backscattering amplitude is given by

$$|\mathbf{f}(-\mathbf{s}_0, \mathbf{s}_0)|^2 \propto V_f k^4 \sin^2 \chi \int dz e^{i2kz} \int d\rho B_{n_1} \left(\sqrt{\rho^2 + z^2} \right) C(\rho)\rho,$$

where V_f is the focal volume. The scattering amplitude in the forward direction becomes

$$|\mathbf{f}(\mathbf{s}_0, \mathbf{s}_0)|^2 \propto V_f k^4 \sin^2 \chi \int dz \int \rho d\rho B_{n_1} \left(\sqrt{\rho^2 + z^2} \right) C(\rho). \tag{3.18}$$

The interpretation of Equation 3.18 becomes clear in the Fourier space. The scattered amplitude is proportional to the Fourier transform of the scattering potential evaluated at a single frequency $k(\mathbf{s} - \mathbf{s}_0)$. The intensity of the scattered light is proportional to the Fourier transform of the refractive index autocorrelation function B at the same frequency. In both cases, these frequencies are points within the Ewald sphere in the Fourier space.

On the other hand, the scattered amplitude and the scattered intensity are integrals of the product of the Fourier transforms of the scattering potential and the correlation function, respectively, and optical transfer functions integrated over a certain range of frequencies lying within the Ewald sphere. The range of these frequencies and the optical transfer functions depend on the illumination and collection geometry. For example, in the case of light recorded by a microscope, the optical transfer functions depend on the source size and numerical aperture of collection [2]. Typically, if light scattered in forward directions is recorded, the integration is performed over low frequencies, while backscattering samples the Ewald sphere around the maximal $2k$ frequencies.

Until now, we have focused on two important quantities: the intensity and the field. In many applications, including microscopy, interferometry, etc., spatial and temporal coherence effects must be considered. A key property that is used to describe these effects is the *mutual intensity*, which is also known as the mutual coherence function. The reader is reminded of the definition of the mutual intensity

$$J(\mathbf{r}_1, \mathbf{r}_2, \tau) = \langle E^*(\mathbf{r}_1, t) E(\mathbf{r}_2, t + \tau) \rangle,$$

where brackets $\langle\ \rangle$ denote the ensemble average. The intensity can be expressed through mutual intensity as

$$I(\mathbf{r}) = J(\mathbf{r}, \mathbf{r}, 0).$$

The first-order Born approximation for the mutual intensity is [2]

$$
\begin{aligned}
J(\mathbf{r}_1, \mathbf{r}_2, 0) = {} & J_i(\mathbf{r}_1, \mathbf{r}_2, \tau) \\
& - k^2 \iint \left[G(\mathbf{r}_1 - \mathbf{r}_1')\big(n_1'^2(\mathbf{r}_1') - 1\big) \delta(\mathbf{r}_2 - \mathbf{r}_2') + G^*(\mathbf{r}_2 - \mathbf{r}_2')\big(n_1'^2(\mathbf{r}_2') - 1\big)^* \delta(\mathbf{r}_1 - \mathbf{r}_1') \right] \\
& \times J_i(\mathbf{r}_1', \mathbf{r}_2', 0) d\mathbf{r}_1' d\mathbf{r}_2',
\end{aligned}
$$

where J_i is the illuminating mutual intensity, and G is a spherical wave Green function. (G satisfies the Helmholtz equation $(\nabla^2 + k^2)G(\mathbf{r} - \mathbf{r}') = \delta(\mathbf{r} - \mathbf{r}')$.)

In principle, the first-order scattered waves can be added up to the incident wave, and the resulting field can be substituted back in the integral (instead of the incident field only) to obtain a better approximation. This is the *second-order Born approximation*. The iterative process can be continued for higher-order terms. This procedure, however, rarely allows for analytically tractable solutions, and the series frequently have poor convergence. Thus, if the first-order approximation is not sufficient and a more accurate solution is needed, it is typically more practical to rely on computational simulations such as the finite-difference time-domain method, as discussed in Chapter 6.

There are many computational methods that are equivalent to the Born approximation. These approximations are discussed in detail in Chapter 6. The first-order Born approximation is equivalent to the *discrete dipole approximation* (DDA) method of solving a scattering problem numerically—the medium is represented as a collection of electrical dipoles, each emitting a scattered wave, with these waves not interacting with the other dipoles. A numerical implementation of the higher-order Born approximation is the *coupled-dipole approximation* (CDA), in which the scattered waves are further traced and their contribution to the total field inside the particle is determined.

The Born approximation can be further simplified if not just the phase shift but also the size of a scattering particle is small compared to the wavelength. Let us consider a particle of radius a such that $ka \ll 1$. (Here we do not necessarily assume that the particle must be spherical; the radius a is defined as half of the linear diameter of a particle that can have an arbitrary shape.) In this case, the exponent in the Born approximation integral can be neglected, and the approximation is further reduced to the approximation of *Rayleigh scattering* [3]

$$R(\mathbf{s},\mathbf{s}_0) = \frac{1}{V} \int_V \left(n_1^2(\mathbf{r}') - 1 \right) d\mathbf{r}'$$

and, thus,

$$I^{(s)} = \frac{1 + \cos^2\theta}{2} \frac{k^4}{r^2} \alpha^2 I_0, \tag{3.19}$$

where $\alpha = (V/4\pi)R$ is the *total electric susceptibility*, and I_0 is the intensity of the incident light. The scattering cross-section equals

$$\sigma_s = \frac{8}{3} \pi k^4 \alpha^2 \propto \frac{a^6}{\lambda^4}. \tag{3.20}$$

This scattering pattern as a function of angle is that of a dipole (i.e., $1 + \cos^2\theta$ dependence). Another important feature of Rayleigh scattering is that its spectral behavior is essentially independent of the particle shape and internal structure: It is an inverse 4th-power dependence on wavelength λ; this is sometimes referred to as a "featureless" scattering pattern. In fact, this is a consequence of the Ewald's sphere principle, which was introduced earlier. The Fourier transform in Equation 3.16 is evaluated for a range of spatial frequencies, with the highest cut-off frequency being $2k$ (backscattering, $\mathbf{s} = -\mathbf{s}_0$) [1], and the scattered field or intensity does not depend on higher frequencies including $\sim 1/a$. It is, however, incorrect to interpret this result in a sense that it is not possible to distinguish the size of a particle based on its scattering pattern (either as a function of scattering angle or wavelength of light λ) for particles smaller than λ (or $\lambda/2$, which is the diffraction limit of the resolution of an optical microscope). In reality, for visible light, $ka = 1$ is achieved for $a \sim 50$ nm, and the Rayleigh regime does not start until the particle is so small that $a < \lambda/20 \sim 20$ nm. Light-scattering spectra $I^{(s)}(k)$ for particles larger than $\lambda/20$ do not exhibit a k^4 dependence; if their spectrum is approximated as a power law k^β, the exponent $\beta < 4$, and it does depend on the particle size.

For particles larger than $1/k$, the first-order Born approximation is also known under a different name: the *Rayleigh–Gans–Debye (RGD) approximation*. As discussed earlier, it is valid when $ka|n_1 - 1| \ll 1$. The name RGD approximation is frequently used in respect to scattering by homogeneous spherical particles. In this specific case,

$$R = \frac{3}{k_s^3 a^3} \left(n_1^2 - 1 \right) (\sin k_s a - k_s a \cos k_s a), \tag{3.21}$$

with $k_s = k|\mathbf{s} - \mathbf{s}_0| = 2k\sin(\theta/2)$. As can be seen from Equation 3.21, the phase function shows more "features" as the particle size increases. The anisotropy coefficient g increases with a. The scattering pattern is predominantly in the forward direction. Both the spectral and the angular patterns exhibit characteristic oscillations in wavelength, with the corresponding frequencies of these oscillations being proportional to a.

Although relationship Equation 3.21 was derived for a spherical particle, the overall trends are pertinent to particles of essentially any shape. Even though the exact form of the scattering pattern depends on the internal structure of a particle and its shape, the Fourier transform relationship between the scattering potential and the scattering amplitude ensures that the main frequency of the spectral or angular oscillatory pattern is primarily determined by the maximal phase shift that is gained by a wave propagating through the particle, which translates into a dependence on the size of the particle.

When does the RGD approximation break down? For the range of spatial refractive index fluctuations that exist in tissue ($n_1 - 1 \approx 0.01 - 0.1$), the largest cutoff a is expected to be on the order of a micron. This covers most cytoplasmic organelles within a cell (other than the cell nucleus), collagen fibers in connective tissue, and many other tissue structures. As we discussed earlier, for all intents and purposes, the validity of the approximation is greater, and the RGD (or Born) approximation is valid even for larger structures. Finally, for backscattering, the Born approximation is typically more accurate than for calculating forward scattering, which further extends the range of the validity of the approximation when only side or backscattering is considered.

If a particle is much larger than the wavelength and the phase shift is not negligible, the first-order Born approximation is no longer valid. In tissue optics, this is pertinent to a case when we consider scattering by large scatterers such as individual cells or large organelles such as nuclei. In these cases, the scattering amplitude and the scattering cross-section can be estimated using the *Wentzel–Kramers–Brillouin approximation* (WKB). Just like the Born approximation, the WKB approximation has been developed and widely applied to solve quantum mechanical scattering problems. In optics, several authors have independently arrived at essentially the same approximation; thus, the method has acquired many different names, including the *Van de Hulst*, the *anomalous diffraction*, or the *interior wave number approximation*.

The WKB approximation is used when two conditions are satisfied: $ka(n_1 - 1) > 1$ and $(n_1 - 1) \ll 1$. The starting point is, again, the integral Equation 3.15, but now we replace the field inside the particle not with the incident field (as in the Born approximation) but with a propagating wave. If the incident light is a plane wave traveling in direction $\mathbf{z}$ ($\mathbf{z} = \mathbf{s}_0$), the scattering amplitude becomes

$$\mathbf{f}(\mathbf{s},\mathbf{s}_0) = -\frac{k^2}{4\pi}(\mathbf{s}\times(\mathbf{s}\times\mathbf{e}^{(i)}(\mathbf{r})))VR(\mathbf{s},\mathbf{s}_0),$$

where

$$R(\mathbf{s},\mathbf{s}_0) = \frac{1}{V}\int_V \left(n_1^2(\mathbf{r}')-1\right)\exp\left(ikz_1' + ik\int_{z_1}^{z'} n_1(\mathbf{r}'')dz'' - i k\mathbf{r}'\mathbf{s}\right)d\mathbf{r}'$$

Here, $z_1 = z_1(\mathbf{x}, \mathbf{y})$ is the z-coordinate of an entry point of the incident light ray into the particle.

The scattering cross-section can be estimated by applying the forward scattering theorem

$$\sigma_s = k\,\mathrm{Im}\int_V \left(n_1^2(\mathbf{r}') - 1\right)\exp^{ik\int_{z_1}^{z'}(n(\mathbf{r}'')-1)dz''}\,d\mathbf{r}'. \tag{3.22}$$

As an important and illustrative example, consider a spherical homogeneous particle of radius a. In this case, the integral in Equation 3.22 can be taken analytically, and the scattering cross section becomes

$$\sigma_s = 2\pi a^2\left\{1 - \frac{\sin 2y}{y} + \frac{\sin^2 y}{y^2}\right\} \tag{3.23}$$

with $y = ka\Delta n$ and $\Delta n \equiv n_1 - 1$. This equation is very accurate only in the limit of negligibly small Δn.

A more accurate equation with an extended range of validity in terms of Δn can be obtained by taking into account surface effects and the elongation of the light path inside the particle due to light refraction at the point of entry into the particle [4]

$$\sigma_s = 2\pi a^2\left\{1 + \left(\frac{\Delta n}{y}\right)^{2/3} - n_1\frac{\sin 2y}{y} + n_1\frac{\sin^2 y}{y^2}\right\}. \tag{3.24}$$

For Equation 3.24 to be valid, Δn does not have to be much smaller than 1.

Elements of the scattering matrix (or scattering amplitude) can also be found for small scattering angles as [5]

$$S_1 = S_2 = \frac{k^2}{2\pi}\iint_A \left(1 - \exp(-i\xi(\mathbf{r}))\right)\exp(-i\delta(\mathbf{r}',\theta))d^2\mathbf{r}, \tag{3.25}$$

where $\mathbf{r}$ is a vector in plane A orthogonal to the direction of propagation of the incident light, ξ is a phase difference between a wave that enters the particle at a position given by $\mathbf{r}$ and passes through the particle along a straight trajectory and a wave that propagates outside the particle, and δ is the phase difference between the rays scattered by different parts of the particle. The integration is performed over the geometrical cross-section of the particle A that is orthogonal to the direction of the propagation of the incident light.

For a spherical particle, Equation 3.25 can be simplified

$$|\mathbf{f}(\theta)|^2 \approx a^2 x^2\left\{\left(\frac{J_1(x\theta)}{x\theta} - \sqrt{\frac{\pi}{2}}\frac{J_{1/2}(\gamma(\theta))}{\sqrt{\gamma(\theta)}}\right)^2 + \left(\frac{2y}{\gamma^2(\theta)}\right)^2\left(\cos\gamma(\theta) - \frac{\sin\gamma(\theta)}{\gamma(\theta)}\right)^2\right\}, \tag{3.26}$$

where $x = ka$ is called the *size parameter*, and $\gamma = \sqrt{x^2\theta^2 + 4y^2}$. In the limit $y \to \infty$ and $\Delta n \to 0$, scattering amplitude Equation 3.26 approaches that of the Fraunhofer diffraction on a disk, as expected

$$|\mathbf{f}(\theta)|^2 \approx a^2 \, \frac{J_1^2(x\theta)}{\theta^2}.$$

The WKB approximation works quite well in the forward direction and provides a good estimate of the scattering cross-section. However, it is frequently not sufficient to describe backscattering and needs to be modified for these purposes. Because the WKB method lacks accuracy for backscattering, attempts have been made to merge it with the Born approximation-based analyses. For example, the WKB approximation or Mie theory (see discussion that follows) was used to model forward scattering from cells and cell nuclei, while Born approximation was used to model backscattering under the assumption that the nuclear structure is fractal [6].

As seen from Equations 3.23 and 3.24, the spectrum of the scattering cross-section exhibits characteristic oscillations in wavelength with frequency proportional to $a\Delta n$. This spectral pattern is called the *interference structure*. It is a result of interference in the far field of two waves: one propagating along the longest diameter of the particle and another that does not interact with the particle. For example, for a spherical particle with a diameter on the order of a few microns and $\Delta n \sim 0.05$, one would count several oscillations within the visible spectral range. In reality, the spectrum is more complicated and shows at least three types of spectral features, with the interference structure having the lowest frequency of the three.

Higher-frequency spectral features are referred to as the *ripple structure*. These oscillations are of a much higher frequency, which is proportional to the interference structure's frequency $a\Delta n$. The ripple structure cannot be described by the WKB approximation. Further, the frequencies of the oscillations in wavelength differ depending on the angle at which the scattered light is recorded. In the forward direction and in the spectrum of the total scattering cross-section, the ripple structure is a result of the interference of surface waves, which explains the lack of dependence on Δn. In backscattering, the ripple structure has a different frequency and can be modeled by the Born approximation. The fact that scattering from such large scatterers can still be modeled by the Born approximation may sound surprising, but, in fact, it agrees with the understanding that in the case of *backscattering*, the validity of the Born approximation greatly exceeds the general validity criterion $ka\Delta n \ll 1$. The frequency of the ripple structure in backscattering is primarily determined by the overall size of the particle. Of course, when scattering from an ensemble of particles is measured rather than scattering by an isolated particle such as a single cell, size distribution tends to wash out some if not most of these oscillations.

The third spectral feature that has the highest frequency of the three is due to the *whispering-gallery-mode* (WGM) resonances. These are extremely narrow peaks with full widths of half maximum that depend on the ratio of the particle size to the wavelength and, for a micron-size sphere, are well under 1 nm. The width of the resonance is related to the so-called quality or Q-factor of a sphere as a resonant cavity. Physically, these are waves that circulate and are confined within the sphere due to total internal reflection at the higher refractive index sphere-surrounding medium interface and focused by the surface. This forms a standing wave corresponding to light that is trapped in circular orbits within the surface of the sphere. The modes propagate along zigzag paths around the sphere. Although observable in experiments with nonbiological perfectly spherical structures, these resonances are not likely to be relevant to scattering from biological objects.

WKB approximation is not limited to modeling light scattering by homogeneous spheres and can also be applied to nonspherical inhomogeneous particles. A scattering cross-section of a heterogeneous and/or nonspherical particle is actually similar to that of a sphere that would produce the same maximal phase shift as the nonspherical or inhomogeneous particle, that is, the so called *equiphase sphere* (*EPS*) [7]. For example, consider a homogenous but nonspherical particle with refractive index n and a maximal length d in the direction of propagation of light. The maximal phase shift is kdn. Thus, the sphere of diameter d and refractive index equal to the average refractive index of the nonspherical particle is the equiphase sphere. The total scattering cross-section of a nonspherical and inhomogeneous particle can then be described by Equation 3.24.

The validity criterion of the EPS approximation for an inhomogeneous particle is given by [7]

$$\beta_n = 4\frac{\sqrt{L_C d}}{\lambda}\delta_n < 1, \tag{3.27}$$

where δ_n is the standard deviation of refractive index variations inside the particle, and L_C is the refractive index correlation length. A similar validity condition for a nonspherical (e.g., irregularly shaped) particle is

$$\beta_r = 2\sqrt{\frac{2}{\pi}}(n'-1)\frac{\Delta}{\lambda}\sqrt{\Gamma} < 1, \tag{3.28}$$

where Δ is the radial standard deviation from its best-fitting sphere, and Γ is the radial–angular correlation angle. We see that the EPS approximation and the scattering cross-section given by the anomalous diffraction approximation work the best for particles with high spatial frequencies (small features) of either the refractive index perturbations (i.e., small L_C) or surface perturbations (i.e., small Γ). On the other hand, large clumps of refractive index and highly elongated shapes make the EPS approximation break down.

An advantage of the approximate methods discussed previously is that they may provide an analytical solution to scattering problems that otherwise cannot be solved from the first principles. There is a class of particles, however, for which the equation of scattering can be solved exactly. These are homogeneous spheres, and the solution is known as the *Mie theory*. Consider a spherical particle of radius a

$$S_1(\theta) = \sum_{n=1}^{\infty}\frac{2n+1}{n(n+1)}(a_n\pi_n(\cos\theta) + b_n\tau_n(\cos\theta)), \tag{3.29}$$

$$S_2(\theta) = \sum_{n=1}^{\infty}\frac{2n+1}{n(n+1)}(a_n\tau_n(\cos\theta) + b_n\pi_n(\cos\theta)),$$

$$\sigma_s = \frac{2\pi a^2}{\alpha}\sum_{n=1}^{\infty}(2n+1)(|a_n|^2 + |b_n|^2),$$

where

$$a_n = \frac{\psi_n(\alpha)\psi_n'(\beta) - n'\psi_n(\beta)\psi_n'(\alpha)}{\varsigma_n(\alpha)\psi_n'(\beta) - n'\psi_n(\beta)\varsigma_n'(\alpha)}, \quad b_n = \frac{n'\psi_n(\alpha)\psi_n'(\beta) - \psi_n(\beta)\psi_n'(\alpha)}{n'\varsigma_n(\alpha)\psi_n'(\beta) - \psi_n(\beta)\varsigma_n'(\alpha)},$$

$$\pi_n(\cos\theta) = \frac{P_n^1(\cos\theta)}{\sin\theta}, \quad \tau_n(\cos\theta) = \frac{d}{d\theta}P_n^1(\cos\theta),$$

$$\psi_n(x) = \sqrt{\frac{\pi x}{2}}J_{n+(1/2)}(x), \quad \varsigma_n(x) = \sqrt{\frac{\pi x}{2}}H_{n+(1/2)}(x),$$

$$\alpha = ka, \quad \beta = k\Delta na,$$

$J_{n+(1/2)}$ and $H_{n+(1/2)}$ are Bessel and Hankel functions, respectively; $P_n^1(\cos\theta)$ is the associate Legendre polynomial; and θ is the scattering angle.

Although precise, the Mie theory provides a solution in the form of an infinite series, which can only be evaluated numerically; thus, it is difficult to gain physical insights into scattering by use of the Mie theory. On the positive side, the solution is widely available, enabled by well-tested computer codes. The Mie theory has been widely used in tissue optics to model light scattering in tissue. For example, the Mie theory can be used to estimate the size distributions of tissue scatterers based on angular or spectral scattering patterns. Because realistic scattering particles in tissue are, generally, neither spherical nor homogeneous and are not located in the far field of each other (the Mie theory is, strictly speaking, valid only for isolated scatterers located in the far field each of each other), this kind of analysis can be used only for qualitative assessment.

Albeit qualitative, analyses based on single-particle approximations such as RGD, Rayleigh, and WKB approximation and the Mie theory do provide some insights into light scattering in tissue, and the main conclusions drawn from these considerations are frequently correct. When applying these models—and the Mie theory in particular, as its applicability is based on quite a few *a priori* unrealistic assumptions—it is important to know which tissue optics questions can and cannot be addressed by this approach. In particular, if one views size distributions recovered by use of these models in the context of *length scales* of refractive index variations rather than real scattering particles, the answer is close to the correct physical picture.

A more rigorous solution than the Mie theory would require taking into account heterogeneous distribution of refractive index including length scales as small as a few tens of nanometers and as large as a few microns and interactions among particles located in the near field of each other. This can be accomplished by either numerically solving Maxwell equations or by employing approximate solutions that treat tissue as a field of continuous refractive index variations.

Some of the most practically significant computational techniques are discussed in Chapter 6. Available methods range from computational analogs of light-scattering theory approximation (e.g., discrete dipole approximation is equivalent to the Born approximation; couple dipole approximation is equivalent to the higher-order Born approximation) to exact solution of Maxwell equations for a specific configuration of scattering particles (e.g., the multiple sphere algorithm calculates scattering by an arbitrary spatial arrangement of spheres of arbitrary sizes) to numerical solution of Maxwell equations from the first principles (e.g., the finite-difference time-domain method and its more computationally efficient but lower spatial resolution approximation, the pseudo time-domain [PSTD] method).

A powerful means of modeling light scattering in spatially continuously varying refractive index media is based on the Born approximation. It is discussed in detail in Chapter 6. For a continuous medium, the concept of a scattering cross-section is replaced by cross-section per unit volume, which, as we saw above, equals the scattering coefficient

$$\mu_s = \frac{\sigma_{s,\delta V}}{\delta V} = \int_{4\pi} \frac{|\mathbf{f}(\mathbf{s},\mathbf{s}_0)|^2}{\delta V} d\Omega,$$

$$\frac{|\mathbf{f}(\mathbf{s},\mathbf{s}_0)|^2}{\delta V} = \frac{k^4}{(2\pi)^2} \sin^2 \chi \int B_{n_1}(\mathbf{r}')e^{-ik(\mathbf{s}-\mathbf{s}_0)\mathbf{r}'}d\mathbf{r}'. \tag{3.30}$$

If the tissue is isotropic, $B_{n_1}(\mathbf{r}) = B_{n_1}(r)$ and

$$\frac{|\mathbf{f}(\theta)|^2}{\delta V} = \frac{k^4}{\pi} \sin^2 \chi \int_0^\infty B_{n_1}(r) \frac{\sin k_s r}{k_s} r dr,$$

where $k_s = 2k \sin(\theta/2)$. These equations allow one to determine the phase function and the scattering coefficient of a continuous medium with refractive fluctuations when the autocorrelation function of the refractive index fluctuations is known. Inversely, as we will see in Chapter 6, this relationship helps us understand what sort of information can be learned about the tissue correlation function when a spectrum of $\mu(\lambda)$ is known.

In summary, light scattering depends on the spatial refractive index distribution, which in turn depends on the spatial distribution of local mass density. Light scattering depends on a wide range of length scales of tissue structures. On the low end, the limit of sensitivity is approached in the Rayleigh scattering regime for length scales a such that $ka = 1$; for optical wavelengths, this corresponds to a few tens of nanometers. On the upper end, refractive index correlation eventually vanishes, which sets the upper limit of light scattering sensitivity. This broadband sensitivity of light scattering to various sizes of tissue structures from length scales as small as a few tens of nanometers to as large as a few microns combined with an extraordinarily complex tissue architecture at these length scales makes light scattering a powerful approach to interrogate tissue and, at the same time, a highly complex process to account for.

Lab Discussion

As described previously, light may interact with tissue via several mechanisms. For example, light may be absorbed by tissue and generate heat. Typically, such an interaction will have a spectral dependence where some colors (wavelengths) are preferentially absorbed while others are absorbed only minimally. As another example, light may interact with tissue via a scattering mechanism. Here, the direction of the light is changed due to variations in the refractive index, that is, the structure of the illuminated tissue. While scattering can also be detected by changes in spectral components, it can also be observed by its angular dependence.

Evidence of light–tissue interaction can take several forms, but all experimental methods for detecting it include common features, such as a light source and a detector. The selection of these experimental components will depend on the desired interaction to be studied. This section introduces several laboratory exercises that advance knowledge in these areas by probing specific interactions. Exercises include: (1) *Measurement of absorption:* Experimental methods will be developed for absorption measurement using hemoglobin concentration as a target; (2) *Biomedical applications of absorption measurement:* Several example experiments of absorption measurements for biomedical applications are given, including glucose detection, alcohol detection, pulse oximetry, and jaundice detection; (3) *Measurement of scattering:* Experiments will measure key parameters that characterize scattering, including measurements of the phase function (angular dependence of scattering), backscattering and forward scattering spectra, and decoupling the effects of scattering and absorption when absorption spectra are known; (4) *Biomedical applications of scattering measurements:* Examples of typical projects that measure scattering to learn about a biomedical sample are given, including assessing cell confluency using reflectance, polarization imaging of skin, and monitoring of cellular processes using scattering.

Laboratory 1: Absorption Measurement of Hemoglobin Concentration

Absorption measurements are typically accomplished using spectrophotometry, where transmission through a sample is compared across several wavelengths. When the absorption properties of a sample are large compared to its scattering properties, the attenuation law described in Equation 3.12 can be reduced to one that describes purely absorption, termed the *Beer–Lambert law*

$$I = I_0 e^{-A} \tag{3.31}$$

where A is the absorbance, defined as

$$A = -\log (I/I_0) = \mu_a L = Ec\, L, \tag{3.32}$$

with the absorption coefficient μ_a, defined as the product of the molar absorbance E and the concentration of the absorber c. Typically, the units of the absorption coefficient are given in cm^{-1} or mm^{-1}, such that the product with the optical path length L gives a unitless quantity. The concentration c is typically given as M (moles/liter) so that in order to yield the correct units, absorbance E is usually defined as cm^{-1}/M. The units of absorbance are usually expressed in terms of *optical density* (OD), a logarithmic measure in powers of 10. Thus, a value of $A = 0$ means no absorption (100% transmission), $A = 1$ indicates 90% absorption, $A = 2$ indicates 99% absorption, etc.

An important aspect of transmission measurements is the normalization by the reference intensity I_0. Since the reference intensity is always larger than the transmitted intensity I, often by several orders of magnitude, the measurement becomes vulnerable to variations in I_0. To compensate for this, most absorbance measurements will incorporate a separate reference channel that will account for variations in source intensity and system throughput that might otherwise introduce unwanted noise.

Laboratory Protocol

Instrumentation

The laboratory experiments in this section require a *spectrophotometer*. While several commercial versions of this instrument are available, they are typically expensive and not readily

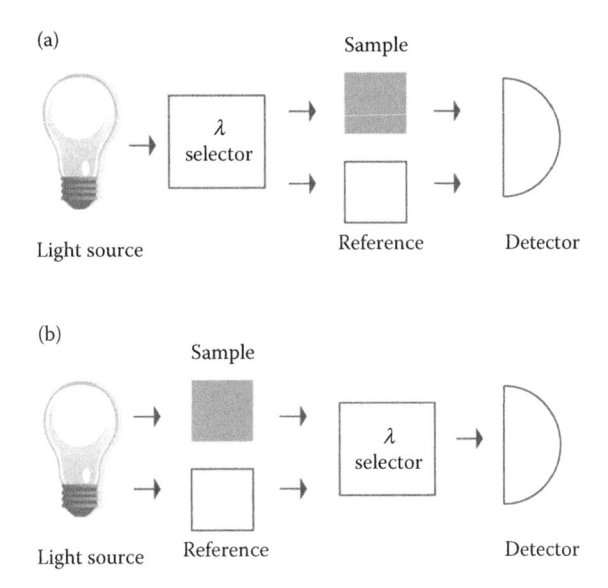

FIGURE 3.2
General schematic of spectrophotometry schemes. Both schemes require a light source, a mechanism for spectral selection, and a detector to convert optical signals to electronic signals. Spectral selection can occur either (a) at the input to the sample or (b) at the output.

available for laboratory exercises. Here, a basic schematic of a low-cost spectrophotometer is presented; however, if a commercial instrument is available, it can be substituted, and one can proceed directly to the investigations of absorption and other light–tissue interactions.

Figure 3.2 shows two basic schemes for implementing a spectrophotometer. Light from a broadband source is directed to transmit through both a sample and reference standard. There are two options for implementing this setup. In the first approach, Figure 3.2a, the light is spectrally filtered prior to incidence on the sample. The filtering can be accomplished using available laboratory equipment, such as a scanning monochromator, as would be implemented in a commercial spectrophotometer. As an alternative to this, a low-cost spectrometer can also be used to provide a filtering mechanism, although its throughput would be expected to be too low for these experiments. In Figure 3.2b, the spectral filtering occurs after the light has been transmitted through the sample. Here, a commercial monochromator or low-cost spectrometer can also be used. In either case, spectral selectivity must be realized by mechanically scanning an aperture across a dispersed spectrum. As an alternative, a more easily manageable system for laboratory experimentation is presented, which uses the approach in Figure 3.2a but achieves spectral selectivity by using electronic control of multiple light-emitting diodes (LEDs).

Figure 3.3 shows a schematic of a spectrophotometer system that can be easily built and used for the laboratory exercises in this chapter as well as other applications. The system uses illumination by one or more LEDs, which can be simply exchanged to provide spectral flexibility for a wide range of uses. The components of this system are now discussed in detail.

LED Sources

A wide range of LEDs are commercially available for fairly low cost, typically a few cents to a few dollars apiece depending on the desired spectral window and output power.

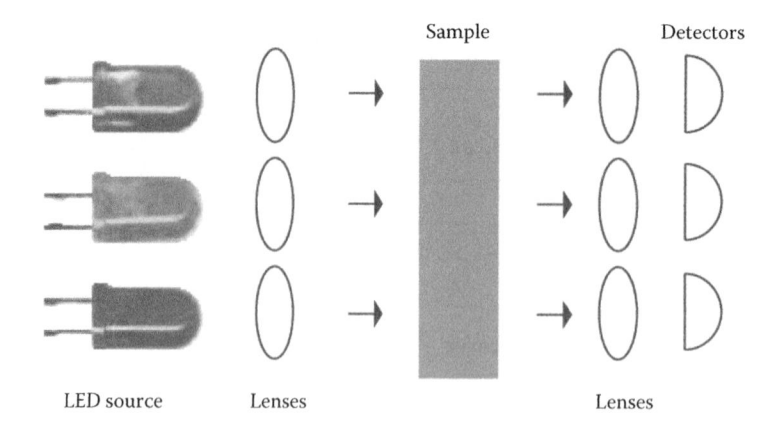

FIGURE 3.3
Schematic of spectrophotometer for laboratory exercises. One or more LEDs is used as a light source. Light from each LED is collimated and transmitted through the sample of interest, and its output is focused onto a separate detector.

Each LED must be provided sufficient current to produce the desired output. Specification sheets provided by the manufacturer can be referenced to determine the optimal circuitry. Parameters to consider here are:

1. Wavelength—the spectral output of the LED should be matched to the absorption feature of interest. A second LED can be used to provide the reference intensity by locating its center wavelength in a spectral region with low absorption.

   Example: For Hb absorption, an LED in the blue–green (400–550 nm) range would match up well with the absorption maxima. However, the absorption in this window may provide too much attenuation to produce readily detectible transmission signals. Instead, Hb absorption is usually measured using a red LED (650–680 nm wavelength) to provide a relatively strong absorption, and an NIR LED (near 830 nm, the isosbestic point) is used as a reference.

2. Forward voltage—this parameter indicates the voltage across the LED when driven by a typical current. Using this voltage and the typical current, the resistance value of a *current limiting resistor* can be determined, which should be inserted in series between the LED and the voltage source.

   Example: A forward voltage of 2.2 V at 20 mA is specified. To produce the desired 20 mA using a 9-V source, a current limiting resistor of value $R = V/I = (9 - 2.2)/0.02 = 340\ \Omega$ is needed.

   The LED can be operated using a mechanical switch, typically available for a few cents, or a transistor may be inserted to control it with a low current control voltage, from a data acquisition (DAQ) board, for example.

3. Optical output—there are two key parameters here that should be carefully considered. The first is the angular divergence of the source. For LEDs, this may be as large as 50 or 60°. To capture this output for the transmission measurement, a collimating lens is needed with a fairly high numerical aperture. Selection of these lenses is discussed later, but as a note, it should be recognized that only a fraction of this angular light may be captured for use by the lens.

The second parameter that is important here is the luminous intensity, usually specified as millicandelas (mcd). However, the typical units for intensity are lumens or watts. Since candelas are equal to lumens/steradians, the angular divergence of the source is needed for the conversion.

Example: The output of the LED is given as 2000 mcd with a 50-degree divergence angle. Upon converting to steradians, this corresponds to an output of 1.177 lumens.

For reference, a standard 100-W light bulb produces a typical output of 1700 lumens. However, both lumens and candelas are referenced to the responsivity of the human eye; thus, IR LEDs will have zero lumens and zero mcd. The lumen-to-watt conversion will depend on the wavelength of the LED. For example, a white LED may provide 92 lumen/watt, but the above-mentioned 100-W bulb will provide only 12 lumen/watt due to less optical output that is observable by the human eye.

Lenses

The role of the lenses is to collimate the output of the LEDs to direct as much of the output power as possible to the sample for transmission measurement. Availability of plastic aspheric lenses with focal lengths in the 2- to 10-mm range allow for a low-cost implementation of the optical system, with a typical cost of $5 to $20 per lens. The numerical aperture of plastic aspheric lenses can be substantial (up to >0.50) and should be matched to maximize the angular output received from the LED but also provide sufficient focal length to mechanically mount the LED–lens combination. Identical lenses can also be used to focus the transmitted light onto the detectors.

Detectors

For this application, PIN silicon photodiodes are recommended for detectors. These can have a substantial active area of a few mm^2 and provide good sensitivity across the spectral range of interest. For the application to detect Hb, PIN photodiodes will provide >0.5 responsivity across the range from 600–850 nm. The acceptance angle of the photodiode should be matched to the numerical aperture of the focusing lenses. Photodiodes will generate a photocurrent when exposed to the transmitted light. A simple circuit can use a resistor to convert the photocurrent to a voltage, which may then be sensed by an oscilloscope or digitized by a data acquisition (DAQ) card on a PC. A more sophisticated approach is to implement a transimpedance amplifier, as shown in Figure 3.4.

The advantage of using the transimpedance amplifier is that the low-output impedance of the operational amplifier (op-amp) will ensure that the voltage-sensing mechanism, such as the oscilloscope, will not draw too much current and affect the photodiode output. A resistor value (R) of 1–10 kΩ is usually sufficient. A further improvement to this circuit is to reduce high-frequency noise by including a small capacitor in parallel with the resistor. This will roll off the frequency response of the circuit according to the RC time constant of the combination. For most biophotonics applications, a response of <1–10 KHz is needed, suggesting a capacitance on the order of 50–500 pF. An additional improvement is to add a resistor between the noninverting input of the op-amp and ground with an equivalent value as the feedback resistor. This will serve to null out any offsets due to nonideal op-amp behavior.

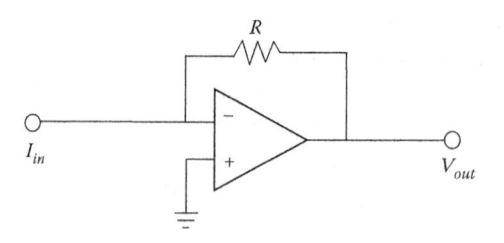

FIGURE 3.4
Transimpedance amplifier. Output current from the photodiode is input to this op-amp circuit. The resistor converts this to an output voltage $V_{out} = I_{in}R$.

Mechanical Design

The final step in implementing the spectrophotometer is the mounting of the components to a robust mechanical base. The components can be hand-machined using wood or Delrin plastic or fabricated using a 3-D printer and computer-aided design (CAD) software. In either case, the key requirement is that the design provide a highly reproducible geometry for adding, holding, and removing a sample chamber. It is recommended that a standardized sample chamber be used for transmission experiments, such as a 10-mm light path plastic cuvette, typically available in large packs at a unit cost of 15–25 cents each. The sample chamber should then be mounted in a holder that grips the cuvette snugly so that motion of the chamber cannot introduce variability in the optical measurement. Figure 3.5 below shows a typical design based on attaching mounts to a rail. Key tolerances here are matching the internal width of the cuvette mount to the cuvette size and correct positioning of the lenses relative to the LED and detector to enable collimation and focusing,

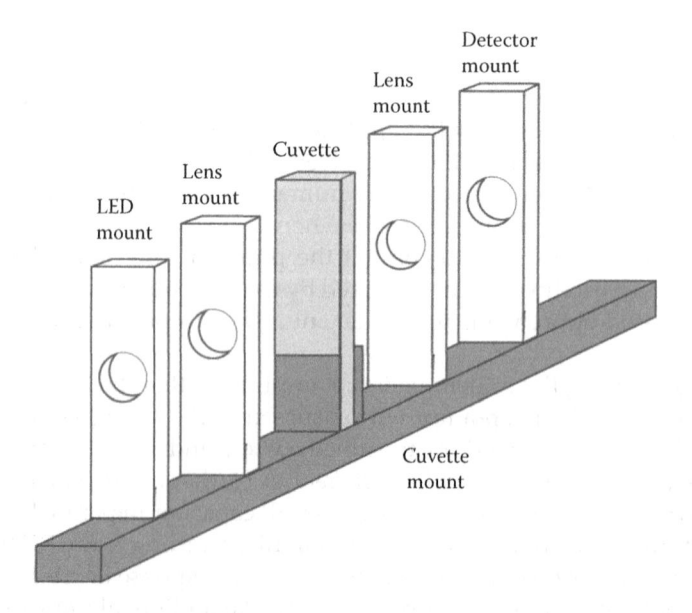

FIGURE 3.5
Mechanical design for spectrophotometer. Rail design holds each mounting piece at fixed positions. Each mount has a drilled hole matched to the diameter of the respective component. The cuvette is held in place by a mount composed of two vertical flanges that provide a repeatable position for the sample.

respectively. The lens-to-LED distance can be determined based on the specifications of the selected components but should be verified at the lab bench prior to attaching components to the rail to help optimize collimation. The lens-to-detector mount distance is not as crucial but should also be experimentally verified to ensure optimal detection.

Sample Preparation

What is needed:

- Human ferrous stable Hb in lyophilized powder form (Sigma-Aldrich, St. Louis, MO)
- Sodium dithionite
- Phosphate buffered saline
- 10-mm light path plastic cuvettes
- Pipettes and glassware for mixing and diluting samples

Hb samples may be prepared by mixing the Hb with deionized (DI) water until the desired concentration is achieved. Exposure to ambient air of this mixture will produce oxy-Hb. To produce deoxy-Hb, a trace amount of sodium dithionite, which removes the oxygen in oxy-Hb, may be introduced to a solution of the Hb powder and phosphate buffered saline (PBS). The use of PBS here is to ensure that reactions that degrade specific absorption do not occur [8].

A stock solution of 300 g/L (4.7 mM) should be prepared for each oxy-Hb and deoxy-Hb. Serial dilutions can be prepared for each as well as mixtures of the two, as resources allow. The below experiment examines the attenuation of oxy-Hb and deoxy-Hb as a function of concentration (samples 1–4, corresponding to 1:1, 1:2, 1:3, and 1:4 dilutions of oxy-Hb or deoxy-Hb stock solution in DI water) and transmission through a 50/50 mixture of the 1:4 dilutions of the two stock mixtures (sample x).

Experiment

1. Using a water-filled cuvette, determine the transmission value for zero concentration by recording the voltage output of the detector for illumination by each of the two LEDs.

2. Remove and replace the water-filled cuvette 10 times, and record the value for transmission each time for each LED.

3. Calculate the average and the standard deviation of these 10 measurements as the I_o value and the uncertainty in that value.

4. For each of the dilutions of oxy-Hb, illuminate the sample with each of the LEDs, and record the voltage output of the detector. Remove and replace the sample 10 times to obtain an average value of I. The uncertainty in this value can be estimated by calculating the standard deviation of the 10 measurements.

5. Repeat these experiments using the series of dilutions of deoxy-Hb, recording 10 transmission measurements for each of the LEDs and calculating the average and the standard deviation.

6. Repeat this experiment using the 50/50 mixture, recording 10 transmission measurements for each of the LEDs.

Data Analysis

1. Determine the absorbance of each oxy-Hb sample by calculating $A = -\ln(I/I_o)$.
2. Plot the absorbance of each sample versus the concentration of oxy-Hb. It may be more direct to construct a plot using a concentration normalized to the stock solution; that is, the 1:1 dilution would be $0.5\ C_o$, where $C_o = 300$ g/L.
3. Compare the slope of this line with the expected value, which is the product of the cuvette path length and molar absorption of oxy-Hb.
4. Repeat the absorbance calculations for each deoxy-Hb sample by using the IR (830 nm) as I_o and the red LED transmission as I.
5. Plot the absorbance of the deoxy-Hb samples versus concentration and determine the slope.
6. Compare the relative accuracy of the two approaches for reference measurements by determining the error in assessing the molar extinction values when compared to theoretical values.
7. For the mixture, the transmitted intensity at each wavelength can be described as

$$I(\lambda) = I_o \exp(-L(E_{oxy}(\lambda)c_{oxy} + E_{deoxy}(\lambda)c_{deoxy})) \tag{3.33}$$

where E_{oxy}, E_{deoxy}, c_{oxy}, and c_{deoxy} are extinction coefficients and concentrations for oxy-deoxy-Hb, respectively, and L is the path length through the cuvette. Upon converting the intensity values to absorbance, the two equations (one for each wavelength) can be inverted to solve for the two unknowns (c_{oxy} and c_{deoxy}). Determine the concentration of each species and the partial oxygen saturation S.

$$S = \frac{c_{oxy}}{c_{oxy} + c_{deoxy}} \tag{3.34}$$

Determine the accuracy of the oxygen saturation measurement compared to the theoretical value.

8. Summarize and discuss results.
   a. Did the absorbance values follow a linear trend? If not, discuss at what concentrations it departed and why.
   b. Compare the error in your calculation of molar extinction values to that due to the variability observed in your transmission measurements.
   c. Compare the two reference measurement methods, and discuss the relative strengths and weaknesses.
   d. Discuss improvements in the instrumental design and experimental measurements that could improve the accuracy of the analysis.

Laboratory 2: Biomedical Application of Absorption Measurements

Although Hb absorption is an important biomedical measurement, there are several other relevant absorption measurements that can be executed as laboratory exercises. A particularly compelling target is **noninvasive assessment of blood glucose concentration** due to its importance in management and care of diabetic patients. This is a challenging measurement to achieve, especially if one seeks to execute it through the skin and to sense glucose in whole blood, since by itself glucose offers very little absorption in the visible spectrum. However, by revising the experiment to focus on measuring glucose concentration in water, transmission measurements can be executed that correlate with concentration. Recent work has shown that there are absorption signatures for glucose in the near-infrared (NIR) to infrared (IR) wavelength regions, for example 2.1–2.5 μm [9]. The spectrophotometry approach described in the previous exercise can be adapted to observe absorption of glucose by using mid-IR LEDs and photodiodes but typically only with concentrations that are significantly higher (10× or more) than biomedically relevant ranges (typically 50–300 g/dL). Mid-IR parts are usually costlier than their visible-spectrum counterparts and are not usually carried in stock by electronics wholesalers. A significant difficulty in working in this spectral range is that IR viewers and sensors are typically needed. Only a few labs will have alternative power meters applicable in this spectral region. As an alternative, a good method of measuring glucose concentration is to exploit its *optical activity* rather than target absorption. With small modifications, the spectrophotometry scheme presented above can be adapted to use polarization to detect optical activity. Two polarizing filters are needed, and these can be purchased or easily made from plastic polarizing sheets. One polarizer is inserted between the light source and the collimating lens to polarize the incident light. A second polarizer is inserted between the lens and the detector to analyze the returning light. By orienting the detectors to allow maximum transmission through a reference cuvette, variations in intensity can be used to measure optical activity. Using Malus's law, $I = I_o \cos \theta^2$, the angle of rotation can be determined by measuring the dip in intensity. As a more advanced measurement, the orientation of the analyzing polarizer can be rotated through 360° to find the angle of maximal transmission. The expectation for this instrument should be to measure tenfold higher concentration than the biomedically relevant range for the optical activity of glucose.

Another compelling target is determining **blood alcohol levels using optical transmission experiments**. As with glucose, there is little absorption by alcohol in the visible spectrum. However, substantial absorption can be seen in the mid-IR, near 3.3–3.4 μm. The spectrophotometer can be adapted for this wavelength range, but the availability and the cost of components in this range put this out of reach for most laboratory courses. An alternative approach is to use a differential absorption measurement where the presence of alcohol is determined by the amount of water that is displaced. By adapting the spectrophotometer to closely monitor an absorption band of water in the 1.6–1.8 μm region, the presence of alcohol can be determined by observing a decrease of absorption in this band [10], as there is comparatively little absorption by alcohol in this range. While light sources and detectors are available in this region of the spectrum, novice engineers may have difficulty working with invisible light output. The accuracy of this approach is more suitable for assessing the alcohol content of a water–alcohol mixture than the relevant concentrations for assessing blood alcohol levels. As another note, the presence of mixed alcohol beverages in the laboratory may provide too tempting a distraction to permit effective investigations.

As an alternative biomedical measurement for the spectrophotometer, bilirubin makes an excellent optical absorption target for **detection of jaundice**. The bilirubin molecule is seen in high concentration in babies with jaundice, where the skin presents a yellow color. Bilirubin appears due to the breakdown of hemoglobin molecules. While it is typically cleared by the liver, if there is dysfunction, the excess bilirubin can spread from the blood to the skin, where it may be detected optically, or to the brain where irreversible damage can occur. The strong absorption in the 400–475 nm range coincides well with light sources and detectors and thus provides an attractive target for absorption measurements. Similar experimental procedures and data analyses as those used in the hemoglobin lab above can be used to assess bilirubin concentration as a function of transmission.

An advanced project that builds on the spectrophotometer design for Hb detection is to develop a **pulse oximetry system**. This instrument also relies on measurement of absorption but isolates the component due to blood flow by taking advantage of its pulsatile volume changes. If this approach is used to measure pulse via volume changes, it is plethysmography, which is the primary building block of pulse oximetry. The most direct approach is to adapt the spectrophotometer to measure transmission through the finger instead of the cuvette. The transmitted optical signal I can be observed to vary in time. The amplitude of this variation is attributed to the pulsatile blood flow, and the concentration of Hb can be determined from Equation 3.35. One difficulty in this analysis is that the optical path length L is not generally known. However, the time-varying nature of the pulsatile flow can be exploited to remove this quantity from the calculation. Taking the derivative of the measured absorbance with time yields

$$\frac{dA}{dt} = -\frac{d}{dt}\ln(I/I_o) = \frac{dL}{dt}(E_{oxy}(\lambda)c_{oxy} + E_{deoxy}(\lambda)c_{deoxy}).$$ (3.35)

This expression can be determined for each wavelength for the two LEDs. Taking the ratio of these two time-varying quantities will remove the pathlength dependence

$$R = \frac{dA_1/dt}{dA_2/dt} = \frac{E_{oxy}(\lambda_1)c_{oxy} + E_{deoxy}(\lambda_1)c_{deoxy}}{E_{oxy}(\lambda_2)c_{oxy} + E_{deoxy}(\lambda_2)c_{deoxy}}.$$ (3.36)

Using the known quantities of the extinction coefficients at each wavelength and for each type of hemoglobin, $E_{oxy,deoxy}(\lambda_{1,2})$, the expression can be solved to yield the oxygen saturation. From Equation 3.36, it can be shown as

$$S = \frac{c_{oxy}}{c_{oxy} + c_{deoxy}} = \frac{RE_{deoxy}(\lambda_2) - E_{deoxy}(\lambda_1)}{RE_{deoxy}(\lambda_2) - E_{deoxy}(\lambda_1) + RE_{oxy}(\lambda_2) - E_{oxy}(\lambda_1)}.$$ (3.37)

One experimental difficulty with this setup is maintaining the transmission geometry with a live finger contained in the cuvette holder. Variations in intensity due to nonpulsatile motion will reduce the ability to isolate the time-varying component of the blood flow and may not produce sufficient quality results for analysis beyond a mere demonstration of the operating principle. Instead, the setup can be adapted to use a commercial pulse oximetry probe, which is well adapted to measure transmission through the finger. An example is the Nellcor Oxisensor® II D-25 probe (Covidien), which typically costs $25. The control circuitry can then be wired up to light the LEDs contained in the sensor, and transimpedance amplifiers can be used to detect the photocurrent from the detections to give a measure of the transmission at each wavelength. To correctly implement this scheme, the LEDs must be pulsed in a controlled manner, and development of a software program suitable for instrument control, such as LabView (National Instruments), is recommended. The advantage of using this type of program is that it can sync data acquisition with the illumination and provide online analysis of the obtained signals.

Laboratory 3: Scattering Measurement of Structure

As discussed previously, elastic scattering of light is the dominant process behind light transport in tissues. Methods for solving the direct scattering problem are essential in developing models for modeling light transport. In addition, analysis of light-scattering patterns as a function of scattering angle and wavelength can provide diagnostically useful information about cell and tissue structure by solving the inverse scattering problem. The focus of this section is on computational implementation of forward-light-scattering models and verifying their predictions with laboratory exercises.

Attenuation and Anisotropy of Scattering

Light propagation through tissue depends on the rate at which it is attenuated, given by the scattering coefficient μ_s as given in Equation 3.11. This parameter describes the rate at which optical power is removed from the incident illumination but not what happens to that light as its transport continues within the tissue. Another key parameter that governs the transport of light is the anisotropy factor g, defined as the average of the cosine of the scattering angle (see Equation 3.13). In the first part of this laboratory exercise, a Mie theory calculator will be used to help understand the effects of scattering anisotropy on light transport relevant to biological scatterers. The calculations will help inform the experimental investigation of light scattering that follows.

Calculations

1. Go to the following website to access an online Mie scattering calculator: http://omlc.ogi.edu/calc/mie_calc.html.

   Alternative Mie scattering calculators can be used such as MiePlot or MieCalc (see Wax and Backman, *Biomedical Applications of Light Scattering*, for more exhaustive discussion of available light-scattering codes).

   Enter the following parameters:

   Sphere diameter: 0.5

   Wavelength in vacuum: 0.633

   Index of refraction in medium: 1.33

   Real index of sphere: 1.59

   Image index of refraction: 0

   Number of angles: 100

   Concentration: 0.01

   Various plots of the differential cross-section as calculated by Mie theory will be generated. Note the polar plot shows a highly forward-peaked scattering pattern. Record the scattering cross-section and anisotropy coefficient.

2. Repeat this calculation with a sphere diameter of 0.05 microns. Again, record the scattering cross-section and anisotropy coefficient.

   Note that this pattern presents more evenly distributed scattering. The parallel polarization shows equal magnitude in the forward and backward directions, while the perpendicular polarization is isotropic, with equal scattering in the forward, backward, and side directions. Compare the anisotropy coefficients between the 0.5- and 0.05-micron spheres.

3. Use Equation 3.10 to estimate the scattering coefficient for these models. Note that the concentration used in this simulation is spheres per cubic micron.

4. Now calculate the transport coefficient (Equation 3.14) for each sphere type.

5. Determine the concentration needed to produce a scattering coefficient of 0.2 mm^{-1} for each type of sphere. Calculate the transport coefficient for the two types of spheres at this concentration. Discuss the difference in magnitude between the two, and make some predictions on how this will affect the transport of light, including changes to light scattered in the forward and side directions.

Instrumentation

This laboratory will use the spectrophotometer developed in the previous laboratory exercise. However, instead of using both the red and NIR LEDs, this laboratory exercise will use only the red LED. If the center wavelength of the red LED significantly varies from the 633 nm used in the above calculations, it may be worthwhile to rerun them using the corrected wavelength. As an alternative, a He-Ne laser ($\lambda = 632.8$ nm) can be substituted for the LED source. In this case, the collimating lens is likely not needed, as most He-Ne lasers emit highly collimated output. In addition, to permit an effective investigation of Mie scattering, the instrument shown in Figure 3.5 should be adapted to include a second detection path that enables detection of scattered light in the side directions. To accomplish this, a second rail is added, oriented 90° and joined at the original rail below the cuvette. Additional mounts will be needed to accommodate a second collection lens and detector. The photocurrent from this detector will require an additional transimpedance amplifier and a second data acquisition or oscilloscope channel for collecting side scattering light signals.

What is needed:

- Spectrophotometer
- 0.05-micron and 0.5-micron–diameter polystyrene ($n = 1.59$) microspheres
- Distilled water ($n = 1.33$)
- 10-mm light path plastic cuvettes

Sample Preparation

Based on the calculations in the first part of this laboratory, prepare samples of each microsphere with a concentration corresponding to an absorbance of 2 for a 10-mm light path. The 0.05-micron–diameter microspheres will likely be available from manufacturers such as Polysciences (Warrington, PA) in concentrations ranging from 1% to 10% w/v. The number of particles per mL can be calculated from this value using the density of polystyrene, 1.05 g/mL. The solution of 0.05-micron–diameter spheres from the manufacturer will likely need to be significantly diluted to produce the desired concentration. It may be more efficient to create a 1:10 dilution as a stock solution and then use this to create the samples for the experiment.

Experiment

1. Using a water-filled cuvette, determine the transmission value for zero concentration by recording the voltage output of the detector for illumination by the red LED.

2. Verify that the signal collected by the side scatter detector is effectively zero.

3. Remove and replace the water-filled cuvette 10 times, and record the value for transmission and side scatter each time.

4. Calculate the average and the standard deviation of these 10 measurements as the I_o value and the uncertainty in that value.

5. Measure the transmission and side scatter for the 0.05-micron–diameter microspheres, repeating each measurement 10 times by removing and replacing the cuvette to obtain an average value of I and Is. The uncertainty in these values can be estimated by calculating the standard deviation of the 10 measurements.

6. Repeat the measurements for the 0.5-micron–diameter microspheres, removing and replacing the sample 10 times to obtain an average value of I and I_s, and their standard deviation.

Data Analysis

1. Compare the magnitude of the transmission and side scatter for each of the two microsphere samples relative to the reference intensity. Qualitatively discuss the differences between the two.

2. Calculate the ratio of transmission to side scatter for each of the two microsphere samples. How does the anisotropy parameter from the Mie theory calculations relate to the relative amounts of side scatter seen for each microsphere sample?

3. Summarize and discuss results.

   a. Did the transmission values agree with the predictions of the Mie theory? If not, discuss possible origins for this discrepancy.

   b. Compare the variability observed in your transmission measurements to assess its role in possible disagreement between data and theory.

   c. Compare the magnitude of the side scatter to the transmission for each of the two microspheres and explain the source of their differences.

   d. Give examples of biological structures that can be modeled using the Mie theory and how measurements of the scattering coefficient, transport coefficient, and anisotropy can be used to detect pathological conditions.

Backscattering Cross-Section Spectra

The spectral dependence of scattering can also be used to assess structural features. As discussed above, *Rayleigh scattering* has a simple spectral dependence, an inverse fourth-power dependence with wavelength λ (see Equation 3.20), but this regime is not relevant until sizes of less than $\lambda/20$ are reached. For visible and NIR wavelengths, this limit corresponds to sizes in the 20–50 nm range. Thus, a wide range of diagnostically useful biological structures lie above this limit and can be probed using spectral reflectance measurements. As the size of scatterers increases, different approximations for analytically expressing the spectrum of scattered light, including the RGD approximation Equation 3.21 and the WKB approximation Equation 3.23, become valid, although each will have a range of parameters over which it is valid. On the other hand, the Mie theory is an exact analytical method for assessing the scattering spectra of spherical particles. With an adequate computational platform, the Mie theory can be used to analyze scattering from a broader range of spherical objects. In this laboratory exercise, a Mie theory calculator will be used to compare the validity of various models often used to explain scattering characteristics of biological scatterers. The calculations will help inform experimental investigations of the spectral dependence of light scattering that follow below.

Calculations

1. For this exercise the MiePlot Mie Scattering calculator will be used. The package may be downloaded from http://www.philiplaven.com/mieplot.htm and installed using the instructions on that page.

   (As an alternative, MieCalc is a purely online Mie calculator, http://www.light-scattering.de/MieCalc/eindex.html.

   This is a versatile tool for executing Mie calculations in a simple environment but does not allow for export of data or saving of plots in the online version.)

2. Start MiePlot and wait for the control screen to load.

3. In the top center of the window, click the drop-down menu that says, "Intensity v. scattering angle" and select "Qext/Qsca/Qabs v. wavelength."

4. Select the remaining parameters as follows:

   Drop size radius (μm) = 5.0

   Select the "disperse" radio button, which will cause additional entries to appear, and make sure 1% Std Dev, and N = 20 are selected.

   Wavelength scale:

   Minimum = 200 nm

   Maximum = 1000 nm

   Number of steps = 200

   Intensity scale = select "Logarithmic"

   Under the "Advanced" menu tab at the top (to the right of "File" and "View"), select "Refractive Index," then "Sphere," then "Fixed." A new area will appear on the control screen titled "Light." In this area, you can enter the refractive index as 1.59 for the real part and leave the imaginary part as 3×10^{-6}.

   Click the "Advanced" menu tab again and now select "Refractive Index," then "Surrounding Medium," then "Water," then "Segelstein."

5. Click the "New Plot" button. This may take a few seconds to run depending on the processor speed and the available memory.

   The output should show a function that oscillates about 2 with increasing magnitude and then decreases toward zero above the 500-nm wavelength. The line that is visible is *Qext*, the extinction efficiency. It overlaps with *Qsca*, the scattering efficiency, a quantity that relates the geometric cross-section to the scattering cross-section, and is defined as $\sigma_s = Q_{sca}\pi(d/2)^2$, where σ_s is defined in Equation 3.6.

   Save this plot as a text file called 10micronspectrum.txt

6. Repeat for scatterer radius of 0.5 micron and save the file as 1micronspectrum.txt. Note that clicking "Overplot" will show these data on the same plot as the original. Clicking "New Plot" will display the data on the same plot without rescaling the *y*-axis.

7. Repeat for scatterer radius of 0.1 micron and save this file as 200nmspectrum.txt. Note the difference in magnitude of this scattering and its spectral shape.

8. For each of these three spectra, determine which is the best fit, Rayleigh scattering Equation 3.20, RGD approximation Equation 3.21, or the WKB approximation Equation 3.23, and plot the analytical form over the numerical data.

Instrumentation

This laboratory exercise requires the use of a **spectrometer** to analyze the spectral dependence of scattering from aqueous solutions of polystyrene microspheres ($n = 1.59$). A wide variety of commercial compact spectrometers are available, such as the USB2000 from Ocean Optics (Dunedin, FL), typically in the $2000–$3000 range. This type of instrumentation can become an important resource for teaching biophotonics, and an investment in this area will reap benefits for many years. In addition to the robust hardware, this type of spectrometer usually comes with fairly intuitive software for observing and recording spectra and may even provide an interface for popular equipment control software platforms such as LabView (National Instruments). As an alternative, the spectrometer that is built for Lab 6.b.i can also be used to acquire spectral data. However, it is important that sufficient spectral resolution be achieved with this setup to enable analysis of the Mie scattering data.

For this experiment, a **reflectance probe** is also needed. A basic schematic of the reflectance probe is shown in Figure 3.6. Light from the source is coupled into one of the inputs of the fiber probe. This light is delivered to the sample via the delivery fiber. The light returned by the sample is collected by the collection fibers and returned by the coupler to the spectrometer for analysis. Fiber optic reflection probes are commercially available but can cost $500–$700. As an alternative, a lower-cost version can be built by purchasing a 1×2 multimode fiber coupler suitable for the visible spectrum. To adapt such a coupler for reflectance measurements, the ends should be connectorized to (a) provide a robust interface with the scattering samples and (b) to connect to the light source and spectrometer. Customization of these parts can be done as part of a design project for students or via a commercial optics prototyping service, although this probably would consume most of the savings.

The final component that is needed is a suitable **light source**. For these experiments, it is desirable to have a spectrum that is flat over the wavelength range of interest (visible, 400–700 nm) and provides sufficiently high power to produce significant backscattering

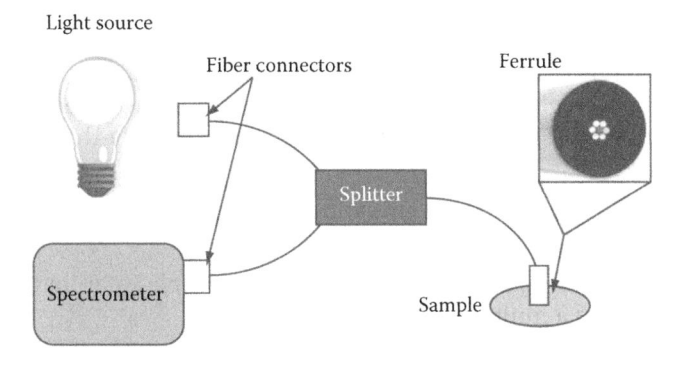

FIGURE 3.6
Schematic of reflectance measurement scheme. Light from the source is coupled into the reflectance probe via a fiber connector. The splitter within the probe channels light from the input path to the ferrule (fiber, inset) which interfaces with the sample. Light returned from the sample is directed by the splitter to the fiber connectorized spectrometer. (Ferrule inset image from www.oceanoptics.com.)

signals. Tungsten lamps are the most economical choices for this laboratory experiment, although X_e arc lamps may also be suitable if available. The output of the light source should be coupled into the input path of the reflectance probe; thus, a proper connector is needed. For multimode fibers, an SMA connector will offer sufficient coupling and reproducibility. Several commercial options are available, such as the light sources for spectroscopy available from Ocean Optics for a few hundred dollars. As an alternative, white LEDs can produce significant amounts of power and can be coupled into a multimode fiber fairly easily. This approach may save up to half the cost of the components but will require students to design the fiber coupling mechanism and optics.

What is needed:
- Spectrometer (visible range (400–1000 nm))
- Reflectance probe
- Light source
- 0.2-micron, 1-micron, and 10-micron–diameter polystyrene ($n = 1.59$) microspheres
- Distilled water ($n = 1.33$)
- Cell culture dishes/well plates or other sample holders

Sample Preparation

Using the methods developed for the previous laboratory exercise, determine the concentration of 0.2-, 1-, and 10-micron–diameter polystyrene microsphere needed to create an absorbance of 2 for a 1-mm light path. The scattering cross-section for each microsphere size may be determined from MiePlot. The smaller microspheres will likely be available from manufacturers such as Polysciences (Warrington, PA) in concentrations ranging from 1% to 10% w/v. The number of particles per ml can be calculated from this value using the density of polystyrene, 1.05 g/mL. The solution of 0.2-micron–diameter spheres from the manufacturer will likely need to be significantly diluted to produce the desired concentration. It may be more efficient to create a 1:10 dilution as a stock solution and then use this to create the samples for the experiment.

Experiment

1. Acquire a reference spectrum by holding the reflectance probe against a reference standard, such as a Spectralon target. If such a target is not available, a bright white piece of paper can be substituted, preferably thicker stock such as a business card.

2. Acquire a "dark spectrum" by initiating an acquisition with the light source disconnected from the reflectance probe.

3. Place a small quantity (a few milliliters) of each sample in an individual culture dish or well within a multiwall plate.

4. Acquire a spectrum for each sample.

5. Notice that the spectra depart from the predicted Mie values due to high-frequency noise. This typically arises due to speckle effects or interference between light scattered by nearby spheres. To overcome this noise source, acquire 10 spectra from each sample and save these data.

6. Calculate the average spectrum for each sample, either using an available function in the data acquisition software or by saving each individual spectrum and then averaging them in a mathematical analysis software environment, such as Excel (Microsoft) or MATLAB (MathWorks).

Data Analysis

1. To isolate the scattering cross-section due to the sample, the measured scattering spectrum must be corrected by the source spectrum. Mathematically, the detected signal $I(\lambda)$ is represented as:

$$I(\lambda) = \mu_s(\lambda) S(\lambda) + D(\lambda),$$

where $S(\lambda)$ is the source spectrum and $D(\lambda)$ is the dark spectrum of the instrument. This correction is applied by subtracting the dark spectrum from the measured average spectrum and then dividing by the source spectrum. For each of the microsphere scattering spectra, apply this correction to produce the scattering coefficient for each.

2. Plot each of the calculated scattering coefficients against the best model, as determined from the calculations in the first part of the lab. It is optimal to convert the scattering coefficient functions to scattering efficiencies and compare to the optimal theoretical model for each size.

3. Calculate the size parameter $x = ka$ and phase shift $\rho = 2ka|n_1 - 1|$ for each of the particles studied here. Based on these values, discuss whether the approximations used to define the validity of each of the theoretical models would still be valid if the relative refractive index went from $n_1 = 1.2$ (1.59/1.33) to 1.03, a more appropriate value for biological scatterers.

4. Summarize and discuss results.

    a. Did the measured scattering values agree with the predictions of the Mie theory? If not, discuss possible origins for this discrepancy.

    b. Compare the magnitude of the observed *backscattering* measurements to the scattering efficiencies predicted by the Mie theory. Discuss what modifications should be made to theoretical analysis to provide a better comparison.

    c. Give examples of biological structures that can be modeled using the Mie theory and how measurements of the scattering spectrum can be used to detect pathological conditions.

Laboratory 4: Biomedical Application of Scattering Measurements

Light scattering can be a powerful means of assessing structure for biomedical applications, as the methods are generally noninvasive and can often provide information on a finer scale than imaging. In addition, while an image of biological structures contains a wealth of information, it may be difficult to analyze those images to produce a simple figure of merit that gives the diagnostic information needed for a particular measurement. Instead, light-scattering measurements can often provide direct access to the structural feature of interest. As an example application, light scattering can be applied to assessing the **confluency of cultured cells using reflectance measurements for tissue engineering applications**. For many cell biology applications, the number of cells that have grown within a space, such as a cell culture dish or flask, is an important indicator of cell viability and growth rates. While cell confluency can be easily assessed using an optical microscope by a trained observer, the need for continual monitoring of a given sample or difficulty in accessing that sample opens the door for a unique biophotonics solution. In this approach, a reflectance probe is used with a light source, similar to the experiments above, but with the spectrometer replaced by a simple photodiode. The reflectance probe is then periodically raster-scanned across the cell culture surface, and modifications in reflectance are used to identify the portions that have been overgrown by proliferating cells. The advantage here is that the measurement has no adverse effect on the sample and thus can be applied repeatedly to monitor its evolution. In addition, this approach can provide quantitative metrics for cell proliferation studies compared to the quantitative assessments of the trained observer. This approach can also provide advantages in studies of cell growth on substrates that are opaque or provide limited access for imaging, such as tissue engineering scaffolds. As an example, engineered blood vessels may be grown in a tubular scaffold composed of an opaque material such as titanium [11]. Rather than destroying the nascent tissue sample to assess its development, a fiber optic reflectance probe, adapted to scan at a right angle to the fiber axis using a prism, can be used to probe the lumen of the titanium tube and assess the confluency of seeded endothelial cells.

A biomedical application of light scattering that has shown particular clinical relevance is the use of **light-scattering spectroscopy to detect cancer**. In this approach, the spectral dependence of light scattering is used to distinguish normal, healthy tissues from precancerous and cancerous cells and tissues. There are several mechanisms used to make this distinction that are relevant to the laboratory exercises above and can form the basis of student design projects. One of the first successful applications of light scattering in biomedicine was the use of the spectral dependence of scattering to determine the size distribution of cell nuclei [12]. This experiment can be repeated by students for a successful project that demonstrates analysis of scattering spectra. The experimental setup is similar to that shown in Figure 4.4. Light is delivered to a cell sample and returned to a spectrometer or other means for spectral analysis. The reflectance probe scheme used in the above laboratory exercise can also be used. Instead of probing a microsphere sample, monolayers of cultured cells, such as T84 human colonic tumor cells, are used. In the original experiments, a diffusing layer of $BaSO_4$ is placed beneath the cell monolayer to mimic the diffuse reflectance from underlying tissue. The light-scattering data may then be analyzed using Equation 3.23 to determine the average size of the cell nuclei in the sample. Another approach for this analysis is to Fourier transform the periodic component of the spectrum to determine the size distribution of the cell nuclei. This approach was used successfully in several trials to detect precancerous cells within intact human tissues [13–15], although later studies identified that absorption due to hemoglobin can confound the information obtained by Fourier analysis. As a control measurement, students can execute a protocol to image cell nuclei using a fluorescent label, such as 4',6-diamidino-2-phenylindole (DAPI), and then quantitatively analyze the size of the cell nuclei using a software program such as ImageJ (NIH).

While the above light-scattering spectroscopy approach uses analysis of scattering spectra to identify the size of specific Mie scatterers, that is, cell nuclei, the method of analyzing the spectral dependence of scattering can also be applied to study light transport through analysis of the transport coefficient μ_s. This approach, termed *elastic scattering spectroscopy*, has shown the ability to distinguish cancerous tissue regions but generally by using advanced statistical methods such as principal component analysis. While such a goal may be too ambitious for a student design project, the basic approach of probing a sample to detect changes in scattering and absorption can form the basis of a successful project. For example, the reflectance probe scheme shown previously can be used to assess the scattering and absorption properties of various tissue phantoms as simulants of normal and healthy tissue. Such phantoms can be composed of polystyrene microsphere suspensions, intralipid samples, or even milk, with changes in scattering simulated by changing the concentration or composition of scatterers and changes in absorption simulated by adding hemoglobin or even something less specific such as coffee or India ink. The reflectance spectra measured from such phantoms can be analyzed to demonstrate the sensitivity in assessing scattering/absorption changes and then compared to literature values for the scattering and absorption coefficients of normal and cancerous tissues. The advantage of using a substance such as hemoglobin is that it has a well-defined spectrum with clear regions of absorption, which can simplify analysis methods that seek to separate the contributions of absorption from scattering [16].

References

1. M. Born and E. Wolf, *Principles of Optics*. Cambridge, UK: Cambridge University Press, 1999.
2. N. Streibl, 3-dimensional imaging by a microscope. *J. Opt. Soc. Am. A Opt. Image Sci. Vis.* 1985;2(2):121–127.
3. A. Ishimaru, *Wave Propagation and Scattering in Random Media*. New York, NY, Oxford: IEEE Press, and Oxford University Press, 1997.
4. Z. Chen, A. Taflove, and V. Backman, Equivalent volume-averaged light scattering behavior of randomly inhomogeneous dielectric spheres in the resonant range. *Opt. Lett.* 2003;28(10):765–767.
5. R. G. Newton, *Scattering Theory of Waves and Particles*. New York: McGraw-Hill Book Company, 1969.
6. T. T. Wu, J. Y. Qu, and M. Xu, Unified Mie and fractal scattering by biological cells and subcellular structures. *Opt. Lett.* 2007;32(16):2324–2326.
7. X. Li, A. Taflove, and V. Backman, Recent progress in exact and reduced-order modeling of light-scattering properties of complex structures. *IEEE J. Sel. Top. Quantum Electron.* 2005;11(4):759–765.
8. K. Dalziel and J. O'Brien, Side reactions in the deoxygenation of dilute oxyhaemoglobin solutions by sodium dithionite. *Biochem. J.* 1957;67(1): 119.
9. A. K. Amerov, J. Chen, G. W. Small, and M. A. Arnold, Scattering and absorption effects in the determination of glucose in whole blood by near-infrared spectroscopy. *Anal. Chem.* 2005;77(14):4587–4594.
10. S. Engelhard, H.-G. Löhmannsröben, and F. Schael, Quantifying ethanol content of beer using interpretive near-infrared spectroscopy. *Appl. Spectrosc.* 2004;58(10):1205–1209.
11. H. E. Achneck, R. M. Jamiolkowski, A. E. Jantzen, J. M. Haseltine, W. O. Lane, J. K. Huang, L. J. Galinat et al., The biocompatibility of titanium cardiovascular devices seeded with autologous blood-derived endothelial progenitor cells: EPC-seeded antithrombotic Ti implants. *Biomaterials* 2011;32(1):10–18.
12. L. T. Perelman, V. Backman, M. Wallace, G. Zonios, R. Manoharan, A. Nusrat, S. Shields et al., Observation of periodic fine structure in reflectance from biological tissue: A new technique for measuring nuclear size distribution. *Phys. Rev. Lett.* 1998;80(3):627–630.
13. V. Backman, M. B. Wallace, L. T. Perelman, J. T. Arendt, R. Gurjar, M. G. Muller, Q. Zhang et al. Detection of preinvasive cancer cells. *Nature* 2000;406(6791):35–36.

14. I. Georgakoudi, B. C. Jacobson, J. Van Dam, V. Backman, M. B. Wallace, M. G. Muller, Q. Zhang et al., Fluorescence, reflectance, and light-scattering spectroscopy for evaluating dysplasia in patients with Barrett's esophagus. *Gastroenterology* 2001;120(7):1620–1629.

15. I. Georgakoudi, E. E. Sheets, M. G. Muller, V. Backman, C. P. Crum, K. Badizadegan, R. R. Dasari, and M. S. Feld, Trimodal spectroscopy for the detection and characterization of cervical precancers *in vivo*. *Am. J. Obstet. Gynecol.* 2002;186(3):374–382.

16. G. Zonios, J. VanDam, L. T. Perelman, V. Backman, R. Manoharan, and M. S. Feld, Quantitative histological analysis of colonic tissue using diffuse reflectance spectroscopy at colonoscopy. *Gastrointest. Endosc.* 1997;45(4):80.

4

Microscopic Tissue Imaging

Principles of Optical Microscopy

This chapter will focus on the optical methods for imaging the microstructure of biological tissues. There are many approaches for optical microscopy that use different imaging methods to observe and record structural features, but they all share the common principle of creating an enlarged image of a small object. In general, there are three requirements for an optical imaging modality. First, a magnified image of the object must be produced. Second, fine details of the object must be resolved. Third, the image must be viewable by either the eye or a recording device. While this last requirement may seem obvious, it is an important aspect of optical microscopy. This chapter will present a brief review of the basics of optical microscopy, but will then shift focus to examining the sources of contrast in various implementations of microscopic imaging.

The **compound microscope** at its heart is a two-lens magnifying system (Figure 4.1). A microscopic object is placed beyond the focal point of the objective lens and a real, inverted image of that object is formed a distance $L + f$ away. The objective lens also functions as the **aperture stop**, the limiting clear aperture of the system that dictates the amount of light that can be collected, as well as the **entrance pupil**, the image of the aperture stop through the optical system. The image formed by the objective lens must fit within the barrel of the device indicated in Figure 4.1 as the **field stop**. A second lens, **the eyepiece**, magnifies the image further such that the total magnification is given by

$$\text{Mag} = M_{obj} \times M_{eye} \tag{4.1}$$

where M_{obj} is the magnification of the objective, defined as the height of the image (S_i) divided by the height of the original object (S_o), with a minus sign included to account for the inverted orientation

$$M_{obj} = -\frac{S_i}{S_o} \tag{4.2}$$

Using the lens maker's equation, the image and object locations can be related to the focal lengths as

$$\frac{1}{L+f} + \frac{1}{O+f} = \frac{1}{f} \tag{4.3}$$

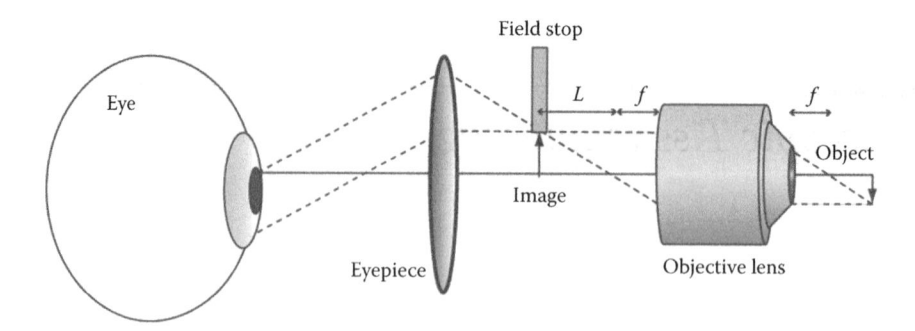

FIGURE 4.1
Schematic of the compound microscope with ray traces. Object is magnified by the objective lens to produce an intermediate image, which may then be viewed through the eyepiece.

where $O + f$ is the distance from the lens to the object, and $L + f$ is the distance from the lens to the image. Using Equations 5.2 and 5.3 and analyzing the triangles in Figure 4.1, the magnification of the objective can be solved in terms L and f as

$$M_{obj} = -\frac{L}{f} \tag{4.4}$$

which is simply the ratio of the length L, called the *tube length*, and the focal length of the objective.

As noted in Equation 5.1, the total magnification also depends upon the properties of the eyepiece. Here, the image formed by the objective is located at the focus of the eyepiece lens, which would place the image formed by the eyepiece at infinity. Thus, the magnification that is achieved is angular and will depend on the distance between the eye and eyepiece

$$M_{eye} = \frac{d_i}{f_{eye}}, \tag{4.5}$$

where d_i is the distance to image, 10 inches for most standard microscopes, and f_{eye} is the focal length of the eyepiece. Thus, the total magnification can be written as

$$\text{Mag} = M_{obj} \times M_{eye} = -\frac{L}{f_{obj}} \times \frac{d_i}{f_{eye}} \tag{4.6}$$

As an example calculation, for a tube length of $L = 160$ mm, a distance d_i of 254 mm (10 in.), and typical focal lengths for a 5× objective ($f = 32$ mm) and a 10× eyepiece ($fe = 25.4$ mm), a total magnification of $M = 5 \times 10 = 50$ is determined.

While the compound microscope was a very useful instrument and a standard laboratory tool for many years, there were practical problems that arose, leading to new designs. In particular, different manufacturers would use different tube lengths, making standardization of magnifications impossible. The Royal Microscopy Society (RMS) standardized the tube length to 160 mm in the nineteenth century to address this issue. (The RMS also standardized the screw threads on objectives, adding the RMS marking to the barrel, commonly known as the "royal screw").

While a standardized tube length improved reproducibility of magnifications, additional problems arose as more advanced microscopy techniques developed. Designers of new microscopy methods sought to insert optical elements such as filters, beamsplitters, and prisms in the beam path to improve imaging properties. These elements resulted in changes to the effective tube length and affected imaging properties. Instead, modern microscope objectives are now **infinity corrected** to solve these problems.

Infinity-corrected objectives (see Figure 4.2) are designed for optimal imaging with the object located at the focal point. Rays from the object emerge from the objective, traveling parallel. Off-axis rays also emerge, traveling parallel but not along the optic axis. The result is an image located at infinity, and the region with parallel rays is called "infinity space." In the compound microscope, a tube lens is used to convert from infinity space to generate an intermediate image, which is then viewed by the user via an eyepiece. Figure 4.3 shows a schematic of this arrangement.

The magnification of the infinity-corrected objective depends on the ratio of its focal length to that of the tube lens used for generating the intermediate image, as in Equation 5.4. Thus, the tube lens focal length (L in Figure 4.3) will set the effective magnification and is referred to as the reference focal length. Typical values for this focal length range from 160 to 200 mm depending on the microscope manufacturer. In general, infinity-corrected objectives cannot be used with finite tube length microscopes, but the reverse can be done at the cost of reduced numerical aperture and magnification.

While infinity-corrected objectives offer practical advantages, such as the ability to insert optical elements in the beam path, there are also imaging advantages. For example,

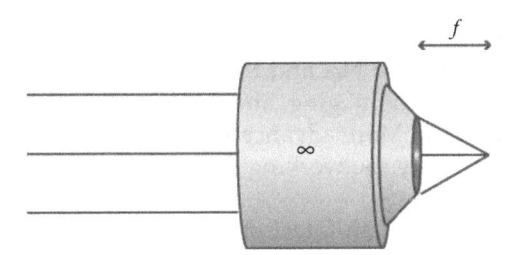

FIGURE 4.2
Illustration of infinity-corrected objective.

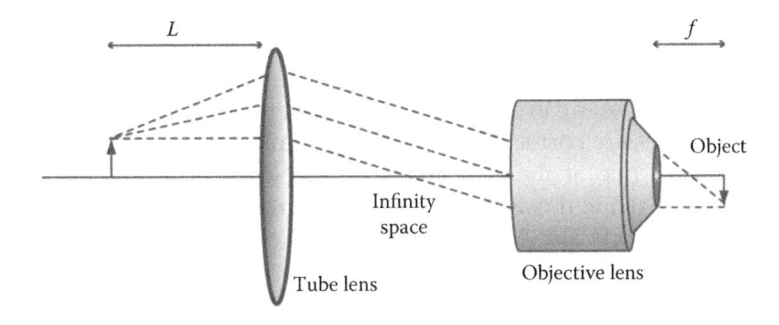

FIGURE 4.3
Imaging with an infinity-corrected objective. The magnification is given by the ratio of the tube lens focal length (L) and the objective focal length (f).

with the longer tube length, the angle for off-axis rays decreases, leading to fewer artifacts. Another advantage arises from the need for an increased focal length of the objective, resulting in a longer parfocal distance, that is, the distance from the top of the objective barrel to the sample (~45 mm), and a longer working distance, defined as the distance from the end of the lens to the sample. As a note, the working distance is rarely equivalent to the focal length of the objective. Objectives with long working distances are often marked LWD.

In general, a simple lens will allow for only a limited range of magnifications and introduce several imaging artifacts. Instead, modern objectives, tube lenses, and eyepieces (in rare cases) include multiple elements to correct for imaging problems such as aberrations and flatness of field. The design of modern objectives is a rich field, and entire texts are dedicated to their design. Rather than giving an in-depth discussion of this approach, it is better to focus instead on critical parameters that users can refer to in order to assess the suitability of an objective for their application.

The resolution of a microscope objective is its ability to discriminate two closely spaced objects. It depends on the wavelength of light used λ, the numerical aperture (NA), and the refractive index n of the material between the object and lens

$$R = \alpha\lambda/n\sin\theta = \alpha\lambda/\text{NA} \tag{4.7}$$

In this expression, the factor α represents a number that depends on the model used to define what criteria are used to determine that two objects are separated. A generally accepted value is to take $\alpha = 0.5$ based on heuristic arguments. However, using the properties of diffraction to analyze resolution will give a value of $\alpha = 0.61$, which is more rigorously correct. The refractive index $n = 1$ for air, but Equation 5.7 shows that increasing this value will improve resolution. To this end, immersion objectives can use water ($n = 1.33$) or oil ($n = 1.45–1.5$) to increase the NA and thus improve resolution. Objectives must be designed to allow for immersion imaging, and the barrels are typically marked w/w for water or OIL for oil immersion. More advanced objectives are designed with a correction collar that can be turned to vary the working distance or allow for a range of immersion media.

Three types of imaging errors can affect objective performance. These include *spherical aberration*, which arises when different off-axis rays are focused to different points, degrading the resolution of the lens. This aberration can be mitigated by using an aperture to limit off-axis rays but at the cost of reducing resolution. It can be corrected via better control of lens grinding and materials used. *Chromatic aberration* results from variations of the refractive index of objective materials with wavelength, resulting in a different focal length for each color. The result is a colored "ghost" in the image. This aberration is corrected by using two different types of glass, each with a different dispersion. For example, using crown and flint glasses can correct chromatic aberration with careful optical design. Only the best objectives are corrected for all colors, while lower-cost lab microscopes will still retain a greenish hue at the best focuses. Color-corrected objectives are typically labeled as achromatic, while those corrected for red, green, and blue will be marked apochromatic. *Field curvature* is also a defining feature of objective performance. In paraxial approximation, it is assumed that the focal plane of a lens is perpendicular to the optic axis. However, the focus lies in a paraboloid surface, called the Petzval surface, with off-axis foci moving closer to the lens. The impact on imaging is that it can be difficult to keep the center and the edges of the image in focus simultaneously. Specially designed objectives correct for this and are labeled "plan" or "plano." For example, a plan achromat will have flattened field curvature and two-color correction. The need for flat field objectives has become more acute for photomicrography, where the whole image is recorded at once. This aberration can also limit axially sensitive techniques such as confocal microscopy. Finally, it should be noted that some manufacturers include corrections in the tube lens and eyepiece, which can limit the use of objectives designed for particular microscopes.

Even with a superior imaging system, microscopes can produce poor images without proper illumination. The goal is to produce glare-free, even illumination that is bright enough to produce contrast. Two forms of illumination are widely used to reach this goal, critical and Kohler illumination. Figure 4.4 shows schematics that compare these two forms of illumination.

In critical illumination, the image of the filament within the bulb is made to coincide with the object plane. For the structure of the illumination not to influence the evenness of the image, a diffusing glass is typically inserted to randomize the incoming light. In this scheme, the condenser diaphragm is used to limit the angle of illuminating rays. This is an important feature and enables the numerical aperture of illumination to match the collection of the objective. If this diaphragm is set too wide, stray light will reduce contrast; if set too narrow, the full resolving capabilities of the objective will not be realized.

In comparison, Figure 4.4 below also shows the scheme for Kohler illumination. First devised by August Kohler, a staff member at Carl Zeiss AG, and published in 1893, the Kohler illumination scheme represents an advance over critical illumination by offering

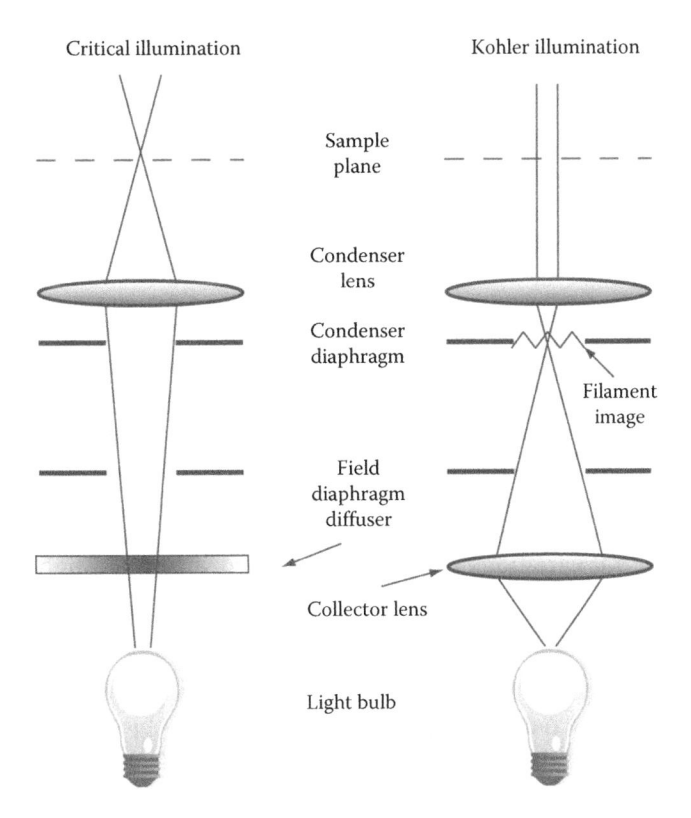

FIGURE 4.4
Comparison of critical and Kohler illumination.

more control and providing more even illumination. The scheme differs from critical illumination by including a collector lens with the light source at its focus. This arrangement now puts the image of the filament before the condenser lens and coincident with the condenser diaphragm. This arrangement allows control of the amount of light incident on the sample by adjusting the condenser diaphragm without changing the uniformity or area of illumination. The field diaphragm (or field aperture) is conjugate to the image plane such that adjusting its width controls the area of illumination. Since the filament is not focused on the sample, it produces uniformly diffuse illumination. Laboratory Exercise 1 will discuss implementation and alignment of Kohler illumination.

Types of Contrast in Microscopy

Phase Contrast

Visualizing the individual features of a given object via microscopy depends on their ability to generate *contrast*, defined as the difference in intensity (and color) that permits them to be distinguished from the background. In general, the eye needs approximately a 2% (0.02) contrast to sense a difference. However, with digital photography, this limit can be artificially extended by a significant degree. The mathematical definition of contrast is

$$Contrast = \frac{(I(s) - I(b))}{I(b)} \times 100\% \tag{4.8}$$

where $I(s)$ is the intensity of the signal and $I(b)$ is that of the background. The observation of contrast due to changes in the total amplitude of transmitted light is generally termed *bright-field microscopy*, with the key mechanism of generating contrast arising from absorption of light.

Unfortunately, many microscopic specimens, in particular biological cells, provide little inherent contrast since they are mostly transparent and absorb only small amounts of light, producing correspondingly small changes in amplitude. Perhaps the easiest method for rendering transparent objects, such as cells, visible is to add a staining agent. For example, many tissue specimens are stained with hematoxylin and eosin (H&E) to create contrast. The hematoxylin stain gives cell nuclei a dark blue coloration, while the eosin stain will impart a pink color to the cell cytoplasm. This approach works by localizing the stains within certain structures and relying on their optical absorption properties to create contrast. Here, spectral variations in the amplitude of transmitted light provide contrast rather than changes in brightness due to modulation of the total transmitted intensity. This approach is quite effective since the human eye is very sensitive to changes in wavelength, that is, color.

An alternative approach to improve the visibility of transparent objects such as living cells is to exploit the wave nature of light. While transparent objects change the amplitude of transmitted light very little, they can impart a substantial change in *phase* due to differential optical thicknesses. This contrast mechanism arises due to changes in *refractive index* n, defined as the ratio of speed of light in that medium compared to its speed in vacuum. The optical path length (OPL) is then given by the product of the object's physical thickness *t* and its refractive index: OPL = *nt*. The corresponding phase shift δ is then found by the ratio of the OPL to the wavelength of light multiplied by 2π.

$$\delta = 2\pi \times \text{OPL}/\lambda \qquad (4.9)$$

Note that this phase shift does not change the amplitude of transmitted light but only imparts a delay relative to light that has not interacted with the sample. Since the human eye can only observe changes in amplitude and not phase, to visualize these delays, a method is needed for converting these phase changes into amplitude changes. Two standard types of interference-based microscopy methods have been developed, phase contrast microscopy and differential interference contrast (DIC) microscopy, each with different approaches to converting phase into amplitude changes.

Phase contrast microscopy was first described by Frits Zernike in 1934, and he was awarded the Nobel Prize in 1953 for this advance. The approach is centered on using interference effects to convert phase variations of the light field that has interacted with a transparent sample into amplitude variations, which can then be readily visualized. To understand how this approach exploits the wave nature of light to generate contrast, let us consider the phase object shown in Figure 4.5. Here, the object is characterized by a refractive index *n′* and thickness *t*, but provides no absorptive features to generate contrast. Light that has interacted with this object will be changed, however, due to the increased optical path length. Using Equation 5.9, we see that the change in refractive index will generate an increase in optical path length equal to $\delta = 2\pi(n-n')t/\lambda$, where *n* is the refractive index of the surrounding medium and λ is the wavelength of illuminating light. To characterize the magnitude of this change, let us consider typical characteristics of a cell, which would be on the order of a 5-μm-thick object with refractive index change of $(n-n') = 0.025$, producing a phase shift of $\pi/2$.

To generate contrast from this phase shift, one must differentiate the light that has interacted with the sample from the light that is undeviated. Fortunately, the optical phenomenon of *diffraction* will cause some of the light that has interacted with the sample to change direction (dashed arrows in Figure 4.5), providing a distinguishing feature that can be exploited to generate contrast. The key innovation in phase microscopy is the approach for causing interference between the diffracted and undeviated light.

Let us consider the interference of the undeviated light with the diffractive wave from the example in Figure 4.5. Given that the object presents only a small change in the index of refraction, the resulting diffracted wave will be significantly smaller that the remaining undeviated light but will offer a phase shift. Figure 4.6 shows these two waves

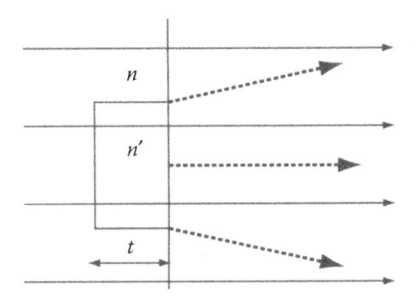

FIGURE 4.5
Light transmission through a phase object. Dashed arrows indicate diffractive light.

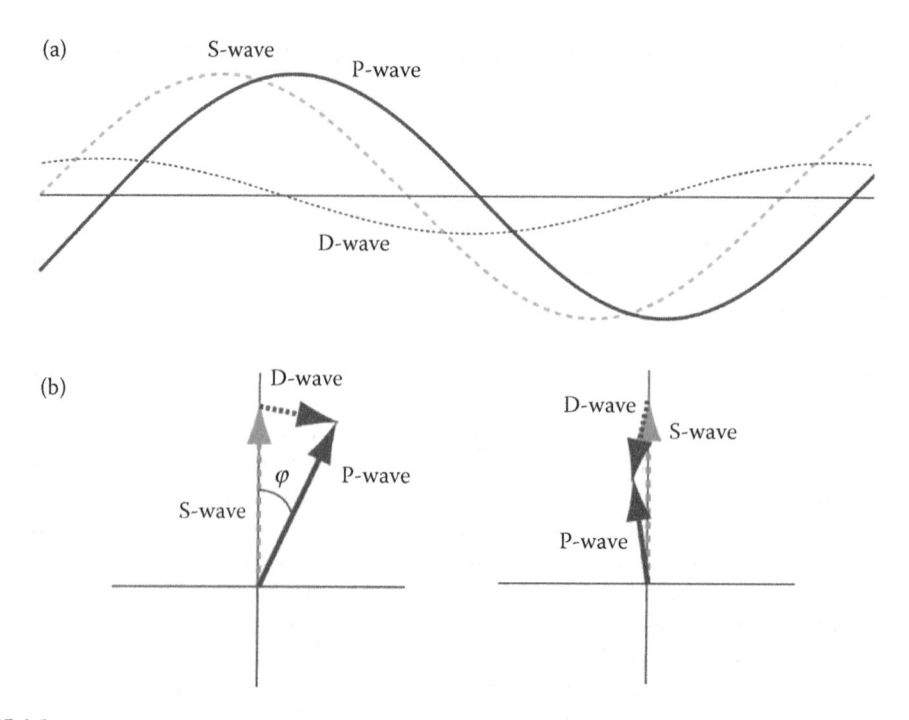

FIGURE 4.6
(a) Interference of surround wave (S-wave) with diffracted light (D-wave), producing the resulting P-wave (particle wave). (b) Bottom left shows the waves in polar coordinates. By shifting the D-wave by 90°, the contrast is increased by maximally altering the amplitude of the P-wave.

(S-wave = "surround" wave for undeviated light and D-wave for diffracted light) and the resulting wave (P-wave = particle wave) due to their interference.

In this example, the interference of the two waves has little effect on the observed contrast. The amplitude of the resulting P-wave is phase shifted, but its amplitude is minimally changed. To create a larger contrast, the amplitude of the P-wave may be changed to a greater degree by shifting the phase of the D-wave. As observed here, a phase shift of 90° of the D-wave causes a greater change in amplitude of the P-wave, improving contrast.

The phase-contrast microscope is based on phase-shifting diffracted light to improve the contrast of the observed image upon interference with the undeviated light. Figure 4.7 shows how this is implemented in a microscopy scheme. The illumination train is modified to include a condenser annulus adjacent to the condenser aperture, located at the back focal plane of the condenser. The annulus masks the illuminating light to create parallel rays incident on the sample. The illumination travels at an angle relative to the optical axis, so the typical description as a "cone of light" is not entirely accurate but is a good analogy to help with intuitive understanding of this scheme. The combination of the condenser and objective image the annulus onto the back focal plane of the object where the phase plate is located. The phase plate is an annular ring of phase-shifting material that imparts a + or − 90° shift to the undeviated light. Schemes with a +90° shift are termed positive phase contrast, while those with a −90° shift are termed negative phase contrast. In practice, the phase-shifting material is usually included in the housing of the objective.

As described above, transparent samples will cause a small amount of light to be diffracted, indicated by the dashed lines in Figure 4.7. The mechanism of contrast here is the conversion of the phase shift due to the sample into an intensity variation by interfering

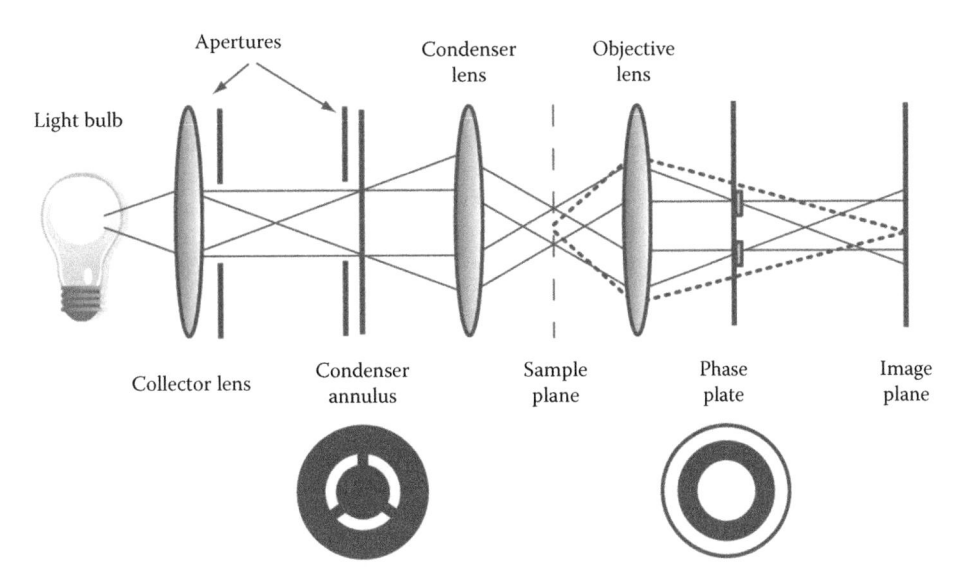

FIGURE 4.7
Optical train of the phase contrast microscope. Solid lines indicate illumination path and undeviated light. The dashed line indicates the path of light diffracted by the sample.

this diffracted light with the undeviated light. Since the illumination through the annulus is imaged onto the phase plate, a complimentary ring of phase-shifting material is used to change the phase of the undeviated light while leaving the diffractive light nearly but not entirely unchanged. A partially absorbing material is usually included in this ring to decrease the amplitude of the undeviated light as well as to increase contrast.

In the image plane, the diffracted wave forms an image of the sample. The dashed lines in Figure 4.7 indicate the path of light from one point on the sample plane to the corresponding point on the image plane. The phase-shifted, undeviated light is evenly spread across the image plane, resulting in interference between the two, causing the phase shifts to become amplitude changes.

The phase-contrast microscope has standardized parts, with the condenser annuli and objectives having rings of matched sizes. The notations Ph1, Ph2, and Ph3 are used to denote annuli of increasing radius. The corresponding phase rings shrink in size with increasing magnification. As noted previously, the ring is deposited on a plate at the back lens of the objective. As the magnification changes, the size of the image changes; thus, different-sized rings are needed to match.

To summarize, phase-contrast microscopy offers the ability to visualize transparent samples, making it well suited for examining live cells. However, the approach requires some degree of diffraction, but not too much, such that the technique is useful only for examining weakly scattering samples. Another limitation is that interpretation of phase-contrast images can be complex due to the coupling of thickness and refractive index in optical path length. Small objects with short path lengths but high refractive indices can appear similar to thicker objects with lower refractive indices. Absorption in the sample can also cause confusing effects that may be misinterpreted as a phase shift.

There are two common artifacts associated with phase-contrast microscopy that one must contend with: halo and shade off. Figure 4.8 shows typical phase-contrast images that illustrate these artifacts. In general, objects imaged with phase contrast tend to show a bright halo around the edges. This is an inherent artifact of the approach, arising from

Halos in phase contrast and DIC microscopy

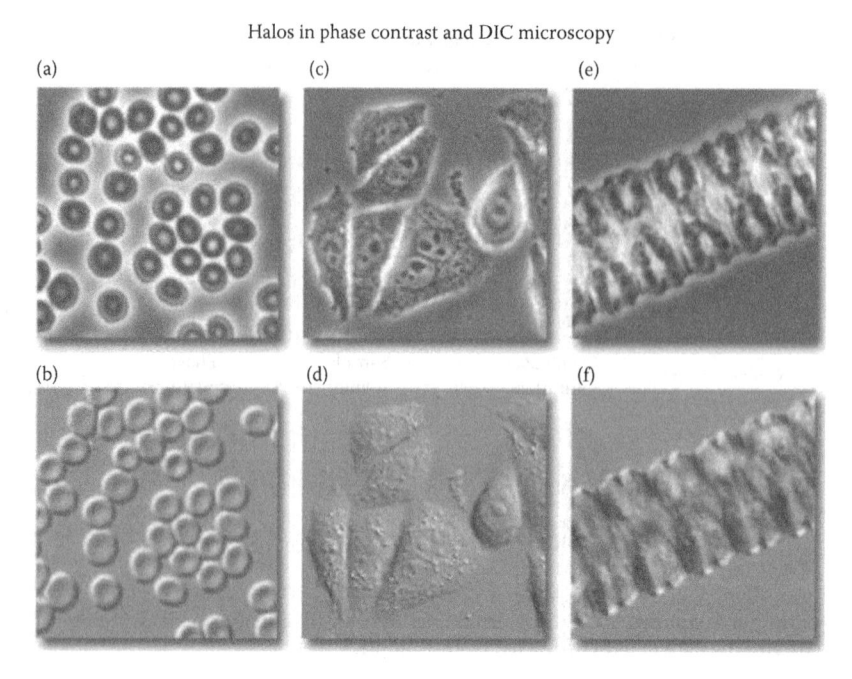

FIGURE 4.8
Comparison of phase contrast and DIC microscopy for red blood cells, HeLa cells, and Zygnema, green algae.

phase-shifting of the diffracted light by the phase ring. Ideally, only the undeviated light would be phase shifted, but since a small portion of the diffracted light must also travel through the phase-shifting ring, the halo artifact is unavoidable. Figure 4.8a–c shows varying degrees of halo. The shade-off artifact occurs in phase-contrast imaging for larger objects. Since only the edges of objects will diffract light, the contrast produced by phase-contrast microscopy can be uneven such that the centers of larger objects will have a different appearance than the edges.

An alternative phase-imaging approach that avoids these artifacts is differential interference contrast or Nomarski microscopy. Although the approach was first devised by Francis Smith in 1955, it was Georges Nomarski who standardized the configuration and for whom it is named. This form of modified polarization microscopy avoids the artifacts peculiar to phase-contrast microscopy but has its own drawbacks to contend with.

Similar to the phase-contrast microscope, the DIC microscope (Figure 4.9) is based on converting phase shifts due to passage of light through a sample into intensity changes that can be directly viewed. In DIC, this conversion is based on the exploitation of polarization, with two orthogonal polarizations, each traversing slightly displaced portions of the sample. This is accomplished by passing polarized light through a Wollaston prism, oriented at 45° relative to the polarization axis, which splits the light into two orthogonally polarized components that are diverging. The condenser lens is placed with its focal plane aligned with the internal face of the prism so that these two components are made to once again travel parallel to the optical axis but slightly displaced from each other when they are incident on the sample. After passing through the sample, these two components are brought back together by the objective lens and a second Wollaston prism. To visualize the phase shift due to the sample, a second polarizer, oriented at 90° to the first, blocks any light that is not phase shifted, creating intensity contrast.

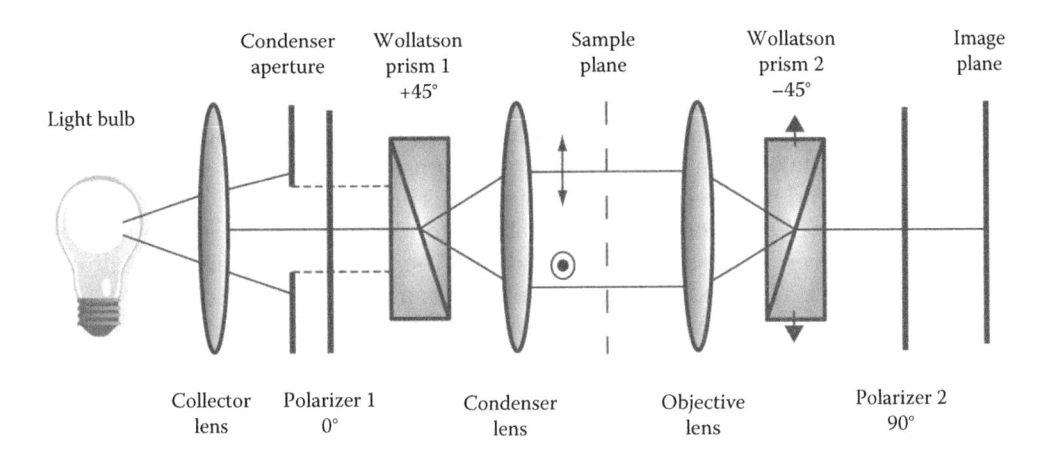

FIGURE 4.9
Optical train for the DIC microscope. Polarized light is split by the first Wollaston prism to generate an offset between two incident polarizations. The second Wollaston prism recombines these two such that only phase-shifted light will pass the second orthogonally oriented polarizer.

The mechanism of contrast here is due to the different phase shifts acquired by each of the two polarizations as it traverses the sample. Figure 4.10 illustrates this effect by considering an object with a small refractive index change compared to the background. The polarization parallel with the page, shown as the arrow, is displaced slightly from the polarization that is perpendicular to the page, shown as the circle. As a result, each polarization experiences different phase shifts at certain transverse positions on the sample. At the center of the sample, both polarizations experience the same phase shift, so there is no net difference between them (these are slightly displaced in the figure for clarity). When these two polarizations are combined by the second Wollaston prism, they again form the same polarization created by the first polarizer. The second polarizer, oriented at 90° to the first, blocks this incident light, resulting in a region of zero intensity. However, when one polarization lies beyond the edge of the object while the second lies within the object, there is a net phase shift. When these two polarizations are recombined, the result is an elliptical polarization that is partially passed through the second polarizer. This produces

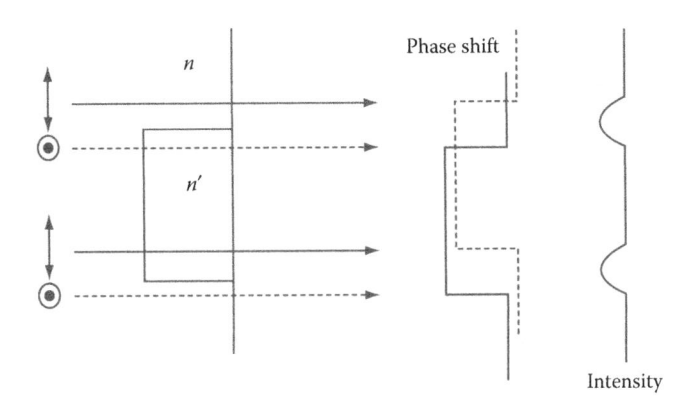

FIGURE 4.10
Illustration of phase shifts acquired by each polarization in DIC imaging and the resulting intensity profile.

a region of nonzero intensity in the image plane, yielding contrast at this point. The DIC image thus appears as a shadowing effect, akin to a relief map, with areas of increased intensity wherever there is a refractive index transition in the sample. In practice, the second Wollaston prism can be translated slightly to induce a bias to the image such that the overall background takes on a grayish hue, positive phase shifts appear as darker regions of lower intensity, and negative phase shifts appear as bright regions.

The clear benefit of DIC microscopy is the ability to visualize transparent samples with some advantages over phase-contrast microscopy. First, the approach enables full use of the condenser aperture, providing higher resolution compared to phase contrast. The DIC approach also avoids the haloing and shade-off problems associated with phase contrast microscopy. Also, since DIC does not require weakly scattering samples, it is possible to examine thicker samples than is possible with phase contrast. The DIC method is not without its own unique shortcomings as well. The main drawback is that contrast is provided in DIC only along the shear axis, that is, the direction of displacement of the two polarizations. While it is possible to rotate the shear axis, it requires a more complex system or sample manipulation. Another significant drawback is incompatibility with birefringent samples. For example, a plastic culture dish will induce its own polarization-dependent phase shifts on the two beams, making it difficult to discern the true contrast due to the sample. Additional limitations include limited light throughput due to the use of polarizers and the fact that the Wollaston prisms, often called DIC sliders, can be expensive.

Spectral Contrast: Scattering

The microscopy methods discussed above generally do not offer spectroscopic contrast, that is, the ability to use different wavelengths of light to discern structure. Exogenous stains can provide colorimetric contrast for certain samples, identifying structures by the addition of color. However, the wavelength-dependent behavior of light scattered by living cells, due to structures on wavelength scales finer than what the eye can see, can also be used to provide contrast. A recent review of optical spectroscopy of cells presents a broad overview of such techniques [1].

Early optical spectroscopic techniques for studying cells, such as light scattering spectroscopy (LSS) [2], used wavelength-dependent scatter to obtain structural information but generally did not provide imaging capabilities. A more recent development is the confocal light scattering and absorption spectroscopy (CLASS) technique, which analyzes the spectra originating from a small confocal volume to infer local structure, providing a unique type of contrast. The principle of confocal microscopy is discussed in detail below as a method for detected light originating from a localized volume. In CLASS, confocal microscopy is combined with spectroscopic detection to obtain contrast at length scales that extend below the diffraction limit. Figure 4.10 shows an example of the technique. While ordinary imaging provides diffraction-limited information (see Equation 5.7), typically on the order of the wavelength of light used, the limit can be circumvented by including spectral information. In Figure 4.10a, a point scatterer appears only as a bright spot but by using spectral information (Figure 4.10b), different-sized subdiffraction scatterers can be discriminated (Figure 4.10c,d).

In partial wave spectroscopy (PWS), spectral information is also used to provide contrast (Figure 4.11). However, rather than ascribing a particular size to a specific structure, in PWS, the spectra are used to describe the statistical properties of intracellular refractive index (RI) fluctuations, quantified by a disorder strength parameter. In this method, spectral information is recorded for a particular region of a cell sample, and the fluctuations in that spectrum are identified with RI fluctuations *within* that sample [5]. The PWS

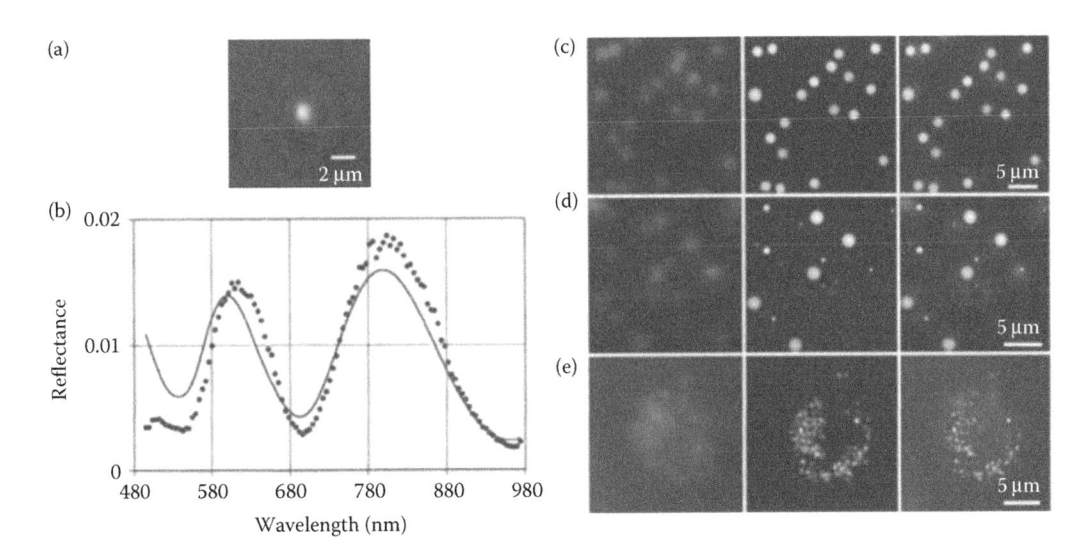

FIGURE 4.11
Confocal microscopy image of a 535-nm bead (a), along with the theoretical (solid line) and measured spectra (points) (b). (Taken from H. Fang et al., *Appl. Opt.* 2007;1760–1769.). (c)–(d) Fluorescent polystyrene beads of various sizes imaged by fluorescence microscopy (left) and CLASS (middle). Right images are an overlay. (e) Human epithelial cells (16HBE14o) with the lysosomes stained with LysoTracker Red DND-99. (Taken from I. Itzkan et al., *PNAS.* 2007;104(44):17255–17260.)

technique also circumvents the diffraction limit, and theoretically there is no limit to the minimum scales to which it is sensitive, providing unique insight into the fundamental building blocks of cells.

Spectral Contrast: Fluorescence

Another approach for exploiting color variations to provide contrast is the use of fluorescence as a source of contrast. Certain molecules, due to their chemical structure, can absorb light and re-emit it at a longer second wavelength. This process can create contrast since imaging schemes can be constructed where fluorescent molecules are the only emitters in a particular spectral range. The difference between the wavelength absorbed by a fluorescent molecule and the emitted wavelength is termed the Stokes shift, for Sir George G. Stokes, who observed the process in the nineteenth century.

The use of fluorescence for imaging contrast requires several key components in the microscopy scheme, including an excitation source, knowledge of the fluorescing molecules, a wavelength filter to separate excitation and emitted light, and a mechanism to detect the emitted light. Figure 4.12 shows these various components.

In fluorescence microscopy, the main criterion for selection of a light source is that sufficient light intensity can be realized for a small spectral band that aligns with the absorption peak of the targeted fluorescent molecule. Coherent light sources, specifically lasers, can serve as excellent sources for fluorescence microscopy since they offer high intensities and are narrow bandwidth. Some laser sources can be expensive, but many are available at a reasonable cost. Most lasers cannot be arbitrarily tuned to meet a specific absorption peak, but with the wide variety of lasers currently available, close matches can usually be achieved. Lasers are also well suited for creating tightly focused excitation spots, which are particularly useful for scanning techniques such as confocal microscopy (see discussion

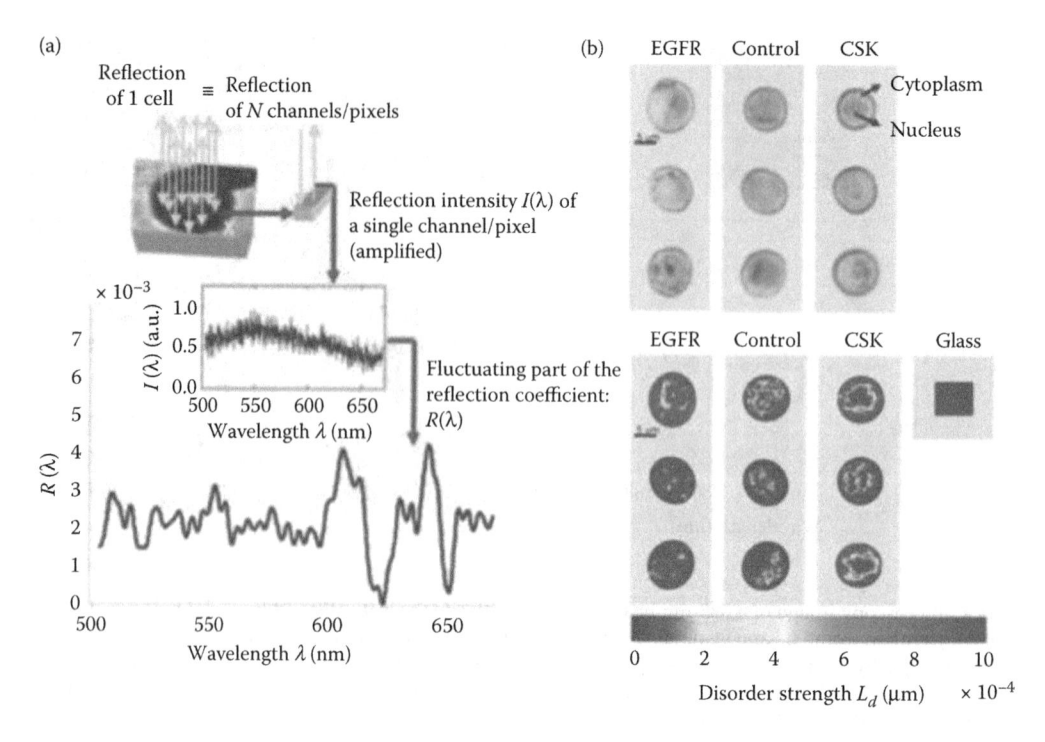

FIGURE 4.12

Example of PWS imaging. (a) Fluctuating component of scattering spectrum is extracted and analyzed. (b) H&E stained (top) and PWS (bottom) images of HT29 human colonic adenocarcinoma cells demonstrate different types of contrast, which can provide complimentary diagnostic information. (Taken from H. Subramanian et al., *Proc. Natl. Acad. Sci.* 2008;105(51):20118–20123.)

that follows). A traditional light source in fluorescence microscopy is a mercury (Hg) or xenon (Xe) arc lamp. These sources are lower cost than most lasers and offer lots of power in the ultraviolet and blue portions of the spectrum, where many fluorescent molecules have their absorption peaks.

Many fluorescent dyes are commercially available today, including a wide array of excitation and emission wavelength parings and an even wider array of targeting moieties. For example, the 4′,6-diamidino-2-phenylindole (DAPI) stain targets cell nuclei, absorbing ultraviolet light and producing blue fluorescent light. On the other hand, the fluorescein molecule has been adapted to many uses in fluorescence microscopy. In particular, fluorescein isothiocyanate (FITC, called "fitsy") absorbs blue light and re-emits green light, with several commercial adaptations that improve its stability and target specific cell structures via antibody labeling. Figure 4.13 shows an example of MCF-7 cells, a breast cancer cell line, with DAPI-stained nuclei and mitochondria stained with a FITC dye.

The utility of fluorescence dyes can be assessed by comparing their quantum efficiency φ, defined as the number of emission photons generated for each excitation photon. This number must be less than 1 and in practice can be a small fraction of a few percent or less. Another limitation of fluorescent molecules is their tendency to photobleach. This is an irreversible process that results in elimination of a molecule's ability to fluoresce and introduces a practical limitation on the length of time a florescence molecule can be used as a contrast agent.

Each fluorescent label requires selection of a filtering scheme to produce high-contrast images. An excitation filter is used to restrict the excitation light to the portion of the spectral

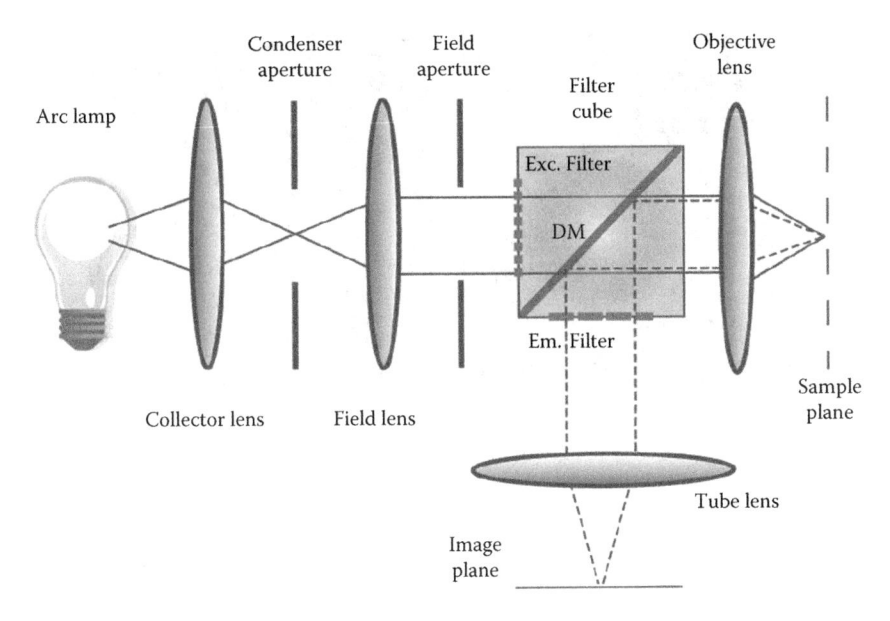

FIGURE 4.13
Schematic of fluorescence microscopy optical train. Excitation light is shown as solid lines, emitted light is shown as dashed lines. Filter cube includes excitation (Exc.) filter, dichroic mirror (DM), and emission (Em.) filter.

region where light is efficiently absorbed by the molecule. A dichroic mirror is used to separate the excitation and emission light to prevent stray excitation light from reducing image contrast. Finally, an emission filter is used to select the range of detected wavelength that corresponds to the emission peak of the molecule. Often, these three filtering elements are incorporated into a filter cube that can be easily exchanged in many fluorescent microscopes. Figure 4.12 shows an example of a filtering scheme installed within such a cube.

The final element in a fluorescence microscopy scheme is selecting an appropriate detector. In general, high numerical aperture objectives will improve fluorescence detection due to their ability gather a large amount of light. To further improve fluorescence detection, a sensor with the ability to provide gain is also desired. Photomultiplier tubes can provide a large amount of gain but are commercially available only as single-channel detectors. This configuration can be used for point-scanning modalities such as confocal microscopy. However, for wide-field imaging of fluorescence, a charge-coupled device (CCD) imager with built-in gain is an optimal choice. CCDs with gain include an image intensifier or electron multiplier that increases the number of counts per detected photon as well as a means to ensure that the signal remains spatially segmented, such as a microchannel plate. Other features that can be useful include blue-enhanced CCDs that incorporate a phosphor sheet to improve response at low wavelengths or the use of a color camera to distinguish multiple fluorescent markers.

Spectral Contrast: Nonlinear Scattering

Inelastic scattering of photons by a sample can serve as a contrast mechanism for microscopic imaging. In this approach, spectral resolution is needed to measure the energy shift of inelastically scattered photons, also termed spontaneous Raman spectroscopy. This approach is highly specific to molecular transitions and can provide contrast based on

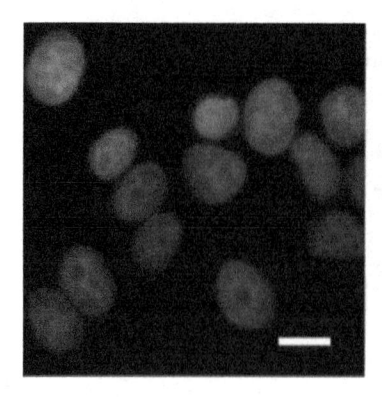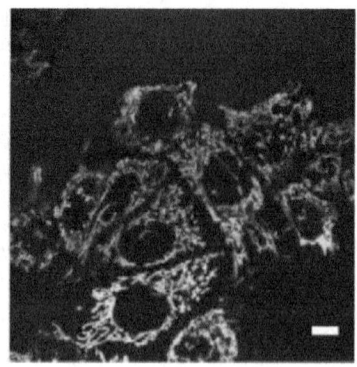

FIGURE 4.14
Examples of fluorescence microscopy images of MCF-7 cells with DAPI-stained nuclei (left) and FITC-stained mitochondria (right). Scale bar is 10 μm. (Adapted from K. J. Chalut et al., *Cancer Res.* 2009;69:1199–1204 [6].)

biochemical composition of a sample. To generate an image using Raman shifts (see Figure 4.14) as the contrast mechanism, raster scanning of a laser spot across the sample is typically required, and spectral data are acquired for each point. Confocal collection (see below) can improve spatial resolution by ensuring that the detected light originates from the same spot as the excitation. The efficiency of Raman scattering is quite low, typically on the order of one photon in 10^6, so acquisition times of a few seconds per pixel are quite common. Methods for improving throughput include parallel collection schemes or use of signal enhancement via metal particles or substrates, termed surface-enhanced Raman scattering (SERS).

A variation on this approach of Raman scattering spectral shifts for microscopic contrast is coherent anti-Stokes Raman scattering (CARS) microscopy. In CARS microscopy, the sample is illuminated with three frequencies of light, ω_{pump}, ω_{probe}, and ω_{Stokes} (see Figure 4.14), although in most CARS schemes, the same frequency is used for the pump and the probe beams. Contrast is generated by matching the energy difference between the pump beam and the frequency-shifted Stokes beam to a Raman scattering transition in the sample. The resulting output is at new frequencies, given by $\omega_{AS} = 2\omega_{pump} - \omega_{Stokes}$ (the anti-Stokes frequency shown in Figure 4.14). Since these new frequencies are blue shifted (anti-Stokes) from the illumination, they can easily be distinguished from any fluorescent background. CARS microscopy was first demonstrated in 1982, imaging cells with contrast based on the 2450 cm^{-1} resonance of D_2O. In recent years, the approach has been widely pursued for microscopy due to its high sensitivity and molecular contrast. As with Raman spectroscopy, CARS can be used to extract the chemical composition of tissue, but because CARS is a nonlinear method, it can provide 3-D optical sectioning in tissue.

Depth Sectioning in Microscopy

In the previous section, various types of contrast in microscopy have been considered based on different interactions between light and biological samples. By tailoring the optical system to exploit a particular type of contrast, powerful microscopy approaches have been developed and become widely available. While it may seem difficult to see a common thread across these modalities, they do all share the aspect of projection imaging

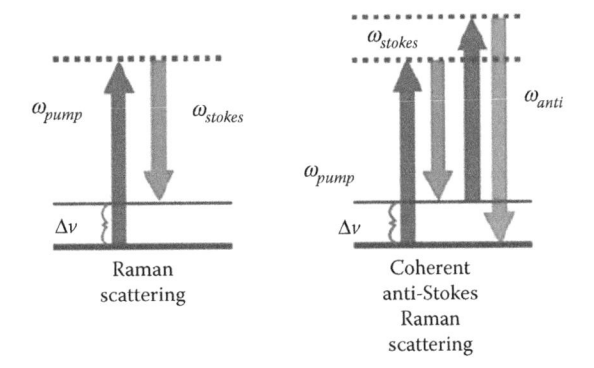

FIGURE 4.15
Energy transitions for Raman scattering and CARS.

where the information is resolved in the transverse dimensions but integrated across the third, axial dimension. We now turn our attention to optical microscopy techniques that can resolve information in the axial dimension via depth sectioning.

The first optical microscopy technique for depth sectioning to become widely practiced was confocal microscopy, invented by Marvin Minsky and patented in 1957. The main principle is illustrated in Figure 4.15. Light from a source is spatially filtered to appear as a point, which is then imaged onto the sample of interest. An alternative is to use a collimated laser beam, which is then brought to a focal point on the sample by the objective. The light returned from that point is imaged onto a confocal plane with a matched pinhole. The amount of light that passes through the second pinhole provides information about the sample properties at that particular location. Light that originates above or below the pinhole, as illustrated in Figure 4.15, is not efficiently passed through the pinhole, resulting in the ability to discriminate by depth. A three-dimensional image can then be realized by moving the illumination spot across the sample. The original design of the confocal microscope used an arc lamp with a pinhole as the light source and required the sample to be translated to generate an image. In contrast, today's modern confocal microscope is a complex integrated system that includes laser illumination, optical scanning of the incident spot, electronic detectors, and computer control and display.

An alternative approach for optical sectioning is coherence gating via low-coherence interferometry (LCI). LCI is the basis of optical coherence tomography (OCT), a biomedical imaging technique used to provide cross-sectional images of tissue *in vivo* [7]. In this approach, depth resolution is achieved by exploiting the low-coherence properties of a broadband light source in an interferometry scheme (Figure 4.16). Based on a Michelson interferometry scheme, light from a low-coherence source is split by a beamsplitter (BS) into a reference beam and an input beam to the sample. The reference beam is reflected by a moving mirror (M), conferring a Doppler frequency shift. The reference beam is combined with the optical field returning from the sample at BS, producing an interference pattern between the two, provided that the path lengths are matched to within the coherence length of the source. The interference pattern oscillates at the heterodyne frequency, given by the reference beam Doppler shift. The photocurrent measured by the detector is demodulated to yield the amplitude of the signal field. By scanning the path length of the reference beam, the sample reflectance is mapped out as a 1-D depth profile.

In OCT, three-dimensional tomographic images are built up from these one-dimensional axial profiles by lateral scanning. When used for biomedical imaging, OCT enables

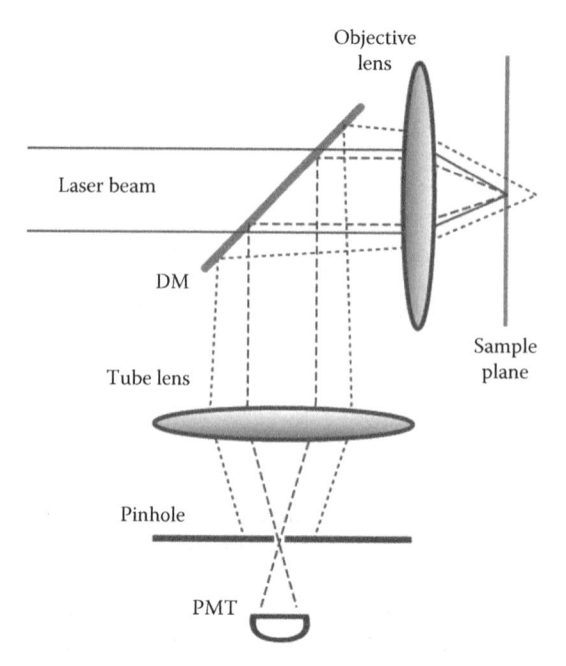

FIGURE 4.16
Optical train for confocal microscope. Dashed line shows light originating in focal plane of objective. Out-of-focus light shown by the dotted line is not efficiently passed through the confocal pinhole to the photomultiplier tube (PMT).

noninvasive cross-sectional imaging with high spatial resolution, typically about 5 to 10 microns in the axial direction and a few microns in the lateral direction. By rapid scanning of the optical path length, high-frame-rate imaging can be achieved, while the high dynamic range of the heterodyne detection scheme permits imaging to depths of a few millimeters in biological tissues.

In recent years, LCI and OCT have shifted to frequency (or Fourier) domain detection instead of the time-domain detection described previously due to increased signal fidelity [8,9]. To understand the difference between these two signal acquisition methods, we again consider a Michelson interferometer but with detection in the Fourier domain (FD), as illustrated by Figure 4.17. Here, a source with a wide bandwidth, $S(\lambda)$, is split into a sample and reference fields. Light scattered by the sample (idealized here as a series of discrete scatterers $m = 1, 2,$ and 3) is recombined with the reference field at the beamsplitter and detected using a spectrometer. Fourier transformation of the spectral data produces a correlation function that describes the optical path length differences between the distance from BS to the scatterers and the distance to the reference mirror, again producing a depth-resolved reflection profile. Figure 4.17 presents the resulting profile for the ideal sample (Figure 4.18).

Another recently developed method for depth-resolved imaging is photoacoustic microscopy, in which wavelength-specific absorption provides the contrast, but detection is based on receiving acoustic signals generated by the absorbed energy. This approach can provide not only the structural distribution of a chemical, such as hemoglobin, but its quantitative concentration may be mapped. The main intrinsic photoacoustic contrasts for biomedical imaging are hemoglobin or melanin absorption, earning the technique major *in vivo* utility for functional imaging of blood vessels, brain hemodynamics, and tumor angiogenesis, among others. By employing tightly focused excitation beams, photoacoustic imaging at

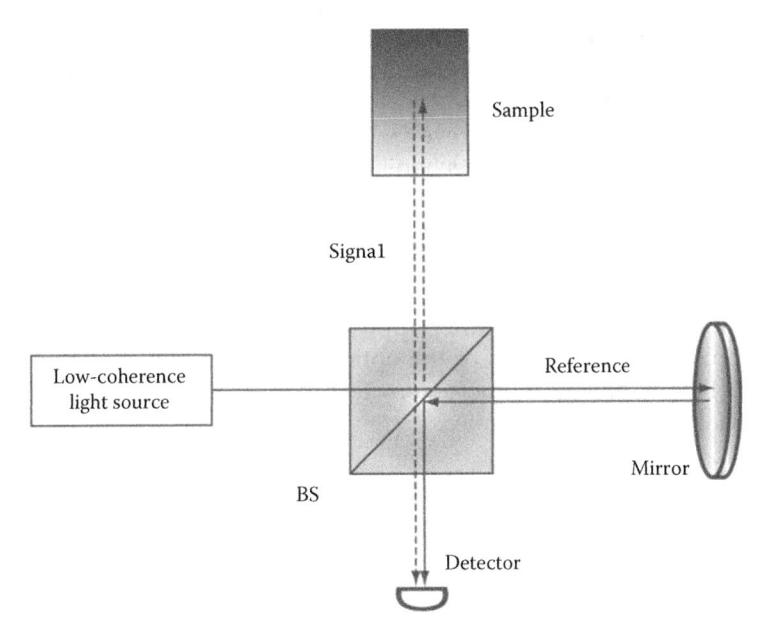

FIGURE 4.17
Low-coherence interferometry scheme.

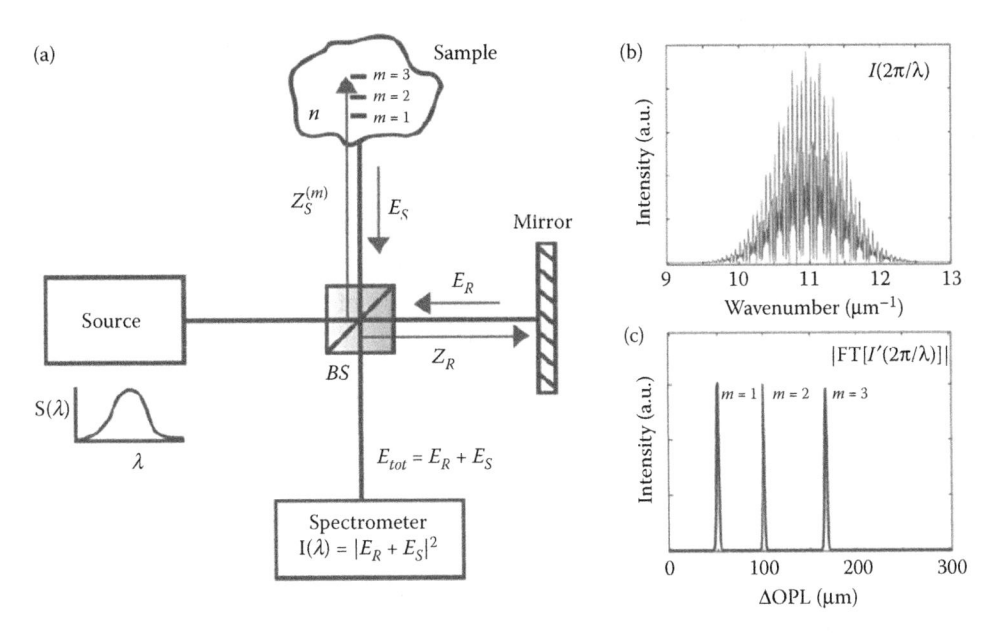

FIGURE 4.18
(a) Schematic of FD-OCT system using a Michelson interferometer geometry and a source with power spectral density $S(\lambda)$. (b) The spectrally detected interferogram can be Fourier-transformed to produce a depth-resolved reflection profile (c). (Adapted from F. E. Robles, *Light Scattering and Absorption Spectroscopy in Three Dimensions Using Quantitative Low Coherence Interferometry for Biomedical Applications*, PhD Thesis in Medical Physics. 2011, Duke University: Durham, NC [10].)

the microscopic scale has also been reported [11,12]. Additional functional information can also be obtained by using exogenous contrast agents, including organic dyes, nanoparticles, reporter genes, or fluorescent proteins [13].

In summary, optical microscopy has evolved to provide several modalities that optimize imaging based on different contrast mechanisms. While the spatial refractive index distribution of the sample provides the basis for generating contrast, techniques such as phase contrast and DIC microscopy employ clever methods to convert these subtle changes into intensity variations that can be easily viewed. Other contrast mechanisms include spectrally resolved methods that use scattering to reveal structures on finer scales than is possible with traditional microscopy. Other approaches use spectral resolution to enhance contrast by employing the richness of the optical spectrum and the color sensitivity of the human eye. Finally, advanced microscopy techniques enable depth-resolved imaging, producing optical sectioning of samples.

Laboratory Exercises

In this section, several laboratory exercises are presented that provide hands-on experience with various microscopy techniques, ranging from basic exploration of microscope characteristics to application of advanced techniques. Exercises include: (1) Overview of the microscope—This laboratory will provide orientation to the microscope while conducting measurements that reinforce basic concepts of microscopy; (2) Contrasts in microscopy—Exercises here compare various sources of contrast with experimental imaging measurements, including phase measurement and fluorescence; (3) Build your own microscope; (4) Interference-based imaging—This section conducts a series of experiments that bridge basic interferometry to optical imaging with coherence gating as a depth-sectioning technique.

Laboratory 1: Overview of the Microscope: Kohler Illumination

As described previously, proper illumination is needed to obtain optimal images from even the highest-quality microscopes. The main focus of this laboratory exercise is to familiarize users with the basic parts and construction of a typical laboratory microscope. The instructions here were originally designed for using the Zeiss Axiovert 200 but have been generalized to be applicable to most modern lab microscopes.

Laboratory Protocol

Instrumentation

This laboratory exercise requires access to a compound light microscope. While the construction and the components of each microscope will vary by manufacturer and model, there all largely contain the same components (Figure 4.19).

> *Eyepiece*—Each microscope will have a method for the user to view the sample. In Figure 4.1, a binocular arrangement is shown, which allows viewing with both eyes. This kind of scheme will generally allow the user to vary the distance between the two eyepieces and independently focus them. Alternatives include

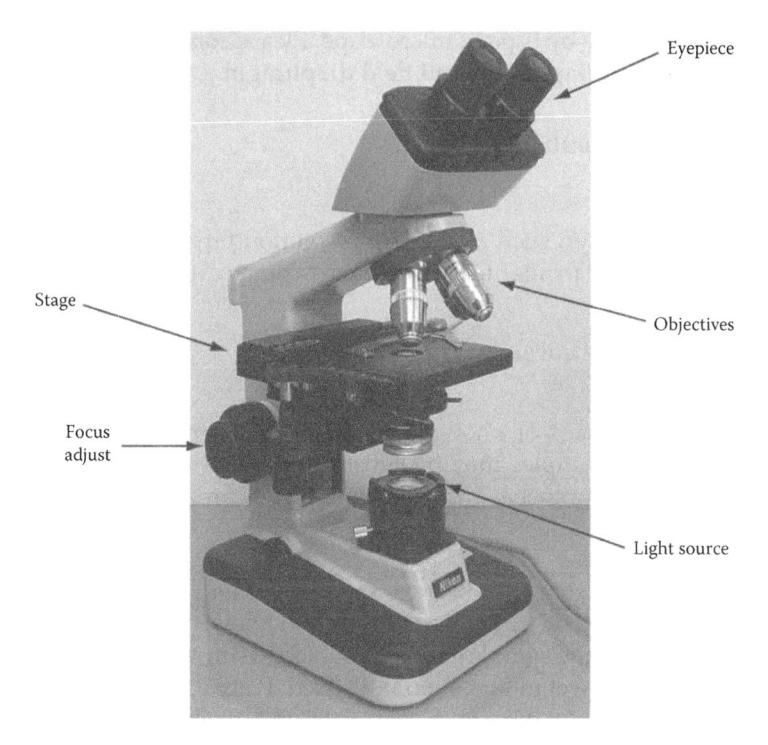

FIGURE 4.19
Typical laboratory compound microscope

the monocular for single-eye viewing and the trinocular, which will offer an additional port for installing a camera.

Objectives—Compound microscopes will typically include multiple objectives installed on the same turret or nosepiece that allow the user to switch between various magnifications.

Light source—Most microscopes will include an internal illuminator that can be activated with an integrated power switch. The first laboratory exercise below will center on proper alignment of this component.

Stage—The stage is the location where samples of interest are placed for examination. Depending on the type of microscope, the stage may include simple clips to hold a sample fixed during imaging, or it may include mechanical translators to enable fine manipulation in one or two directions.

Focus adjust—Microscopes contain a knob that translates the objectives up and down to enable the user to find the optimal focus for imaging the sample. Some will contain separate knobs for coarse and fine adjustments or may integrate both controls within the same knob. Proper use of the focus adjust is to move the objective as close to the sample as possible prior to imaging and then backing the objective away from the sample to find the best focus. This will ensure that the objective is not damaged by unintentional contact with the sample.

In addition to these components, microscopes will also contain variable apertures or diaphragms that control the light incident on the sample. The location and controls of these diaphragms will vary by type of microscope. Please consult the microscope manual to identify the **condenser diaphragm and field diaphragm**.

Materials and Sample Preparation

What Is Needed

- High-contrast specimen such as a USAF resolution target. An alternative can be made by using a laser printer to place a microscopic pattern on a transparency.
- Ruler.
- Microscope slides and coverglasses.
- Plastic pipette.
- Samples with structures of known size. Polystyrene microspheres are selected here, but biological samples may be substituted when available. Pondwater can offer paramecium or amoeba, and aquarium algae can provide interesting imaging targets.

For technical samples, such as polystyrene spheres, a small drop of solution containing the spheres should be placed on a microscope slide and allowed to dry. A coverglass should be placed on top to protect the sample and held in place using transparent tape at the edge.

For biological samples, a wet mount should be used. Place a small drop from the sample on to the slide. Try and select a drop with material that is visible by eye. Place a coverglass over the sample, and observe with the microscope.

Experiment

1. Kohler alignment
   a. Select the objective with the lowest magnification (e.g., 10× objective) on the nosepiece and ensure the correct stop position.
   b. Open the **field diaphragm** completely by adjusting the control until stop.
   c. Open the **aperture diaphragm** completely by adjusting the control until stop.
   d. For microscopes with multiple integrated modalities, set the illumination train for the bright-field setting, and ensure that no filters or other elements are in the imaging path.
   e. Place the high-contrast specimen on the microscope stage.
   f. Optimize focus on the selected detail of the specimen.
   g. If the microscope offers electronic control of the light intensity, adjust the setting to a light intensity with a comfortable brightness for viewing.
   h. Close the **field diaphragm** until it is visible in the field of view, even if not in focus.
   i. Focus on the edge of the field diaphragm by moving the condenser vertically.
   j. Center **field diaphragm** using centering screws, and open it until the edge of the diaphragm just disappears from the field of view. *Note: Not all microscopes will enable the position of the diaphragm to be adjusted.*
   k. For **aperture diaphragm** setting, remove one eyepiece from the eyepiece tube and set aperture diaphragm to approximately two-thirds of the diameter of

the objective exit pupil. The optimum contrast setting is dependent on the respective specimen.

   l. Insert eyepiece again, and, if required, refocus on the specimen via fine adjust.

   m. Adjust the light intensity via the electronic control to recover a comfortable illumination level.

Note: The field size and aperture change after switching objectives. The diaphragm settings must be repeated to ensure optimal imaging conditions.

2. Virtual image location

   a. Using the 10× objective, image the technical sample (resolution target), and find a set of bars (or other structure) that fills the field of view (FOV).

   b. Take a ruler and move it back and forth until the size of the expected image as seen on the ruler matches its appearance in the microscope.

   c. Use the 40× objective to verify the FOV and virtual image scaling.

3. Resolution

   a. Examine the slide with the 1-micron beads using the 40× objective. Find two lone beads and center them in your FOV.

   b. Switch to 20× and 10× objectives and examine the same two beads.

   c. Note at which magnifications the beads are distinct. For biological specimens, compare observations with expected sizes of structures.

   d. Try the 0.5- and 0.1-micron beads and repeat.

Data Analysis

Part 1

Questions:

1. What effect does adjusting the focus of the condenser have on the illumination?
2. What effect does adjusting the field diaphragm have?
3. Repeat alignment with the 20× and 40× objectives. If any difficulty arises, explain why this occurs by considering the NA of the objectives and condenser.

Part 2

1. Estimate the FOV by finding the true size of the technical sample.
2. For 10× objective, what is the total magnification? Thus, what is the size of the expected image?
3. What is the distance from your eye to the virtual image? (What is the ruler distance?)
4. Since the 40× has four times the magnification of the 10× objective, what do you expect the FOV would be?

Part 3

1. Estimate the resolution of the 40× objective (use Equation 5.7 with the NA on objective barrel and a wavelength of 500 nm).
2. Repeat the calculation for the 10× and 20× objectives.

3. Compare these predictions with experimental observations. Under which magnifications can the beads be resolved?

4. For the additional sizes of beads examined, list which objectives are capable of resolving bead pairs and whether this agrees with predictions.

Laboratory 2: Contrasts in Microscopy

This laboratory exercise will explore the different sources of contrast across the several imaging modalities discussed in Section 5.2. Recall that phase contrast and DIC microscopy are two methods of converting phase shifts due to refractive index variations into changes in light intensity that can be viewed by eye. Phase contrast is based on phase-shifting undeviated light to produce maximum changes in phase when it is differentiated from diffraction light. On the other hand, DIC is based on using polarized light that is split into two closely spaced components by a Wollaston prism when incident on a weakly absorbing or scattering sample. The two components are then recombined with a second prism located at the base of the objective. A final polarizer creates contrast by converting phase variations into amplitude variations. A third modality is also examined here, fluorescence microscopy, which provides spectral contrast by using molecules that absorb at one wavelength and reemit at a second, longer wavelength to label-specific structures.

Instrumentation

This laboratory exercise requires a compound light microscope with capabilities for phase contrast (part 1), DIC (part 2), and fluorescence microscopy (part 3). Use of a microscope with a digital camera installed on the video port for documentation is recommended.

Phase Contrast Settings

Microscopes equipped for phase contrast will enable changes in the illumination train via a wheel or slider in the condenser assembly. The mechanism will insert annular diaphragms that correspond to matching phase rings inside the phase objectives. For proper phase-contrast imaging, it is important that the condenser setting match that of the objective. Note the appropriate phase setting that is etched on the barrel of the objective lens (e.g., Ph1, Ph2), and set the condenser to match. For example, if the 20× objective has Ph2 inscribed on it, then the Ph2 slot in the condenser should be used.

DIC Settings

For DIC microscopy, several components are required and must be properly aligned to achieve the desired effect. The DIC slider should be inserted beneath the objective, and the complementary slider should be installed in the condenser. Note that some microscopes may offer more than one DIC setting, so be sure to match the condenser and objective sliders. The illumination train for DIC includes a polarizer, which should be set at 90°. This should provide a zero-bias image (black background); if this is not the case, the slider should be adjusted to achieve zero bias. The lab procedure below will vary from this setting to explore the properties of DIC.

Fluorescence Settings

For fluorescence microscopy, the microscope should offer a suitable light source and filter sets matched to the excitation and emission properties of the specific fluorescent molecules to be imaged. Fine alignment of the fluorescence illuminator may be required as per the manufacturer specifications. In epifluorescence (excitation through the objective), proper settings of the condenser aperture and field diaphragm may be more difficult to implement than in bright field.

Each microscope equipped for fluorescence imaging will have its own set of filters, depending on its usage. Standard filter sets include DAPI/nuclear stain (UV excitation/ blue emission), FITC/boron-dipyrromethene (BODIPY; blue excitation/green emission), and Rhodamine (green excitation/red emission). Green fluorescent protein (GFP) fluorescence can be observed with the FITC filter but provides better contrast if a UV excitation filter is used. The spectral characteristics of these filters are generally available from the manufacturers and should be provided to students as a resource for this laboratory exercise.

Digital Camera Settings

Optimal imaging with a digital camera can require tinkering with the camera's exposure settings. The goal is to capture an image that represents what is seen through the eyepiece. The exposure time should be adjusted to provide good image contrast while avoiding saturation. The specific setting will depend on the brightness of the illumination, the light-gathering ability of the objective (numerical aperture), and the type of sample that is used. Advanced cameras will also include the ability for color and white balance. It may take some experimentation to find the optimal settings.

Materials and Sample Preparation

What Is Needed

- High-contrast specimen such as a USAF resolution target.
- Polystyrene microspheres (10-micron diameter) or other sample that offers a range of path lengths.
- Sample with birefringence (rayon or other type of fibers).
- Sample with multiple fluorophores (constellation microspheres from Invitrogen make an ideal sample).
- Microscope slides and coverglasses.
- Plastic pipette.
- Cotton swab or toothpick.
- PBS (phosphate buffered saline).

For technical samples, such as polystyrene spheres, a small drop of solution containing the spheres should be placed on a microscope slide and allowed to dry. A coverglass should be placed on top to protect the sample and held in place using transparent tape at the edge.

For cheek cell samples, a wet mount should be used. Scrape the inside of your cheek with the flat end of a sterile toothpick or the wooden end of a cotton swab. Transfer the scraping to a drop of PBS on a microscope slide. Place a coverglass over the sample and observe with the microscope.

For fluorescent samples, students can be provided with a sample of fluorescent beads or experiment with labeling cells with common fluorophores.

Experiment

1. Phase contrast

   a. Set the microscope to Kohler illumination.

   b. Place the cheek cell slide on the stage, and examine under phase contrast.

   c. Select the 10× objective. The phase setting (e.g., Ph1) will be inscribed on it. The condenser must be at the same setting to achieve phase contrast.

   d. Adjust the focus to provide optimal contrast of the sample, adjusting the light intensity as needed.

   e. Without moving the stage, switch to the 20× objective. Adjust the condenser wheel to match the correct phase contrast (PH) setting marked on the objective barrel.

   f. Adjust focus using fine adjustment to put sample in focus.

   g. Finally, switch to 40× and adjust the condenser setting and focus.

   h. Acquire a digital phase-contrast image of your cheek cells under 40× magnification for inclusion in your lab report. If a digital camera is not available, a sketch of the observed field can substitute.

   i. Try switching back to bright field with the aperture diaphragm opened wide. Also, try imaging the cells upon closing the aperture.

2. DIC

   a. Place microsphere sample on the stage, view with the 10× objective, and adjust focus. The microspheres tend to focus light on their own. An optimally focused image will put the edges of the microsphere in focus.

   b. If necessary, adjust slider to ensure it is at zero bias (adjust slider to get black background; both edges of the spheres are white).

   c. Remove the eyepiece and note the black X that is visible.

   d. Record the image using a digital camera or make a sketch for your lab report.

   e. Adjust the slider to get a nonzero bias image (gray background, one sphere edge white, the other black).

   f. Again, remove the eyepiece and note that the black X has moved. Move the slider back and forth to observe the X come in and out of the field of view

   g. Record the nonzero bias image using the digital camera, or make a sketch for your lab report.

   h. Switch to the slide with the birefringent fibers, locate a piece of fiber, and adjust focus.

   i. Place the condenser in the bright-field setting, and remove the slider but keep the polarizer near 90 degrees. The birefringent properties of the fiber should be apparent.

   j. Record a nice color image (if available), or make a sketch/provide a description for your lab notebook.

   k. Replace the slider and return the condenser to DIC setting.

   l. Adjust the slider to produce a nonzero bias image (gray background and colors within the fiber).

   m. Record this image with the digital camera, or make a sketch/provide a description for your lab notebook.

3. Fluorescence microscopy
   a. Set the microscope to Kohler illumination.
   b. Switch to 20× and examine the fluorescent microsphere sample under bright-field imaging. You will see microspheres of various colors and sizes.
   c. Turn on the fluorescence illuminator, and insert the DAPI filter cube. Several colors of microspheres should be visible.
   d. For each of the colors that are visible, we will reconstruct the absorption and emission spectra by noting the observed coloration under each of the available filter sets.
   e. For example, create a column labeled "DAPI filter," and write down each of the five colors you see. Then, switch to the next filter cube and note the colors of the same beads. Finally, close the shutter to the fluorescence illuminator and note the colors seen in bright field.
   f. Use the digital camera to record an image of the sample using the DAPI filter cube for your lab notebook.
   g. If available, image labeled cellular structures and record a copy with the digital camera, or make sketches for your lab notebook.

Analysis

Part 1

1. Estimate the size of the cheek cells based on what fraction of the field of view they represent. Can you identify the nucleus? How big is it?
2. Can you still see the cells clearly under bright field? Explain.

Part 2

1. Compare the properties of phase contrast and DIC by sketching the following objects as they would appear in each. Assume that the objects are 5 microns in height at their thickest point and vary from the background by $\Delta n = 0.03$.

   You may use a mathematics program to calculate the results.

Part 3

1. Based on the bright-field and fluorescence measurements and the online filter characteristics, sketch the absorption/emission characteristics for each type of microsphere or fluorescent molecule in your sample.
2. Compare with the known properties of these fluorescent molecules and note any discrepancies. What additional information would help you correct these?

Laboratory 3: Build Your Own Microscope

Discussion

Although commercial microscopes offer advantages in terms of alignment, ease of use, and imaging modalities, many research projects are at their core a simple microscope. For such projects, adapting a commercial microscope system is not always possible, and researchers often choose to build an instrument from scratch. However, in doing so, it is important to understand the fundamentals of microscope design and how alignment, stop positions, conjugate planes, and illumination affect aberrations and image quality. In this lab, a basic microscope is constructed using minimal components, and the effects of alignment, stop position, and illumination are explored.

Materials

Microscope objective—The choice of objective is not critical and could range from a short focal length lens to a high-quality, well-corrected commercial microscope objective. If using a simple lens, the NA that can be achieved will be limited, and reasonable magnification may require very long image distances. If using an infinite conjugate objective, a tube lens should be used to ensure that the objective is used at the designed conjugate plane, which maximizes image quality. Low-NA objectives are less prone to image degradation from aberrations. A moderately high-powered objective with NA greater than 0.5 is recommended to explore the effects of NA and alignment. Suggested part: Edmund Optics $40\times$ 0.60 NA finite conjugate objective.

Camera—Images can be observed directly using an eyepiece, but quantitative comparison is possible if images can be captured digitally. A low-cost webcam or digital single-lens reflex (SLR) camera works well. The lens should be removed to allow the image to be formed directly on the sensor. Suggested part: Canon digital Rebel series SLR camera. If using a digital SLR, images captured in RAW mode have the advantage of improved linearity for quantitative analysis.

Resolution target (optional)—Qualitative comparisons can be very enlightening, but a resolution target provides a standardized means to simply and quantitatively evaluate microscope resolution and is readily available from many vendors.

Tube lens (optional)—Most high-end modern microscopes are designed to use infinite conjugate objectives in combination with a tube lens allowing for filters, polarizers, and beamsplitters to be inserted in the collimated light path without needing to refocus. Older or cheaper finite conjugate objectives are designed to form an image at a specific back focal distance behind the objective according to the tube length (usually 160 mm). Suggested part: Edmund Optics 150-mm achromatic doublet.

Iris or stop—An adjustable iris ($\times$2) can be used as an aperture stop and eases adjustment of the numerical apertures. Alternatively, holes of various sizes may be cut out of foil. Suggested part: Edmund Optics mounted iris.

Light source—Almost any light source can pressed into service, but a low-cost white light LED or flashlight works well. Suggested part: white LED flashlight. Additional lenses and mounts may also be used to explore various illumination configurations.

Mounts and rails—Some method of mounting and aligning the components will be needed. A rail system is convenient and allows easy adjustment of component positions along the optical axis. Suggested part: one 50-cm rail, one v-mount or optical claw for source, two slide or filter mounts, one microscope objective threaded mount, two iris mounts, six posts and post holders, and six rail mounts.

Translation stage (optional)—Mounting the sample holder on a micrometer stage makes focusing easier. Without an adjustable stage, the sample can be focused by carefully sliding the sample mount on the rail; however, adjusting the position to within the depth of focus requires patience and a steady hand.

Miscellaneous—Scotch tape, microscope slides, etc.

Exercises

Assemble and Align the Microscope

Assemble the components as in Figure 4.1. The distance from the back of the microscope objective to the camera sensor should be equal to the tube length specification of the objective. Common tube lengths are 160 mm (DIN) or 170 mm (JIS) for finite conjugate objectives (Figure 4.20).

Use a slide with a piece of Scotch tape as a diffuser to illuminate the resolution target and focus the image on the camera by adjusting the sample holder position. To avoid scratching the objective, focus by moving the sample very close to the objective (within the working distance) and then slowly move the sample away while watching for the image to come into focus.

Ray Diagram

Use the specifications of the objective to draw a ray diagram of the optical system using a thin lens approximation. An example ray diagram is shown with the chief ray (black) and marginal ray (gray). The chief ray is the ray that passes through the center of the aperture stop and originates at the edge of the object. A marginal ray is a ray that passes through the edge of the aperture stop and intersects the optical axis at the object plane and image plane. Additional rays are shown as dashed lines (Figure 4.21).

When the image plane is moved farther away, what happens to the working distance? In the figure, the stop is located at the back focal plane of the lens. What effect does this have on the chief ray angle in object space? What would happen to the chief ray angle if the stop were moved toward the image or toward the lens? The numerical aperture (NA) is defined as $NA = n \sin(\theta)$, where n is the refractive index, and θ is the angle between the optical axis and the marginal ray. How is the NA in image space related to the NA in object space? The

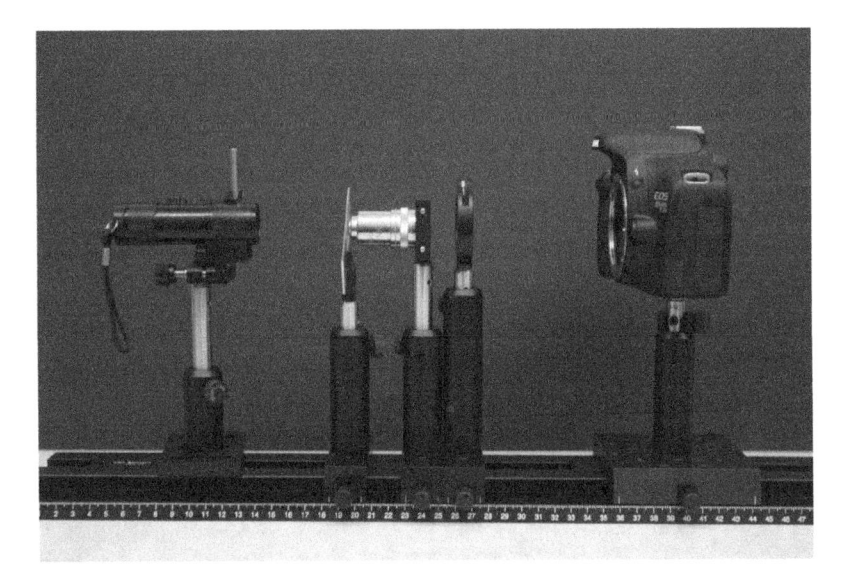

FIGURE 4.20
Simple microscope configuration. From left to right: LED flashlight, sample slide, DIN 40× objective, iris (optional, helps block stray light), digital SLR camera body.

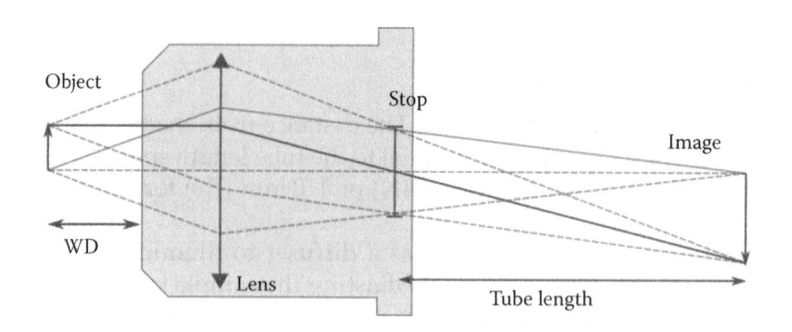

FIGURE 4.21
Microscope objective ray diagram. The chief ray (bold solid) passes through the edge of the object and the center of the aperture stop. The marginal ray (light solid) passes through the center of the object and the edge of the aperture stop.

lateral resolution of a diffraction-limited microscope is inversely proportional to *NA*. If the image plane is doubled, how does the resolution change?

Magnification

Magnification is the ratio of image to object size. The resolution target provides a precise object size, and the image size can be calculated by multiplying the pixel size by the number of pixels across the image of the target. How does the magnification compare to the specification for the objective? Move the image sensor away from the objective to double the image distance and refocus the object. Measure the magnification. How does the image quality compare? Repeat with the image sensor as close to the objective as possible (Figure 4.22).

Stop Shifts and Telecentricity

An aperture stop is the limiting aperture of an optical system. In microscope objectives, the aperture stop is located at the rear focal plane of the objective and can be seen at the back of the objective near the threads. An entrance pupil is the image of the stop in object space, and an exit pupil is the image of the stop in image space. The stop in an objective is typically located in image space (with no lens element between the stop and the image), so the stop *is* the exit pupil. Since the stop is located at the rear focal plane of the objective, the image of the stop in object space, or entrance pupil, is located at infinity. When a pupil

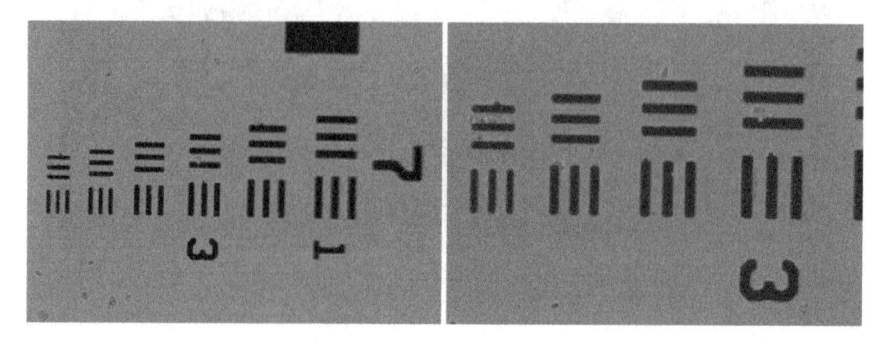

FIGURE 4.22
Comparison of image quality for different magnifications. Left is the resolution target taken at the designed magnification of 20×. Right is the image taken with the detector plane 2× farther away.

is located at infinity, chief rays (rays that pass through the center of the stop) are parallel, and the optical system is said to be telecentric. Microscope objectives are almost always object-space telecentric.

With the image sensor at the nominal position (tube length) and the target in focus, capture an image. Adjust the target position slightly to defocus the target in both directions around focus and capture images. How does the magnification change? Next, place an iris or stop between the objective and the image. Close the iris until it becomes the limiting aperture in the system. Repeat the measurement of magnification through focus. How does the stop position affect focusing? With the new stop position, is the objective telecentric? Does telecentricity make focusing easier?

Measure the distance from the original stop to the new stop. Use the effective focal length from the objective specifications, and calculate the new entrance pupil position.

With the iris in place, adjust focus and capture an image. Close the iris down to the smallest possible size and capture another image. How does the resolution or image quality compare to the image quality without the iris in place? How is the stop size related to the numerical aperture? Measure the stop diameter and distance from the stop to the image sensor. Calculate the image space numerical aperture. How is object-space NA related to image-space NA? How does resolution depend on NA? Compare the observed resolution and calculated theoretical resolution.

Critical and Köhler Illumination

Illumination of the object or sample in a microscope is an important factor for image quality. In an ideal microscope system, the object would be illuminated uniformly with respect to both position and angle. However, most real sources have variations in intensity; for example, a tungsten filament is a complex helical emitter. Light from such a source can be homogenized by a diffuser, but the cost is poor efficiency, and most of the light emitted from the source is not delivered to the sample. A more efficient way to deliver light to the sample is through the use of a lens called a condenser lens that focuses light onto the object.

Critical illumination is the term used to describe the case where the condenser lens is used to image the emitter onto the sample. The problem with such a scheme is that the intensity profile of illumination varies across the object, creating bright and dark regions within the field of view. A better method developed by August Köhler and referred to as Köhler illumination is to image the source onto the front focal plane of the condenser using a collector lens. Each point in the source is then collimated at a different angle within the numerical aperture of the condenser. All these collimated beams of varying intensity overlap at the sample, resulting in very uniform illumination within the field of view.

Using additional lenses, try both critical and Köhler illumination of the sample and observe the image. How does uniformity compare? How does the brightness compare? How do these illumination geometries compare to the use of a diffuser behind the sample?

Illumination NA

The well-known formula for diffraction limited resolution is $r = 1.22\lambda/(2NA)$, which assumes that the illumination or condenser numerical aperture is greater than or equal to the collection or objective numerical aperture: $NA_{con} \geq NA_{obj}$. When this condition is not satisfied, the more general form is used: $r = 1.22\lambda/(NA_{con} + Na_{obj})$. In the simple microscope configuration, NA_{con} can be adjusted by illuminating a diffuser (Scotch tape on a microscope slide) through a fixed aperture size and then moving the diffuser toward or away from the sample. Use this configuration to observe a weakly scattering object such as a buccal cell (easily obtained by swabbing the inside of a cheek and smearing the saliva onto a slide). How does the resolution change as NA_{con} is reduced? How does the contrast change as NA_{con} is reduced?

Dark-Field Illumination

In many biological applications, the object is weakly scattering, and with unstained samples, the contrast is very low. One way to increase the contrast is to eliminate the unscattered light so that the field surrounding an object appears dark. This is referred to as dark-field microscopy and is achieved by illuminating the object at a large angle with respect to the optical axis greater than NA_{obj}. Rays that pass unscattered through the slide will not be collected by the microscope objective. A simple way to implement this is to move the illumination beam to the side of the slide. Observe the buccal cell using a dark field. Move the illumination around to different positions around the optical axis. How does the image change? With this simple illumination geometry, there is an inherent asymmetry to the illumination. Commercial dark-field microscopes use illumination from a ring or cone of angles just greater than the objective NA. This retains better symmetry, eliminating azimuthal dependence in the image as well as increasing the brightness (Figure 4.23).

Laboratory 4: Interference-Based Imaging

This laboratory exercise investigates the use of low-coherence interferometry for optical imaging. Optical coherence tomography makes use of LCI interferometry to perform tomographic biomedical imaging with spatial resolution approaching the wavelength of light. OCT is now widely used to perform imaging in the eye and retina with high spatial resolution, making it very useful for ophthalmic diagnosis.

In the clinic, OCT uses a fiber optic interferometer to perform depth ranging analogous to pulse-echo imaging techniques such as ultrasound. Tomographic images of tissue microstructure can be built up from longitudinal scans of tissue reflectivity versus depth. In any pulse-echo imaging technique, the spatial resolution is determined by the length of the incident pulse of radiation, which is in turn limited by its wavelength. In OCT, the use of light rather than sound waves enables imaging resolution on the order of optical

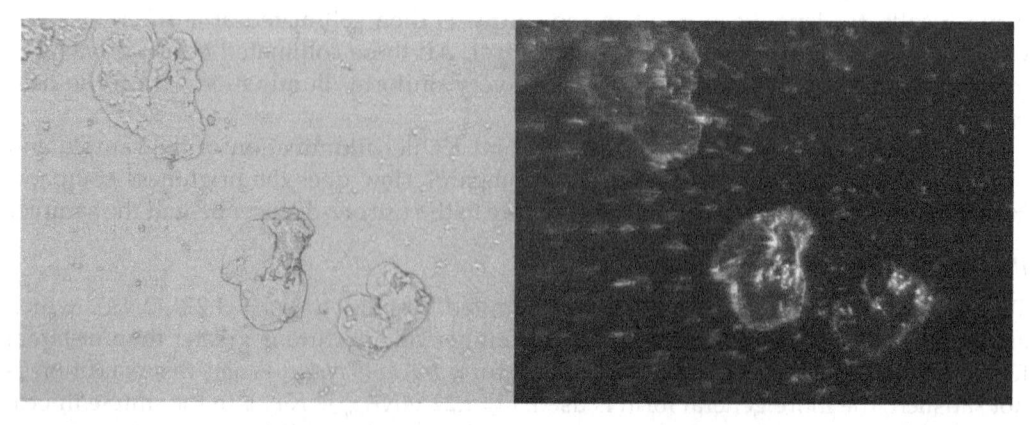

FIGURE 4.23
Images of buccal cells in bright field (left) and dark field (right) using a DIN 40× objective.

wavelengths (<1 μm). However, rather than using ultrashort pulses of light, which can be costly to create and difficult to maintain, LCI makes use of interferometry with resolution dictated by the *coherence length l_c* of a CW light source

$$l_c = \frac{2\ln 2}{\pi n} \frac{\lambda_0^2}{\Delta\lambda} \tag{4.10}$$

Here, λ_0 is the center wavelength of the source, $\Delta\lambda$ is the full-width at half maximum (FWHM) bandwidth of the source, and n is the index of refraction of the material in which l_c is being measured.

The first part of this exercise concentrates on the basic principles of low-coherence interferometry and provides insight into the concept of coherence length. Once this concept is demonstrated and understood, you will use a working OCT scanner to perform high-resolution imaging on real tissue samples.

Instrumentation

For this laboratory exercise, a fiber optic Michelson interferometer is required. It can be constructed as part of the lab exercise if time and resources permit or implemented beforehand by staff for student use. As a precursor to this work, an exercise with a free-space Michelson interferometer is also described. The scheme in Figure 4.16 should be used for this with a coherent light source such as helium-neon laser and a micrometer-driven stage to translate the reference mirror (Figure 4.24).

Fiber coupler—The central component of the interferometer is the 2 × 2 fiber coupler, also known as a fiber optic beamsplitter (FOBS). This component looks like a small metal cylinder with two optical fibers coming out of each end. Like the cube beamsplitter used in the Michelson interferometer, you can think of the fiber coupler as a 50/50 beamsplitter with one input port, two intermediate ports leading to sample and reference arms of the interferometer, and one output port that is directed to the detector. These can be purchased from many suppliers for $100–$400, depending on their characteristics.

Light source—Connected to the input port is a fiber leading from a low-coherence light source. Superluminescent diodes (SLDs), which are similar to laser diodes but emit over a broader range of wavelengths, make excellent sources for LCI and OCT. These are relatively

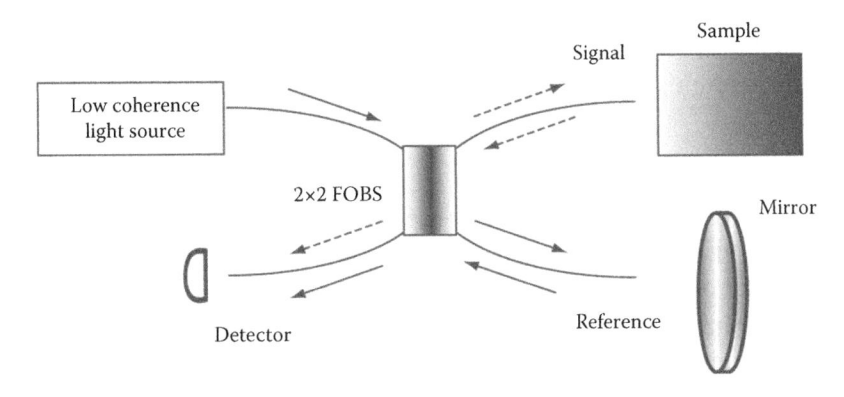

FIGURE 4.24
Fiber optic Michelson interferometer. FOBS—fiber optic beamsplitter.

inexpensive as an individual component, costing around $1,000–$1,500, but require a stable current source and often temperature control.

Collimators—The reference arm output of the fiber coupler needs to be retroreflected using a mirror. An integrated collimator lens is recommended, which is a specially coated and mounted lens that collimates (makes parallel) the light expanding from the fiber tip. When imaging biological samples, it may be desirable to differentially attenuate the reference beam using a neutral density (ND) filter, which is a piece of dark glass of calibrated transmission. Light from the sample arm fiber is also collimated using a fiber collimator either for interferometry or as input to a scanning system for imaging.

IR detector card—This is a highly recommended piece of lab equipment for working with infrared beams. This pink-colored card glows red where infrared light shines on it and can help students trace beams path as well as helping with alignment.

Micrometer-driven translation stage—For low-coherence interferometry, it is essential that the pathlengths of the signal and reference paths can be matched at the micro scale. To achieve this, a micrometer-driven stage that gives you fine control over the length of the reference arm is important.

Detector—The output port of the fiber coupler is input to a detector that can convert the optical signal to an electronic one. A basic photodiode can be used here for simpler systems, such as the free-space Michelson interferometer used in the pre-lab exercise. In this case, the reference arm can be scanned in time by axially translating the position of the mirror in the reference arm and observing the signal on an oscilloscope or transferring the data by a digitizer to a computer. The preferred method for the full set of experiments is to use an infrared spectrometer as the detector to enable Fourier domain detection. While a low-coherence spectrometer can be used for examining spectral features, a high-resolution spectrometer is needed for LCI and OCT. Commercial compact spectrometers for OCT are available from several vendors with high-spectral resolution, good data acquisition speed, and even software for easy readout. If software for spectrometer readout is not available with the spectrometer, a short program should be written that allows students to visualize and save recorded spectra.

Materials and Sample Preparation

What Is Needed

- Free-space Michelson interferometer
- Fiber optic interferometer
- Glass microscope slide to use as sample
- Dual-axis scanning galvanometer

Experiment

1. Pre-lab with free-space Michelson interferometer
   a. Observe the interference pattern when you slowly change the length of the adjustable arm by turning its micrometer positioner.
   b. Collect data for a measurement of the He-Ne laser wavelength. To do this, turn the micrometer positioner a preset distance (try 20 microns) while counting the number of optical fringes that go across the combined spot during the motion or can be observed on the oscilloscope.

2. Fiber optic low-coherence interferometry

   a. Make sure you have identified all parts of the OCT apparatus and understand their basic function. You may wish to do the first part of the analysis, drawing a schematic of the LCI setup in your lab notebook, before beginning.

   b. The main goal of this part of this experiment is to obtain enough spectral interferogram data to measure the spectral properties and coherence length of the light source and estimate the power reflectivity of the slide surface.

   c. To do this, you will need to calibrate the spectrogram plot, that is, the actual number of nm (the units of wavelength) per pixel on the detector array. The *x*-axis of the data obtained from the spectrometer is the pixel number of the detector. The total number will depend on the type of the spectrometer. Note that this is not necessarily the center wavelength of the source but rather the center of the spectral range of the spectrometer.

   d. Create a known path length difference between the sample and the reference arms by turning the reference arm micrometer until the number of interference fringes across the spectrum reduces to zero. The spectrum will seem to oscillate a lot in this position; that is because the whole spectrum is either constructively or destructively interfering at once and is not stable. This is the zero path length position.

   e. Turn the micrometer a known distance (use the calibration marks on the micrometer) until you get a fringe pattern with a significant number of fringes across the spectrum (20+).

   f. Save the spectrogram for later analysis using the file save function in the software. It is recommended that multiple spectrograms be recorded to ensure good data are saved.

   g. Once the spectrogram has been saved for the single reflector with a known path length difference, record one with only the reference arm power (block the light to the sample). This gives the light source spectrum without any fringes, which can be used to estimate λ_0 and $\Delta\lambda$, the characteristics of the light source.

   h. To observe the depth-resolved reflection profile, the data must be processed via Fourier transform. A simple analysis routine can be written in a signal processing language such as MATLAB by the students or provided as a stand-alone program.

3. Optical coherence tomography

   a. In this final part, scanning optics are used to acquire OCT images of biological tissue microstructure. Students working on an extended project can adapt the low-coherence scheme to include a dual-axis scanning mirror that scans the OCT beam across the surface of the sample. However, to allow greater time for exploring the properties of the imaging device, it is recommended that an instructor or teaching assistant set this system up and adapt the software to synchronize the data acquisition with the scanning mechanism.

   b. Students can acquire two- and three-dimensional images of biological samples such as their fingertip. Two-dimensional data can be easily uploaded into signal processing software such as MATLAB and used to generate cross-sectional images of the fingertip. Power 3-D rendering software is available on the

Internet to enable students to use the data they have acquired to make stunning 3-D images.

Data Analysis

Part I

1. Does it the interference pattern go completely dark at its minimum value?
2. Draw a schematic of the Michelson interferometer and label all the components.
3. Calculate the wavelength of He-Ne laser light, given your measurement of the number of optical fringes associated with a change in relative arm lengths. Show all the details of your calculation, including how far you translated, how many fringes you observed, and your wavelength calculation. Don't forget that when you move a mirror, the round-trip path length to that mirror and back changes by twice that amount.

Part II

1. Draw a schematic of the LCI and label all the components.
2. Calculate the central wavelength λ_0, the FWHM wavelength bandwidth $\Delta\lambda$, and the coherence length l_c from your measurements. Report all these measurements in μm.
3. In order to use Fourier transform techniques to convert spectral data into depth-resolved reflection profiles, the data must be represented as a function of *wave number* rather than wavelength. The formula that relates the two is $k = 2\pi/\lambda$. Using the information on the spectrometer characteristics (center wavelength and span), calculate the wave number of each of the pixels in the array, assuming that the pixels are linearly spaced in wave number. Develop a short software program that converts your data from intensity as a function of wavelength to intensity as a function of wave number.
4. Fourier-transform this data to produce an A-scan (amplitude scan = depth-resolved reflection profile).
5. Plot the A-scan you produced. Label all features and artifacts that may be present. Estimate the coherence length from the FWHM of the mirror peak. How does it compare from your predicted result in question 2?
6. Use your saved spectrogram data with known path length difference to calibrate the axis of the spectrogram. Plot the calibrated spectrogram data both with and without the sample reflector. Label the x-axis in radians/μm. Estimate k_0 and Δk from your measurement, and label them on the reference arm–only plot.
7. Estimate the power reflectivity of the slide surface from your data.

Part III

1. Plot, from your saved data, your best B-scan images of human skin. Include distance scales, and note the imaging depth achieved. Point out any anatomical features in your image that you can identify.
2. Use 3DView to plot an interesting 3-D-rendered view of something to include in your report.

Additional Questions

1. What are some methods that can increase the accuracy of the OCT image and decrease the various artifacts? Can you think of a method to remove the complex conjugate artifact, that is, the mirror image that appears?

2. Hard question: In Fourier-domain OCT, images are brightest when near zero path length difference between reference and sample arms. Can you think of a reason this would be so? Hint: Think about the fringe frequency when the path length difference is large and how these fringes are detected.

Confocal Light Absorption and Scattering Spectroscopic Microscopy

Vladimir Turzhitsky and Lev T. Perelman

Introduction

Since the discovery of light microscopy, the detailed understanding of biological function has seen an increasing demand for higher-resolution instruments as well as the capability of obtaining *in vivo* measurements. Unfortunately, most biological subcellular organelles do not have adequate contrast to be easily observed noninvasively without staining. Fluorescence staining is commonly used to obtain biologically specific information; however, most fluorescence staining procedures, although capable of observing living cells, disrupt the cellular processes as part of the procedure. An ideal microscopy technique would possess the resolution of electron microscopy while being entirely noninvasive and not altering the cellular environment and processes.

Optical microscopy has the desired properties of being noninvasive but has fundamental barriers in resolution that are determined by the diffraction limit. Confocal microscopy has the notable feature of being more effective at eliminating out-of-focus contributions and, therefore, has unique depth-sectioning capabilities. However, confocal microscopy often requires staining with fluorophores to obtain an adequate contrast. Confocal light absorption and scattering spectroscopic microscopy is a modality that combines confocal microscopy with light scattering spectroscopy. CLASS microscopy uses a spectroscopic wavelength analysis that allows for improvements in both the contrast and resolution of the image of interest, without the use of any staining agents. Although the fundamental LSS technique of quantifying particle size using the backscattering spectrum was first discovered many years ago [14], the method was not applied to microscopy until recently [15]. The technique is similar to standard confocal microscopy with the exception that every pixel in the image is composed of a spectrum, which is then analyzed to obtain accurate size information about the scattering particle that resides within the confocal volume. The CLASS microscope typically makes use of a supercontinuum laser, which possesses desirable coherence properties, a large intensity, and a large spectral bandwidth. As both confocal microscopes and supercontinuum lasers are prohibitively expensive, this lab will focus on the principles of CLASS microscopy by studying (a) the Mie theory–based analysis used in LSS, and (b) the effect of a lens with a finite numerical aperture on the LSS analysis.

Light Scattering Spectroscopy for the Determination of Particle Size

LSS uses the fundamental characteristics of how the light-scattering spectrum depends on the characteristics of the particles, such as the size and refractive index. LSS typically uses the convenient approximation of assuming that the particles can be adequately modeled as spheres of an unknown size and refractive index. The exact solution of light scattering by spheres can be solved and is known as Mie theory, with the results depending on the diameter of the sphere, the relative refractive index $m = n_s/n_m$, the wavelength of light λ, the polarization, observation angle, and distance to the particle. For the purposes of this lab, we will describe unpolarized light. First, we will explore the behavior of the LSS spectrum for small angles in the backward direction. Then, we will describe the required modifications in modeling a larger numerical aperture such as in a microscope geometry.

A schematic for a typical LSS experiment is shown in Figure 4.1. The most important parameters in this geometry are the angular separations of the fibers and the distance to the sample [16]. The angular separation affects the backscattering spectrum, and the distance affects the signal intensity (Figure 4.25).

If we restrict our attention to refractive indices that are typical for the range of intracellular organelles, then we can estimate m to be within the range of 1.03–1.06. Mie scattering is typically described in terms of the size parameter $x = \pi n_m \delta/\lambda$, where δ is the particle diameter. For small values of m, we can use the Rayleigh–Gans approximation to obtain some insight into the results that we would expect

$$\frac{I_s}{I_0} = \frac{\delta^2}{L^2}(m-1)^2 f\left(x\sin\frac{\theta}{2}\right)\frac{1+\cos^2\theta}{(1-\cos\theta)^2} \tag{4.11}$$

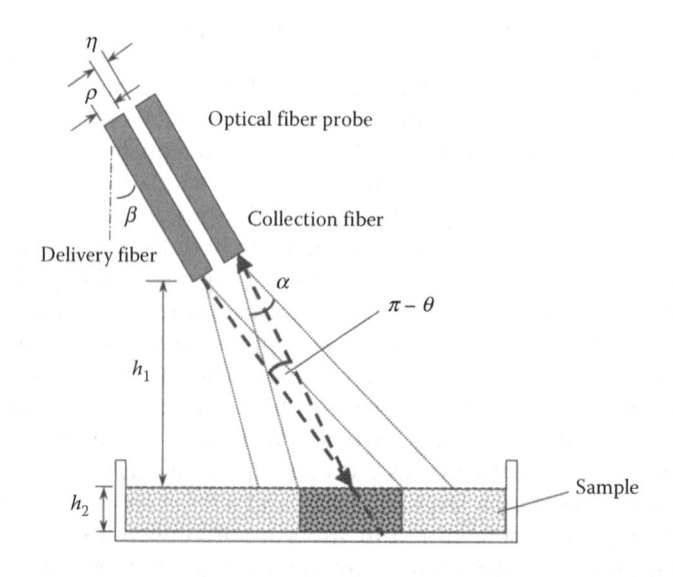

FIGURE 4.25

Geometry for two-fiber LSS experiment. The delivery fiber is connected to a broadband light source, and the collection fiber is connected to a spectrometer/CCD, from which the data are transferred to a computer. The angle θ is determined by height h_1, fiber diameter ρ, interfiber distance η, and tilt angle of the fibers.

where I_s is the scattered intensity, I_0 is the incident intensity, and L is the distance from the particle to the detector. The only wavelength-dependent term in Equation 4.11 is the form factor f, which depends on the size parameter and scattering angle. This form factor can be used to determine the size of a particle from an unknown sample. The exact dependence of the backscattering spectrum can be calculated using Mie theory. Again, if we restrict our attention to small θ and small m, we obtain results such as in Figure 4.2. Regions 1, 2, and 3 in Figure 4.2 show the behavior of the differential scattering coefficient in the wavelength range of 400–800 nm for particle sizes of 900 nm, 300 nm, and 100 nm, respectively. Note that the number of oscillations seen in each region is considerably different. Therefore, by simply observing the number of oscillations that are present within a wavelength region, the approximate particle size can be estimated. A more exact fit can be obtained by using one of the many publicly available Mie codes (Figure 4.26).

Effect of Objective Lens on LSS

The application of LSS to CLASS microscopy involves using confocal pinholes and a high-NA objective lens. These two components serve to isolate a confocal volume. However, a high-NA objective also results in the collection and averaging of a large range of backscattering angles. If the illuminating light is obtained from an incoherent source, the result is an averaging or integration of many LSS spectra from each of the backscattering angles.

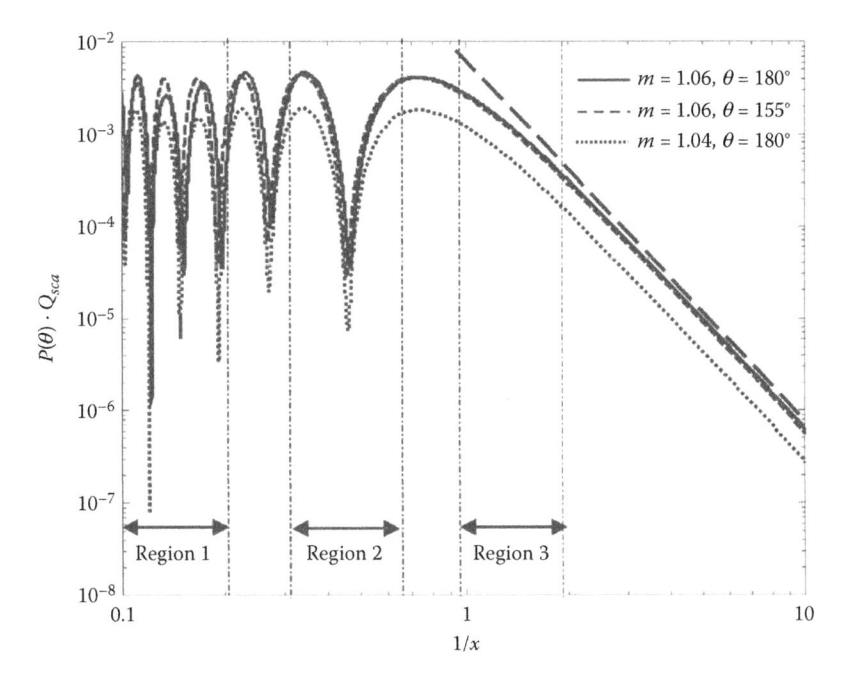

FIGURE 4.26
Plot of scattering cross-section from Mie theory for two relative refractive indices and two backscattering angles. For small-size parameters, the scattering is predominantly Rayleigh with a wavelength dependence of $1/\lambda^4$ (dashed line shows this asymptote). The three regions represent the backscattering signal for the wavelength range of 400–800 nm due to scattering from three particle sizes (region 1: 900-nm particle, region 2: 300-nm particle, region 3: 100-nm particle).

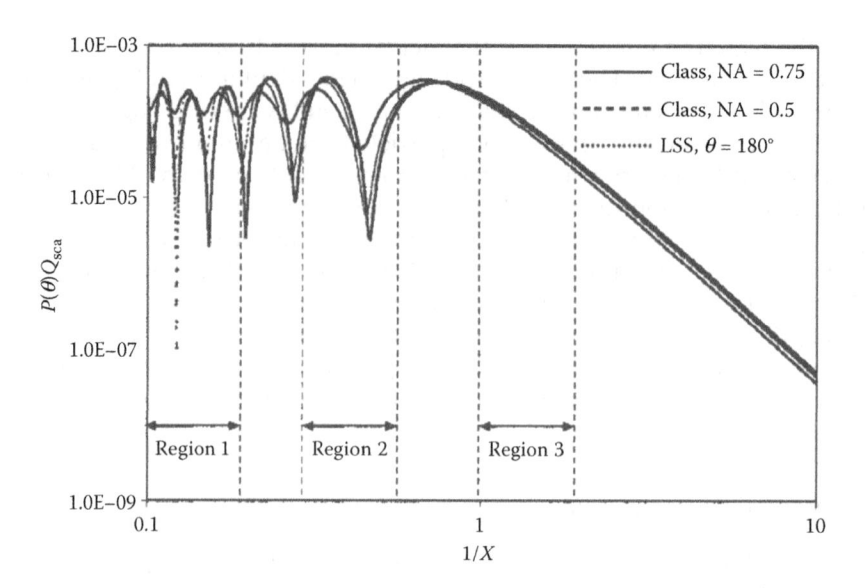

FIGURE 4.27
Comparison of LSS and incoherent CLASS spectra for $NA = 0.75$ (solid line) and $NA = 0.5$ dashed line. Regions 1, 2, and 3 again represent spectra from 900-, 300-, and 100-nm particles, respectively, in a wavelength range of 400–800 nm.

This can be described by integrating the scattering amplitude over the incident and collection NA [17]

$$A(\lambda) = \iint_\Omega \iint_\Omega P(-\hat{k})P(\hat{k}')\exp\left[i\frac{2\pi}{\lambda}R(\hat{k}'-\hat{k})\right]f\left(\frac{\delta}{\lambda},n,\hat{k},\hat{k}'\right)d\hat{k}d\hat{k}', \qquad (4.12)$$

where $\hat{k}'$ is a unit vector in the propagation direction of incident light, $\hat{k}$ is a unit vector in the propagation direction of scattered light, and P is the pupil function of the lens. For simplicity, we use the same pupil function for the exit and entrance of the light. Figure 4.3 shows the effect of varying the pupil function on the LSS spectrum obtained from an incoherent source. It is important to note that this analysis does not directly translate to a CLASS spectrum obtained by using a coherent laser source, in which case the interference structure is conserved. For the purposes of this lab, we refer the reader to an analysis presented as coherent CLASS [18] (Figure 4.27).

Laboratory 5: Experimental Observation of LSS and Effect of Finite NA on LSS Spectrum

1. Equipment and supplies required
    a. USB fiber spectrometer
    b. Optical fiber
    c. Broadband light source (e.g., Tungsten lamp or Xenon lamp)
    d. Three solutions of polystyrene microspheres with an unknown size
    f. Objective or lens

g. Computer with software for controlling USB spectrometer

h. Access to publicly available Mie theory software (optional)

2. Protocol/exercises

a. Part 1: LSS for particle size evaluation.

b. First, a standard optical spectroscopy setup must be aligned. An optical fiber is coupled to a broadband light source, and the light emerging from the fiber is directed toward a sample. A second collection fiber is placed near the illumination fiber and is connected to a USB spectrometer. The distance is then adjusted to obtain an adequate signal level from a reflectance standard such as white paper. The background and reflectance standard signals are saved for normalizing the sample data. Reflectance signals are then obtained from three unknown samples, and the data are stored for processing.

c. Part 2: LSS data collection with a lens.

d. The system is first aligned as in Part 1, but now a lens is added to focus the emerging light onto the sample. The acquisition time is adjusted to obtain an adequate signal, and new background, reflectance standards, and sample signals are acquired.

3. Analysis

a. The data should be processed to account for the wavelength distribution of the source and transmission profiles of the optical components. This is accomplished by subtracting the background (I_{BG}) and normalizing by the reflectance standard (I_{RS})

$$A(\lambda) = \frac{I_{Sample}(\lambda) - I_{BG}(\lambda)}{I_{RS}(\lambda) - I_{BG}(\lambda)}$$

b. $A(\lambda)$ can then be compared to the reference data in Figures 4.2 and 4.3 or a Mie theory calculation.

4. Questions for students

a. What are the sizes of the particles in the measured solutions?

b. What happens to the LSS spectrum when a focusing lens is used and why?

c. What would happen if a wide distribution of sizes is present in the sample, and how does CLASS microscopy deal with this?

Laboratory 6: Spectroscopic Microscopy and Thin-Film Interference

Spectroscopic microscopy merges capabilities of spectroscopy and microscopy, with microscopy providing spatial resolution while spectroscopy enables recovering additional information about cell or tissue structure or chemical composition. The spectroscopic contrast can be generated by elastic scattering, absorption, fluorescence, or inelastic scattering. In this laboratory, we will learn the principles and applications of spectroscopic microscopy. In particular, the exercise will focus on techniques that are based on interference—as we saw in Laboratories 4 and 5, light interference is the underlying basis of a number of optical microscopy techniques, and spectroscopic microscopy is a no exception.

Background

Microscopic imaging has led to major advances in visualizing biological materials at the micro scale. The spatial resolution of current optical imaging techniques such as microscopy is, however, limited by diffraction of light. That is, the best spatial resolution achieved so far (confocal microscopy, phase-contrast techniques) is on the order of $\lambda/2NA$, where λ is the wavelength of the light and NA is the numerical aperture of the imaging optics. This implies that intracellular structures such as organelles, chromatin fibers, proteins, lipids, etc., which are less than or around 100 nm, cannot be imaged by means of optical microscopy. Nevertheless, the light scattered from these organelles can still be detected.

As a branch of optical imaging, spectroscopic microscopy takes advantage of the spectroscopic content in the detected light in addition to the imaging benefits of an optical microscope to obtain information beyond the resolution limit. For example, the recently developed confocal light scattering and absorption spectroscopy unites the imaging resolution of a confocal microscope with the ability to recover size, refractive index, and shape of particles smaller than the wavelength of light based on the principles of light-scattering spectroscopy. High-NA optics is used here to maximize the lateral and axial imaging resolution as well as the collection of broad-angle backscattering from small particles.

Light scattering is not the only spectroscopic technique that can characterize the particles. Alternatively, interferometry-based techniques can also be combined with microscopy to recover the phase information of the scatterers inside the biological samples. One such technique is spectroscopic optical coherence microscopy (SOCM), which combines the use of confocal gated optical sectioning of microscopy with coherent-gated optical ranging. The broad spectroscopic content of the detected light is then used to analyze scattering particles inside biological cells. In additional to elastic light scattering and interferometric principles, inelastic scattering can also be combined with microscopy, as in the case of "integrated Raman and angular scattering microscope (IRAM)." IRAM combines the power of elastic scattering to characterize the morphology and refractive index of a sample and Raman scattering to characterize the chemical composition of the sample. The major difference of IRAM compared to other angle-scattering microscope modalities is that it contains a tightly focused beam ($\sim$7 μm) stemming from the requirement of high-photon flux for Raman microspectroscopy.

A recent approach to extracting intracellular spatial refractive index fluctuations is "partial wave spectroscopic (PWS) microscopy", which measures the backscattering spectrum to assess the refractive index fluctuations within isolated single cells. In particular, PWS microscopy combines the interferometric elastic light-scattering principles to obtain nanoscale refractive index information in the axial plane with microscopic imaging capability in the spatial direction. In PWS microscopy, the sample is illuminated by a broadband, low spatially coherent light source with a low-NA objective (for interferometry principles). The backscattering from the sample is collected using a high-NA objective. The spectral fluctuations at each resolution element are further analyzed to extract refractive index fluctuations at each subcellular location. Although the spatial resolution in a PWS microscope is limited by diffraction, it is sensitive to axial structural fluctuations that are on the order of 10 to 100 nm.

As discussed previously, a class of spectroscopic techniques is based on the phenomenon of the coherent superposition of scattered electromagnetic waves, a phenomenon known as interference. The resultant wave carries information about the physical and

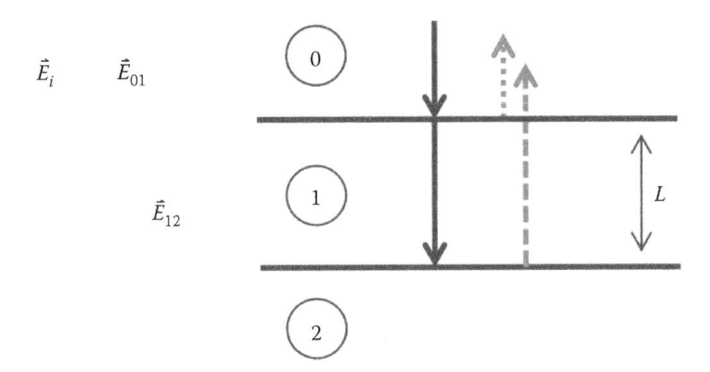

FIGURE 4.28
Interference of reflections from a homogeneous slab.

optical properties of the scattering medium such as its size, refractive index, etc. This chapter describes how the properties of a thin slab with (a) homogeneous and (b) inhomogeneous refractive index distribution can be recovered using light interference. For simplicity, only the case of normal light incidence and case of real refractive indices are considered (Figure 4.28).

Thin-Film Interference Basics: Recovering Film Properties from Spectrum (Normal Incidence)

Consider a three-layered model where light is incident normally from medium 0 onto a thin film (or thin slab) 1 of thickness L. At the interface (0-1) the light will be split into transmitted light and reflected light. The ratio between the amplitudes of reflected light and that of the incident light at a given interface $(x–y)$ is determined by a Fresnel coefficient Γ_{xy}: $\Gamma_{xy} = E_r/E_i = (n_x - n_y)/(n_x + n_y)$. The intensity of reflection or reflectance is defined as $R_{xy} = |\Gamma_{xy}|^2$. The nonreflected propagating wave will be split again at the interface (1-2) when the transmitted light reaches the bottom surface of the film. The superposition of waves reflected from top and bottom surfaces of the middle layer is known as thin-film interference (blue arrows in Figure 4.1). When the thickness of the film is on the order of the wavelength of light, a colorful pattern on the film emerges, which is otherwise called thin-film fringes. For simplicity and subsequent experimental relevance, here we consider the case where media 0 and 2 do not absorb light (therefore, their refractive indices are purely real numbers), and the film's refractive index is $n_1 = n_1' + jn_1''$. The resultant reflectance intensity due to thin film can be written as

$$R(k) = A_1^2 + A_2^2 \exp(-4n_1'' kL) + 2A_1 A_2 \cos(2n_1' kL + \beta) \exp(-2n_1'' kL),$$

where A_1 and A_2 are the Fresnel coefficients of the light waves due to the interfaces (0-1) and (1-2), respectively, and are defined as

$$A_1^2 = |\Gamma_{01}|^2$$
$$A_2^2 = |\Gamma_{01}|^2 |T_{01}|^2 |\Gamma_{12}|^2 |T_{10}|^2, \quad \text{where } |T_{xy}|^2 = 1 - |\Gamma_{xy}|^2$$

It is important to note that when one or both refractive indices of the medium at an interface are complex, the resulting Fresnel coefficients are also complex. For instance, here, Γ_{01} is

$$\Gamma_{01} = \frac{n_0^2 - (n_1'^2) - 2in_0n_1''}{(n_0 + n_1)^2 + n_1''^2},$$

(4.13)

which implies a phase shift between the incident and reflected waves.

To derive the physical or optical properties of the thin film, a combination of the following parameters of the reflectance spectrum can be used (Figure 4.2)

- Baseline light intensity: $A_1^2 + A_2^2 \exp(-4n_1''kL)$
- Spectral oscillation amplitude: $2A_1A_2 \exp(-2n_1''kL)$
- Spectral oscillation frequency: $1/2n'L$
- Phase: β:

$$\beta = \pi - \tan^{-1}\left[2n_0n_1'' / \left(n_1'^2 + n_1''^2 - n_0^2\right)\right]; \quad \text{when} \left(n_1'^2 + n_1''^2\right) > n_0^2,$$

and

$$\beta = \tan^{-1}\left[2n_0n_1'' / \left(n_1'^2 + n_1''^2 - n_0^2\right)\right] \quad \text{otherwise.}$$

Interferometry uses this thin-film interference phenomenon (Figure 4.29). Applications range from calculating the thickness of optical coatings and filters to material characterization (commercially available filmetrics, ellipsometry, optical profilometry) to biosensing (reflectometric interference spectroscopy, RifS).

Slab with Inhomogeneous Refractive Index

For all derivations in this section, we consider only one-dimensional propagation of light through a slab with weak and random refractive index fluctuations ($\Delta n \ll 1$) that is, under the Born approximation.

First, we consider a model of wave propagation in a one-dimensional random medium with the average refractive index n_0 equal to that of the surrounding medium. An electromagnetic wave with electric field Ei is incident on the random medium, resulting in the reflected electric field E_r (Figure 4.3a). All the scattering events in this case happen only

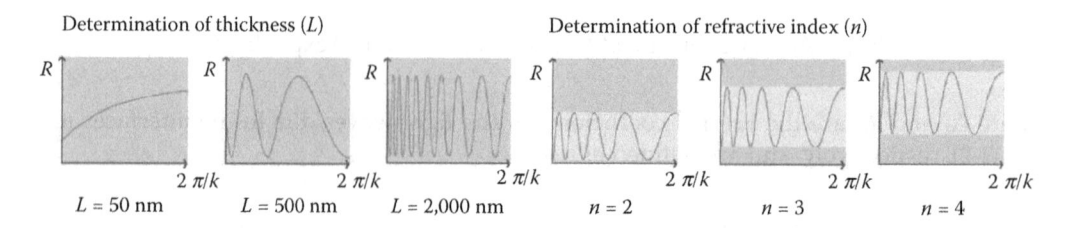

FIGURE 4.29

Determination of thin-film properties from the reflectance interference spectrum.

due to the random fluctuation Δn, and the refractive index of the medium can be defined as $\tilde{n}(z) = n_0(1 + \Delta n(z))$.

Let $B_{\Delta n}(z' - z'')$ denote the autocorrelation function of refractive index variation

$$B_{\Delta n}(z' - z'') = E\{\Delta n(z')\Delta n(z'')\} \tag{4.14}$$

Since the medium has a random refractive index distribution, it can only be described as a statistical entity. Hence, we focus on the statistical properties of the reflected light. Typically, for a particular $B_{\Delta n}(z' - z'')$, the mean and the standard deviation of reflectance at a particular wavelength over all possible realizations of refractive index geometry can be studied ([19,20]).

For a thick slab, compared to the wavelength of light ($2kl \gg 1$), the mean and standard deviation among all possible reflectance intensities coming from a random slab are equal and are described as

$$\langle I \rangle = \sigma_I = (k^2 L)S(2k) \tag{4.15}$$

where $S(k)$ is the power-spectral density of the random medium, defined by the Fourier transform of the autocorrelation function $B_{\Delta n}(z)$

$$S(k) = \int_{z=-\infty}^{+\infty} B_{\Delta n}(z)e^{jkz}dz \tag{4.16}$$

Regardless of the specific correlation function $B_{\Delta n}(z)$ of the random medium, $S(2k)$ depends linearly on the variance of the refractive index fluctuations σ_n^2, and for correlation lengths much smaller than the wavelength $l_c \ll \lambda$, it also depends linearly on l_c

$$\langle I \rangle = \sigma_I \infty (k^2 L)\sigma_n^2 l_c \tag{4.17}$$

It can be shown that instead of calculating the standard deviation of reflectance intensities σ_I over the realizations of a refractive index medium, the standard deviation of the reflected spectrum (σ_k) (amplitude fluctuations as a function of wavelength) can be taken, and it will have the same functional dependence as σ_I. This is an advantage in applications where multiple realizations of the same medium cannot be obtained (Figure 4.30).

The assumptions that went into this result are

- The correlation length is small compared to the thickness: $l_c \ll L$.
- The thickness is large compared to the wavelength: $2kL \gg 1$.

Now, we add refractive index differences at the top and bottom of the slab: the uppermost medium has refractive index n_0, the middle layer n_1, and the lowermost medium n_2. (Figure 4.3b). This geometry implies a big refractive-index mismatch on one interface, causing reflections at the boundary of the layers that are usually larger than those caused by the random component inside the film. For simplicity, a smaller mismatch is assumed at the bottom (this will not affect the primary conclusions but will make expressions simpler). The refractive index fluctuation $\Delta n(z)$ is now defined with respect to n_1: $\tilde{n}(z) = n_1(1 + \Delta n(z))$; $\Delta n \ll 1$.

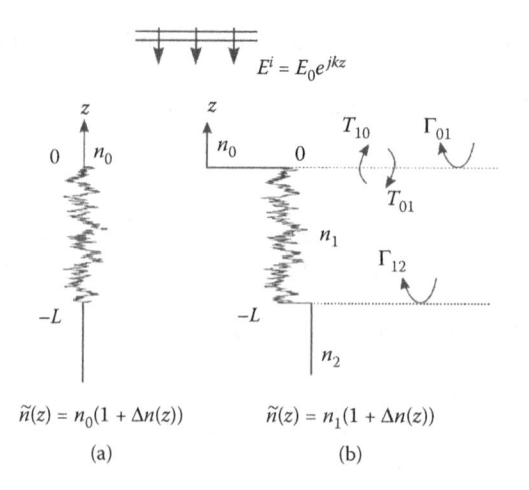

FIGURE 4.30
Wave propagation through an inhomogeneous slab with (a) average refractive index matching that of surroundings and (b) a refractive index mismatch on the top and at the bottom.

The resulting reflectance in this system at a specific wavelength consists of a deterministic (nonrandom) component defined by average values of the refractive indices of the three layers and the random component defined by the properties of the middle layer. That is

$$I = |\Gamma_{01}|^2 + 2\mathrm{Im}\{(\Gamma_{01}\,T_{10}\,T_{01})k_1 \int_{z'=-L}^{0} \Delta n(z')e^{j2k_1 z'}dz'\} = I_d + I_r, \tag{4.18}$$

where Im { } denotes the imaginary part of a variable, I_d is the "deterministic reflection" determined solely by the contrast between average refractive index of the slab with its boundaries, and I_r is the intensity reflectance due to the random distribution inside the slab. The Fresnel coefficients Γ_{01}, T_{01}, and T_{10} are defined as

$$\Gamma_{01} = \frac{n_0 - n_1}{n_0 + n_1}; T_{01} = \frac{2n_0}{n_0 + n_1}; T_{10} = \frac{2n_1}{n_0 + n_1}$$

Since the refractive index fluctuation $\Delta n(z)$ is zero-mean by definition, the **mean** reflectance intensity is equal to the deterministic part, while its **standard deviation** depends on random fluctuations' properties and is given by the following expression

$$\sigma_I = \sqrt{2}\Gamma\sqrt{\sigma_I^{match}} \tag{4.19}$$

where σ_I^{match} is the standard deviation of the reflectance obtained in the previous case (1A). Γ is defined by the Fresnel coefficients: $\Gamma = \Gamma_{01}T_{10}T_{01}$. Combined with the expression for σ_I^{match} we obtain

$$\sigma_I = \sqrt{2}\Gamma\sqrt{(k^2 L)S(2k)} \tag{4.20}$$

The above generalized equation underlines the basic concept behind the detection of weak refractive index fluctuations inside a slab with an average refractive index that is different from its surroundings. That is, the standard deviation of reflected light intensity quantifies the distribution of random refractive index fluctuations. Its low-power dependence ensures high contribution from structures of small sizes and weak contrast, thus preventing self-averaging in the presence of larger structures. Finally, its absolute value is enhanced by a factor of $\sqrt{2}\Gamma$, providing a better signal-to-noise ratio for potential experimental applications.

As it is impractical to experimentally study multiple realizations of the same refractive index distribution, the spectral variation of intensity reflectance ($\sigma_{\delta I}^{\Delta k}$) can be used to characterize the inhomogeneous medium inside a slab. Since the deterministic intensity reflectance is constant (the dispersion of average refractive indices is neglected), the spectral standard deviation of intensity reflectance in the visible range of wave numbers can be expressed as

$$\left(\sigma_{\delta I}^{\Delta k}\right)^2 = \frac{1}{\Delta k} \int_{\Delta k} |I_r(k)|^2 dk = \frac{1}{\Delta k} \int_{\infty} |T_{\Delta k}I_r(k)|^2 dk,$$

where $T_{\Delta k}$ is a top-hat function that is a unity for wave numbers within the experimental bandwidth Δk and zero everywhere else.

The random part of intensity reflectance when expressed through $F(k)$—the Fourier transform of $n(z)$ has the form $I_r(k) = 2\Gamma \text{Im}\{k_1 F(2k_1)\}$, where $F(k) = \int_{z'=-\infty}^{\infty} \Delta n(z')e^{jkz'}dz'$. Thus, its spectral standard deviation according to Plancherel's theorem is equivalent to

$$\left(\sigma_{Ir}^{\Delta k}\right)^2 = \frac{1}{2\pi\Delta k} \int_{z=-\infty}^{\infty} |FT(T_{\Delta k}I_r(k))|^2 dk, \qquad (4.21)$$

where FT denotes Fourier transform. On the other hand, it can be shown that $|FT(T_{\Delta k}\delta I)| = \pi\Gamma|(d\{F(T_{\Delta k}) \otimes \Delta n(z)\}/dz|$, which signifies that the spectral variance of intensity reflectance is explained by the effective refractive index distribution $\Delta n_1(z)$, which is a convolution of the actual $\Delta n(z)$ with a sinc function due to the temporal coherence of light (finite bandwidth) $\Delta n_1(z) = \sin c(\Delta kz) \otimes \Delta n(z)n_1(z) = \sin c(\Delta kz) \otimes \Delta n(z)$. When a narrow spectral interval is analyzed ($\Delta k \ll k$) according to the previously discussed behavior of random refractive index correlation functions as a function of l_c and σ_n (when $kl_c \ll 1$), the spectral standard deviation of intensity reflectance depends on the properties of random refractive index distribution inside the slab as

$$\sigma_I^{\Delta k} \propto \sqrt{(k^2 L)(\sigma_{n_1}^2 l_c)} \qquad (4.22)$$

Backscattering in Three-Dimensional Media

We described a model for one-dimensional wave propagation. However, in a three-dimensional light-scattering reality, although the basic idea behind the phenomenon stays the same, other aspects must be accounted for. Due to spatial coherence of light, all light-scattering events within a diffraction-limited area along the same phase front will contribute to the resultant reflectance. Therefore, to transform the 3-D problem into the previously described 1-D, scattering within the spatially coherent area needs to be averaged and projected onto the direction of wave propagation. In particular, for plane wave propagation along the Z direction, contributions from an Airy disc in the perpendicular X–Y plane have to be accounted for. As a consequence, the spectral variance of intensity reflectance from a three-dimensional inhomogeneous slab includes effective averaging of the actual refractive index within a coherence volume: temporal coherence along the direction of propagation and spatial coherence along the phase front. Nevertheless, the concept of sensitivity to optically unresolvable random refractive index fluctuations (including $\Delta n \ll 1$ and $kl_c < 1$) stays unchanged.

Laboratory Protocol

Instrumentation

Experiments in this section are performed using a conventional reflected-light illumination microscope and projecting the magnified sample image onto a spectrometer. This microscope-spectrometer configuration provides wavelength-dependent amplitude information for the reflected signal from a sample located in the sample plane of the microscope objective.

The above microscope–spectrometer combination achieves two goals:

1. To discern and study the fine detail of the sample using microscope magnification.
2. To detect spectral features resulting from the interference of waves reflected and scattered from within the sample using spectrometer.

Figure 4.3 shows a schematic diagram of a microscope–spectrometer instrument. Spatially and temporally incoherent light radiated from a broadband white light source (here: a xenon arc lamp) is collimated by a condenser lens, passed through a beamsplitter (BS), and then focused onto a sample stage (SS) by a microscope objective lens (OBJ) (here: **60X, N.A. = 0.6**). The sample stage is a three-axis stage that moves in x, y, and z directions. The reflected light from the sample is collected by the objective lens and subsequently transmitted through a beamsplitter and a tunable filter (here: liquid crystal tunable filter working in the visible wavelength range) and then projected onto a charge-coupled device (CCD) detector array that digitizes the intensity signal and transfers it to a computer (not shown). In the above configuration, the objective lens serves as both a collimator lens when the light is being focused onto the sample and an image-forming lens for the backscattered light (Figure 4.31).

Measurement Protocol

1. Turn on the light source, tunable filter, and detector camera. Wait for the devices to stabilize as required (~20–30 minutes).
2. Place the sample on the microscope sample stage.
3. Tune the filter to a wavelength in the middle of the spectral range (around 600 nm), and acquire the image of the sample in Live mode of the CCD camera.

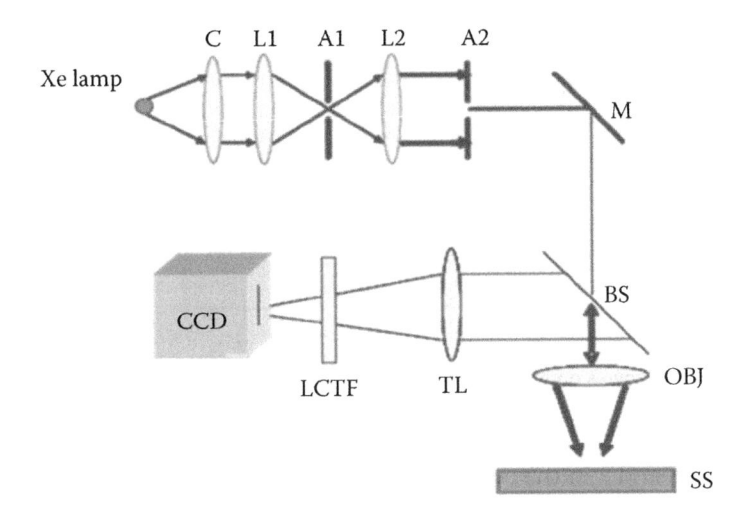

FIGURE 4.31
Schematic of an example microscope–spectrometer configuration. A1, A2, apertures [D(A1) 1/4 2 mm, D(A2) 1/4 4 mm]; BS, beamsplitter; C, condenser; L1, L2, lenses (f 1/4 150 mm); F, flipper mirror; LCTF, liquid crystal tunable filter; M, mirror; OBJ, objective lens; SS, sample stage; TL, tube lens (f¼ 450 mm).

4. Adjust the focus of the image by changing the vertical position of SS to achieve image crispness.

5. Chose object of interest as applicable: change x and y axis controller of SS until you see it.

6. Open a computer file where the data are to be saved.

7. Open the control interface for your system.

8. Acquire the spectrum of light intensity reflected from the sample by tuning the filter to different wavelengths and acquiring the corresponding CCD image.

9. Record the exposure time of the CCD camera.

10. Before moving to the next measurement, check the collected spectra. If the registered counts of CCD camera are above 80% of its maximum, lower the light exposure time.

Experiment 1: Thin-Film Interference Observed from Microspheres

What Is Needed

- Plain glass microscope slides (Fisher Scientific, VWR, etc.)
- Polystyrene spheres (3- and 5-μm diameter, Thermo Scientific)

Sample Preparation

1. Take a clean microscopy glass slide and one of the sphere solutions (Figure 4.32).

2. Put a drop of sphere solution on the slide and let it disperse and dry. If a fume hood is available, put the slide in it to accelerate the drying process.

3. Repeat steps 1 and 2 using spheres of a different size. Keep different-sized spheres on different glass slides.

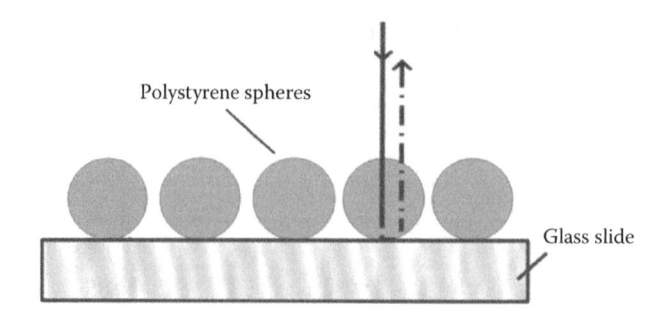

FIGURE 4.32
Sample layout: dry microspheres on microscope glass slide.

Experimental Procedure

1. Follow steps 1 and 2 in the measurement procedure.
2. Place one of the prepared slides with dry microspheres on the microscope sample stage.
3. Follow steps 3 to 10 in the measurement procedure.
4. Record reflection spectra from two different areas in the sample slide.
5. Repeat the above steps using the sample slide containing different-sized spheres.
6. Collect reflectance spectrum from mirror for normalization.
7. measure spectra from a dyed sphere.

Data Analysis (Should Be Performed for the Data from Both Samples)

1. Normalization by incident light: Divide sample reflectance spectrum by the corresponding reflectance spectrum from the mirror. Take into account the difference in camera exposure times.
2. Compare the reflection spectra obtained from a glass slide from the top and sides of spheres. What are the characteristics of the spectra? Explain.
3. Take the brightest point from the top of a sphere. Plot the reflection intensity as a function of wavenumber.
4. Calculate the amplitude and frequency of oscillations. How do they compare between the two samples (light paths a and b in Figure 4.5)? Explain.
5. What is the refractive index of spheres according to spectra collected?
6. Find the absorption coefficient as a function of k and imaginary refractive index of a dyed sphere?

Experiment 2: Interference Spectrum from an Inhomogeneous Film of Nanospheres

What Is Needed

- Plain glass microscope slides (Fisher Scientific, VWR, etc.)
- Polystyrene spheres (20- and 100-nm diameter, Thermo Scientific)

Sample Preparation

This is the same as in the previous section, but 20- and 100-nm bead solutions are used instead. Here, in the microscope field of view, spheres will not be distinguishable (why?); you will see a film of aggregated spheres with characteristic cracks.

Experimental Procedure

1. Follow steps 1 and 2 in the measurement procedure.
2. Place one of the prepared slides with dry nanospheres on the microscope sample stage.
3. Follow steps 3 to 10 in the measurement procedure.
4. Record reflection spectra from an area of interest. Check reflection spectrum for its oscillation frequency. When the film is too thick, the top and bottom reflections become incoherent and do not interfere; hence, no oscillations will be seen. Similarly, if the film is too thin, then no oscillations will be visible either. It is recommended to have ~6 periods in 300-nm bandwidth.
5. Collect four measurements per sample from areas with similar thicknesses for best compatibility.
6. Collect reflectance spectrum from mirror for normalization.

Data Analysis (Should Be Performed for Both Data Sets)

1. Normalization by incident light: Divide sample reflectance spectrum by the corresponding reflectance spectrum from the mirror. Take into account the difference in camera exposure times.
2. Notice that the oscillations will not look as uniform as they did for homogeneous films. What is different between these thin-film oscillations and the ones seen in experiment 1?
3. What sample properties define the L, Δn, and l_c of your thin slabs?
4. Calculate the frequency of oscillations: What is the thickness (L) of the film?
5. Calculate the average amplitude of oscillations.
6. Compare the average amplitude of oscillations from the two samples. Explain the results.

Experiment 3: Interference Spectrum from a Biological Cell as an Inhomogeneous Film

What Is Needed

- 100-nm aluminum-coated microscope slides (Deposition Research Lab, Inc) (Figure 4.33)
- Cytobrush (CooperSurgical, Inc.)
- 70% ethanol

Sample Preparation

1. Obtain *in vivo* cytology squamous epithelial cell brushings as below.
2. Brush the inside of a cheek, and smear the brush onto the metallic surface of aluminum-coated microscope slide (Figure. 4.6).
3. Fix the cells by immersing the slide in 70% ethanol immediately after the smear is obtained.
4. Keep the sample at a room temperature for 20 minutes.
5. Take out the slide, put in a fume hood (if available) with the side containing cells facing up, and let them dry.

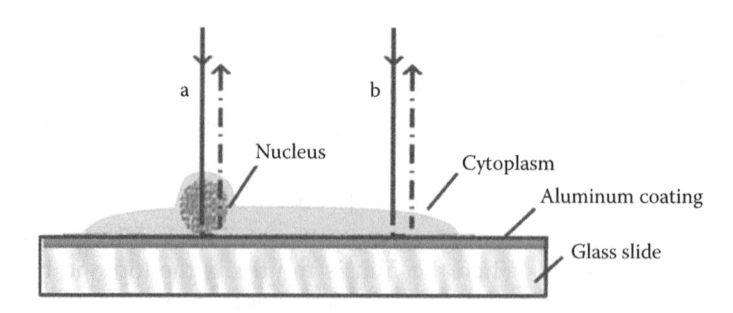

FIGURE 4.33
Squamous epithelial cell fixed on aluminum-coated glass slide.

Experimental Procedure

1. Follow steps 1 and 2 in the measurement procedure.
2. Place the sample on the microscope sample stage, cells facing up.
3. Follow steps 3 to 10 in the measurement procedure.
4. Record reflection spectrum from four areas with individual cells in the center. Avoid folded or overlapped cells. Note that the camera counts saturate at lower exposure times than in the two previous experiments. Why?
5. Collect reflectance spectrum from mirror for normalization.

Data Analysis

1. Normalization by incident light: Divide sample reflectance spectrum by the corresponding reflectance spectrum from the mirror. Take into account the difference in camera exposure times.
2. Compare reflection spectra with those observed from polystyrene spheres. Explain differences in optical properties.
3. Assume refractive index of a cell to be 1.5. Analyze five different spectra from nucleus (Figure 4.5a) and cytoplasm (Figure 4.5b). Calculate their thicknesses.

Answers and Hints

Experiment 1: Data Analysis

1. The characteristic spectra from the brightest spot on a sphere correspond to the interference pattern from a homogeneous slab.
2. The amplitude of oscillation is defined by refractive index contrast above and below the polystyrene spheres; the frequency is determined by sphere diameter— it will be higher for the larger spheres.
3. The refractive index of polystyrene averaged over the visible range of wavelengths is 1.59.

Experiment 2: Sample Preparation

Nanosphere diameters are well below the diffraction limit of the microscope, and that is why spheres cannot be resolved.

Data Analysis

1. The sample represents an inhomogeneous slab: Unlike the constant amplitude of oscillations within the wavelength range seen in the Experiment 1, these oscillations are random.

2. L is the total thickness of the nanosphere slab; Δn depends on the density of spheres within the slab and represents contrast between the refractive indices of air ($n = 1$) and polystyrene ($n = 1.59$); l_c is majorly defined by sphere size and density of aggregation.

3. Since both samples have a similar variance of refractive (both are composed of air and polystyrene), the average amplitude of spectral oscillations from areas with the same thickness as follows from Equation 4.22 will increase with sphere size (l_c).

Data Analysis

1. A biological cell can be represented as a slab with an inhomogeneous refractive index distribution. Therefore, the observed spectra will resemble interference spectra observed in Experiment 2.

Laboratory 7: Measurement of Real and Imaginary Refractive Index of Cells

The refractive index is the major source of contrast in microscopic imaging of biological cells based on elastic processes (e.g., scattering, phase delay, interference) and absorption. The real part of the refractive index causes phase delay, which is the contrast mechanism in phase-contrast microscopy. The spatial variations of the real part of the refractive index lead to elastic light scattering, which is the underlying mechanism of dark-field microscopy. The imaginary part leads to light absorption and is employed in bright-field microscopy of cells stained with contrast dyes. Thus, knowing intracellular refractive index is critically important. The objective of this laboratory is to learn how interference-based spectroscopic microscopy can be used to measure real and imaginary parts of the refractive index.

Background

Since endogenous absorption is weak in biological samples, numerous absorbing dyes have been synthesized to provide artificial contrast for the visualization of tissues, cells, and cellular compartments. Staining dyes are routinely used in basic biology, pathology, and medical diagnostics; the process of staining is easy to perform, and colorful images are obtained without the need of complicated optics. However, while massive attention is focused on the induced absorption, the dramatic changes it brings to phase and scattering properties of stained biological samples are often overlooked.

First, due to the principle of causality, real and imaginary parts of physical entities are interdependent. That is, when absorption is initiated (which is equivalent to introduction of imaginary refractive index), the real refractive index changes in response. Known as Kramers–Kronig relation, this interconnection is discussed in detail in the following didactic section. In addition, regardless of the effect it has on real refractive index, a dye delivers an imaginary refractive index, and that by itself boosts the scattering cross-section of a targeted particle since it increases the spatial variation of the absolute refractive index. This change in scattering properties is most remarkable when other sources of scatter contrast such as size and real refractive index are weak. For example, intracellular particles with characteristic sizes smaller than the wavelength and their unstained refractive index contrast on the order of 1.01 will experience scattering cross-section growth by orders of magnitudes with the introduction of absorption [1]. Unlike the innate scattering contrast of unstained biological samples, staining-induced contrast possesses manifold specificity: (i) it preferably enhances weak scattering events; (ii) the chemical affinity of a certain dye allows targeting the change to macromolecules or organelles of interest; and (iii) dye color offers control of the direction of change in real refractive index; it is decreased at shorter wavelengths to the absorption peak and increased at longer. With the currently available selection of highly specific dyes with various colors, histological staining offers a perfect tool for specific enhancement of scattering in the realm of spectroscopic microscopy techniques.

Relationship between the Real and Imaginary Parts of the Refractive Index

In the beginning of the twentieth century, Hans Kramers [2] demonstrated that the absorption spectrum of a medium is sufficient for the calculations of its real refractive index. Around the same time, Ralph Kronig [3] showed that this dispersion relation is a necessary and sufficient condition for relativistic causality to hold. The bidirectional mathematical relation connecting the spectral profiles of real or imaginary parts of the complex electric susceptibility is known as the Kramers–Kronig relation

$$\text{Re}\{\chi(k)\} = \frac{2}{\pi} P \int_0^\infty \frac{k' \, \text{Im}\{\chi(k')\}}{k'^2 - k^2} dk'$$

$$\text{Im}\{\chi(k)\} = -\frac{2k}{\pi} P \int_0^\infty \frac{\text{Re}\{\chi(k')\}}{k'^2 - k^2} dk'$$

(4.23)

where P is the Cauchy principal value (Figure 4.34).

Since the electric susceptibility of a nonmagnetic material defines its refractive index as $n = \sqrt{1 + \chi}$, the real and imaginary parts of the latter are also interlinked as shown in Figure 4.1. The Kramers–Kronig relation intrinsically implies that for a nonzero real part of refractive index to exist, the medium must have a nonzero imaginary part (absorption) of at least one frequency. Most materials absorb at the frequencies of dipole resonances, which explains the direction of normal dispersion: The real part of the refractive index decays toward lower frequencies (as seen at longer wavelengths to the resonant peak in Figure 4.1).

In a case where absorbing dye particles are embedded into a nonabsorbing host medium, the refractive index of the new compound medium is $n = \sqrt{1 + \chi_{host} + \chi_{dye}}$. Decomposing n into its real (n') and imaginary (n'') components, we get

$$n' = \sqrt{n_{host}^2 + \chi'_{dye} + n''^2}$$

$$n'' = \frac{\chi''}{2n'}$$

(4.24)

That is, when histological dyes spread within transparent biological materials, this set of equations determines how the spectral profile of resultant real refractive index is related to absorption of the stain.

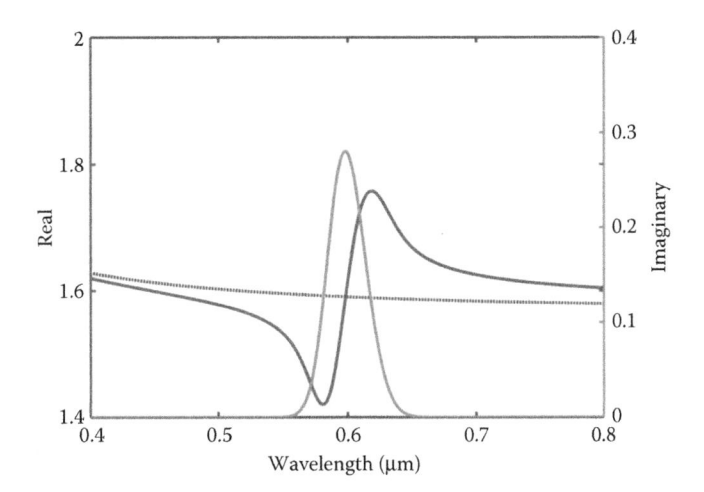

FIGURE 4.34
Refractive index in the proximity of absorption resonance. The dotted line represents refractive index of clear polystyrene, the solid lines represent the real (blue) and imaginary (green) refractive indices of polystyrene that contain dye particles, with an absorption peak at 0.6 μm.

Laboratory Protocol

Instrumentation

This laboratory requires the same instrumentation as the preceding lab. Note that a reflection microscope setup can be substituted with a transmission alignment. In that case, the light will travel only once through the sample, and no aluminum-coated slide will be replaced with a regular glass slide.

What Is Needed

- Plain glass microscope slides (Fisher Scientific, VWR, etc.).
- 100-nm aluminum-coated microscope slides (Deposition Research Lab, Inc).
- Cytobrush (CooperSurgical, Inc.).
- Histological staining supplies for technique of choice. Multicolor staining (e.g., hematoxylin and eosin [H&E]) is recommended.
- 70% ethanol for fixation.
- Mounting medium with a refractive index between 1.5 and 1.55 (e.g., Permaslip [Alban Scientific]). Most fixed tissues and cells have a refractive index in that range, and mounting media are meant to match it to prevent reflections and ensure maximum visual clarity.
- Computing environment for data analysis (MATLAB [MathWorks] recommended).

Sample Preparation

1. Obtain and fix cytology squamous epithelial cell brushings following the sample preparation technique for the Experiment #3 of Laboratory 7 (Figure 4.35).

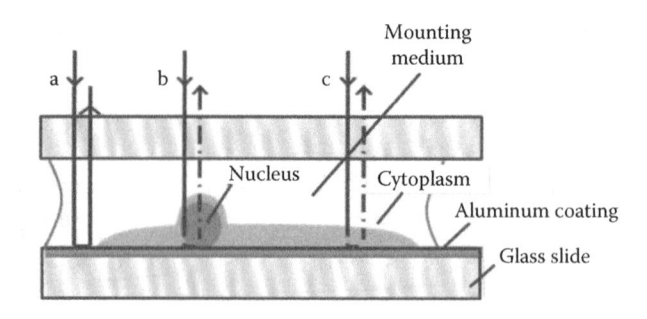

FIGURE 4.35
Sample layout: A biological cell is embedded in a refractive-index matching medium. Colors correspond to H&E staining. If a transmission microscope system is used, aluminum coating is not needed.

2. Follow the staining technique provided by the manufacturer. When H&E is used, stain with hematoxylin prior to eosin to minimize eosin content in the nucleus and enhance color contrast with the cytoplasm.

3. Let the stained slides fully dry, and cover them with the mounting medium first and with a clean regular glass slide next. (Figure 4.2).

Experiment

1. Follow steps 1 and 2 in the measurement procedure of Laboratory 7.

2. Place the sample on the microscope sample stage, clear slide facing up.

3. Follow steps 3 to 10 in the measurement procedure of Laboratory 7.

4. Record reflection spectrum from four areas with individual cells in the center. Choose three thinner-looking cells and one thick. Avoid folded or overlapped cells.

Data Analysis

1. Check reflectance spectra from empty areas (Figure 4.2a). Explain the spectral shape.

2. Biological cell content is best represented as a water solution of (in order of decreasing content): proteins, nucleic acids, lipids, and polysaccharides. Alcohol fixation fully dehydrates biological cells and tissues. To find the refractive index of a fixed cell, use Gladstone–Dale's relation $n_{solution} = n_{solvent} + \alpha_{solute}C_{solute}$, where C_{solute} is solute density in g/mL, α_{solute} is approximated as 0.18 mL/g [4], and the solvent is water. Approximate C as the cell dry density for stratum mucosum: 1.15 g/mL, [4] and you will obtain your n_{host}.

3. Calculate the absorption constant $\alpha(x, y, \lambda)$ using Beer–Lambert's law $I(x, y, \lambda) = I_0(x, y)\exp(-\alpha(\lambda)L)$. You will need:

   a. L: Assume the thickness of cell cytoplasm is 500 nm and the nucleus is 1 micron. Note that the collected light has passed through the sample twice.

   b. I_0: To avoid confusion of light extinction due to absorption with that of scattering or any other losses, consider I_0 equal to the value of intensity you detected at a wavelength at which none of your stains absorb (for H&E, use $I_0(x, y) = I(x, y, \lambda = 700$ nm)). Explain why I_0 varies with voxel location.

4. Find imaginary refractive index $n''(x, y, \lambda)$. How is it related to the absorption constant?

5. Obtain absorption spectra for the stains you used from the manufacturer, and compare with the obtained $n''(x, y, \lambda)$ profile. Discuss the similarity and/or difference observed.

6. As a first iteration of imaginary electric susceptibility $\chi_1''(x, y, \lambda)$ from Equation 4.24, assume $\chi_1''(x, y, \lambda) = 2n_{host}(\lambda)n''(x, y, \lambda)$.

7. Using Kramers–Kronig relation, calculate the $\chi'(x, y, \lambda)$. You can either write your own program or download publicly available code (for MATLAB, use [5]).

8. Calculate the real part refractive index of the compound stained medium $n'(x, y, \lambda)$ using Equation 4.24.

9. The second iteration of $\chi_2''(x, y, \lambda) : \chi_2''(x, y, \lambda) = 2n'(x, y, \lambda)n''(x, y, \lambda)$. Repeat steps 1 and 2. Did the result for $n'(x, y, \lambda)$ change? Are further iterations needed?

10. Summarize results:

   a. Prepare graphs of real and imaginary refractive indices for voxels in cytoplasmic and nuclear areas. Explain spectral differences.

   b. At which wavelength is the change in refractive indices the most significant?

   c. List all assumptions made in data analysis. Which of them were made for simplicity and were unnecessary? Propose experiments to avoid those. Discuss validity of the ones that were necessary.

References

1. A. Wax, M. G. Giacomelli, T. E. Matthews, M. T. Rinehart, F. E. Robles, and Y. Zhu, Optical spectroscopy of biological cells. *Adv. Opt. Photon.* 2012;4(3):322–378.
2. L. Perelman, V. Backman, M. Wallace, G. Zonios, R. Manoharan, A. Nusrat, S. Shields et al., Observation of periodic fine structure in reflectance from biological tissue: A new technique for measuring nuclear size distribution. *Phys. Rev. Lett.* 1998;80(3):627–630.
3. H. Fang, L. Qiu, E. Vitkin, M. M. Zaman, C. Andersson, S. Salahuddin, L. M. Kimerer et al., Confocal light absorption and scattering spectroscopic microscopy. *Appl. Opt.* 2007;46(10):1760–1769.
4. I. Itzkan, L. Qiu, H. Fang, M. M. Zaman, E. Vitkin, I. C. Ghiran, S. Salahuddin et al., Confocal light absorption and scattering spectroscopic microscopy monitors organelles in live cells with no exogenous labels. *PNAS* 2007;104(44):17255–17260.
5. H. Subramanian, P. Pradhan, Y. Liu, I. R. Capoglu, X. Li, J. D. Rogers, A. Heifetz et al., Optical methodology for detecting histologically unapparent nanoscale consequences of genetic alterations in biological cells. *Proc. Natl. Acad. Sci.* 2008;105(51):20118–20123.
6. K. J. Chalut, J. H. Ostrander, M. G. Giacomelli, and A. Wax, Light scattering measurements of subcellular structure provide noninvasive early detection of chemotherapy-induced apoptosis. *Cancer Res.* 2009;69:1199–1204.
7. D. Huang, E. A. Swanson, C. P. Lin, J. S. Schuman, W. G. Stinson, W. Chang, M. R. Hee et al., Optical coherence tomography. *Science* 1991;254(5035):1178–1181.
8. M. A. Choma, M. V. Sarunic, C. H. Yang, and J. A. Izatt, Sensitivity advantage of swept source and Fourier domain optical coherence tomography. *Opt. Express* 2003;11(18):2183–2189.
9. R. Leitgeb, C. K. Hitzenberger, and A. F. Fercher, Performance of Fourier domain vs. time domain optical coherence tomography. *Opt. Express* 2003;11(8):889–894.

10. F. E. Robles, *Light Scattering and Absorption Spectroscopy in Three Dimensions Using Quantitative Low Coherence Interferometry for Biomedical Applications*, PhD Thesis in Medical Physics. 2011, Duke University: Durham, NC.

11. Y. Wang, K. Maslov, Y. Zhang, S. Hu, L. M. Yang, Y. N. Xia, J. A. Liu, and L. H. V. Wang, Fiber-laser-based photoacoustic microscopy and melanoma cell detection. *J. Biomed. Opt.* 2011;16(1):011014.

12. C. Zhang, K. Maslov, and L. H. V. Wang, Subwavelength-resolution label-free photoacoustic microscopy of optical absorption *in vivo. Opt. Lett.* 2010;35(19):3195–3197.

13. C. Kim, C. Favazza, and L. H. V. Wang, *In vivo* photoacoustic tomography of chemicals: High-resolution functional and molecular optical imaging at new depths. *Chem. Rev.* 2010;110(5):2756–2782.

14. L. T. Perelman, V. Backman, M. Wallace, G. Zonios, R. Manoharan, A. Nusrat, S. Shields et al., Observation of periodic fine structure in reflectance from biological tissue: A new technique for measuring nuclear size distribution. *Phys. Rev. Lett.* 1998;80:627–30.

15. I. Itzkan, L. Qui, H. Fang, M. M. Zaman, E. Vitkin, I. C. Ghiran, S. Salahuddin et al., Confocal light absorption & scattering spectroscopic (CLASS) microscopy monitors organelles in live cells with no exogenous labels, *Proc. Nat. Acad. Sci. USA* 2007;104:17255–17260.

16. H. Fang, M. Ollero, E. Vitkin, L. M. Kimerer, P. B. Cipolloni, M. M. Zaman, S. D. Freedman et al., Noninvasive sizing of subcellular organelles with light scattering spectroscopy. *IEEE J. Sel. Top. Quant. Elect.* 2003;9:267–276.

17. H. Fang, L. Qiu, M. M. Zaman, E. Vitkin, S. Salahuddin, C. Andersson, L. M. Kimerer et al., Confocal light absorption and scattering spectroscopic (CLASS) microscopy, *Appl. Opt.* 2007;46:1760–1769.

18. L. Qiu, V. Turzhitsky, L. Guo, E. Vitkin, I. Itzkan, E. B. Hanlon, and L. T. Perelman, Early cancer detection with scanning light scattering spectroscopy, *IEEE J. Sel. Top. Quant. Elect.* 2011;99:1–11.

19. S. B. Haley and P. Erdos, Wave propagation in one-dimensional disordered structures, *Phys. Rev. B Condens Matter* 1992;45:8572–8584.

20. A. Ishimaru, *Wave Propagation and Scattering in Random Media*. New York: Academic Press, 1978.

5

Tissue Spectroscopy

The objective of this chapter is to provide students with exposure and hands-on laboratory experience with the main modalities of biological tissue spectroscopy. Tissue spectroscopy is one of the vibrant research areas within biophotonics. Its significance stems from the practical applications of spectroscopy in tissue diagnostics as well as the wealth of information it may provide to better our understanding of fundamental biology.

The term spectroscopy simply means that a particular type of light–matter interaction is observed as a function of the wavelength of light. The key concept of tissue spectroscopy is that, at least in principle, essentially any type of light–matter interaction can be translated into a spectroscopic modality. Depending on the modality, different properties of tissue can be elucidated. The general strategy is to measure the spectrum of light interacting in a certain way with tissue and analyze the spectrum in order to extract quantitative information about particular tissue properties, whether it is tissue's structure or chemical composition.

In Chapter 3, we discussed in detail the two most prevalent types of light interaction with tissue: elastic scattering and absorption. These translate into several spectroscopic modalities, including elastic scattering spectroscopy, diffuse reflectance spectroscopy, polarization gated spectroscopy, enhanced backscattering spectroscopy, spectroscopic optical coherence tomography, etc., depending on the mode of observation of scattered light as well as the number of scattering events, specific light paths, or path lengths, and other characteristics of scattered light that is being recorded. Depending on these characteristics, the depth of penetration of scattered light into tissue and the information that can be extracted from the signal may vary. The spectroscopic techniques based on scattered and absorbed light are typically used to provide quantitative information about tissue morphology (e.g., refractive index correlation function at nanometer-to-micron length scales, sizes of tissue structures, tissue birefringence) and blood perfusion (e.g., blood concentration, oxygen saturation [a.k.a. tissue oxygenation], average diameter of blood vessels).

The detection of endogenous tissue fluorescence translates into fluorescence spectroscopy. The spectrum of an endogenous fluorescence signal that is recorded from tissue is affected by absorption and scattering—both processes can change the intensity of the emitted fluorescence signal as well as excitation light by either absorbing photons at particular wavelengths or redistributing the light due to scattering. However, it is still possible to identify endogenous fluorescence spectrum after postprocessing of the recorded fluorescence. Collagen and elastin in the connective tissue matrix and nicotinamide adenine dinucleotide reduced form (NADH) and nicotinamide adenine dinucleotide (NAD) in cells are examples of endogenous fluorophores—the intracellular fluorescence is typically considerably weaker than the fluorescence due to the extracellular matrix. Fluorescence spectroscopy is widely used to probe tissue metabolic activity (e.g., NADH, NAD) and the status of extracellular matrices.

Multiphoton fluorescence such as second harmonic generation has been also implemented as a type of spectroscopy. An example of the information that can be obtained from the spectra is the structure and composition of extracellular matrix.

Weaker inelastic scattering phenomena such as Raman scattering and its modalities (stimulated Raman scattering [SRS], coherent anti-Stokes Raman spectroscopy [CARS], and inverse Raman scattering) have been developed into Raman spectroscopy, SRS spectroscopy, and CARS spectroscopy. These techniques typically have very high molecular specificity and are used for "molecular fingerprinting." For example, tissue content of proteins, nucleic acids, and small molecules such as glucose can be measured.

Laboratory exercises in this chapter focus on tissue spectroscopy techniques that are based on elastically scattered light and endogenous fluorescence.

Light Transport in a Medium with Continuously Varying Refractive Index

Discussion

Spectral measurements of scattering properties of tissue typically exhibit a decrease in scattering cross-section per unit volume as a function of wavelength. The exact functional dependence of this decrease varies but often follows a power law. For example, the reduced scattering coefficient is usually proportional to λ^α, where α ranges from -4 to 0. For media composed of discrete particles, $\alpha \sim -4$ indicates very small particles relative to the wavelength (Rayleigh scattering), while flatter wavelength dependence indicates a greater proportion of particles larger than the wavelength. Tissue, however, is not composed solely of discrete isolated scattering particles and is often described as a continuous medium.

Conceptually, it is easy to imagine light transport in a medium composed of discrete scatterers: A wave or ray propagates until it encounters a particle; the shape and material properties of the particle dictate the scattering function; the newly scattered wave or ray continues on until scattered again. The process of scattering when the medium is not made up of discrete isolated particles is a bit subtler. The term "continuous media" is used to describe materials where the refractive index is a continuous random function of position: $n = n(r)$. Biological tissue is an example of a continuous random medium. For the purposes of this discussion, the refractive index will be assumed to be real, and only elastic light scattering is considered.

If the refractive index function $n(r)$ for a particular material is known, light transport can be calculated by methods such as numerically solving Maxwell's equations using finite-difference time-domain (FDTD). However, a random medium may also be described by statistical properties even when the particular refractive index function $n(r)$ is unknown. In other words, even when the exact position of each particle is unknown, the scattering properties of the medium can still be described. For example, a discrete medium such as a suspension of spherical particles may be described statistically by the refractive index contrast, the particle size, and the concentration. Likewise, a continuous medium may be described by the refractive index variance σ_n^2 and the index correlation function. Since scattering arises from fluctuations in refractive index, the excess relative refractive index $n_1(r)$ is considered

$$n_1(r) = n(r)/ <n(r)> -1 \qquad (5.1)$$

where $<n(r)>$ represents the mean refractive index. The correlation function of $n_1(r)$ is then written as $B_n(r)$ following the notation of Ishimaru [1].

In a discrete medium, scattering properties can be calculated for a representative particle. In a continuous medium, the scattering properties must be calculated per unit volume. For this purpose, a volume element dV is introduced. This differential volume element must be representative of the medium and must be large enough to contain the refractive index correlation function. When the fluctuations in refractive index are small, the medium is said to be weakly scattering. Light transport in weakly scattering media can be described using the first Born approximation [2]. This assumes that the incident wave is essentially unmodified as it traverses the volume element, and only single or first-order scattering needs to be considered within dV. However, this does not mean that multiple scattering cannot be modeled within this context. Multiple scattering will certainly occur in any large enough volume. What is required here is that it be possible to define a volume large enough to represent the statistical variation and yet small enough to only weakly alter the incident wave. This is equivalent to requiring that the total scattering cross-section per unit volume can actually be defined.

Differential Scattering Cross-Section

Following Reference 1, the differential scattering cross-section per unit volume can be calculated within the Born approximation by taking the Fourier transform of $B_n(r)$ to get spectral density $\Phi(k_s)$

$$\Phi(k_s) = FT(B_n(r)) = 1/(2\pi)^3 \int B_n(r)\, 4\pi\, r \sin(k_s r)/k_s dr \qquad (5.2)$$

The differential scattering cross-section is then related to the spectral density by

$$\sigma(\theta,\varphi) = 2\pi k^4 (1 - \sin^2(\theta)\cos^2(\varphi))\Phi(2k\sin(\theta/2)). \qquad (5.3)$$

The experimentally measurable scattering properties that are used to characterize a medium can readily be computed by integrating $\sigma(\theta,\varphi)$ to obtain the scattering coefficient μ_s, the anisotropy factor $g = <\cos(\theta)>$, and reduced scattering coefficient μ_s'.

$$\mu_s = \iint \sigma(\theta,\varphi) \sin(\theta)\, d\theta\, d\varphi \qquad (5.4)$$

$$g = 1/\mu_s \iint \sigma(\theta,\varphi) \cos(\theta) \sin(\theta)\, d\theta\, d\varphi \qquad (5.5)$$

$$\mu_s' = \mu_s(1-g) \qquad (5.6)$$

Modeling Continuous Media with a Whittle–Matérn Correlation Function

When the correlation function for a random medium is unknown, a convenient family of curves known as the Whittle–Matérn correlation family may be used [3]. This family of curves is composed of a power law multiplied by a modified Bessel function of the second kind and has the property of describing many realistic functional shapes, depending on the parameter D.

$$B_n(r) = A_n \, (r/L_n)^{(D-3)/2} K_{(D-3)/2}(r/L_n)/K_{(D-3)/2} \tag{5.7}$$

A_n is the value at L_n, L_n is the normalization length, and K is the modified Bessel function of the second kind with order $(D-3)/2$. Parameter D controls the shape of the function: when $D < 3$, the function approaches a power law for small r; when $3 < D < 4$, the shape is a stretched exponential; when $D = 4$, the function is an exponential with correlation length $l_c = L_n$; and as $D \to \infty$, the function approaches a normal distribution.

When the value of D is greater than 3 (Figure 5.1), the function is bounded at the origin, and the variance of refractive index is given in terms of A_n as

$$\sigma_n^2 = A_n 2^{(D-5)/2} |\Gamma((D-3)/2)| \quad \text{for } D > 3 \tag{5.8}$$

where $|\Gamma(\,)|$ represents the absolute value of the gamma function. When $D < 3$, the function approaches a power law of the form r^{D-3} as r approaches zero. Power-law correlation functions indicate that a medium is fractal.

A fractal is defined as a system that exhibits self-similarity at different length scales [4]. A weak form of this definition is statistical self-similarity, which applies to random fractals. The autocorrelation function of a random fractal follows a power law. Conceptually, this is because a power law is scale invariant. In other words, the medium looks statistically similar at different scales or "zoom levels." Autocorrelation functions of real materials can follow a power law only over some finite range of length scales from a small inner to a large outer scale [4]. Obviously, the refractive index correlation cannot continue beyond the length scale of the body, nor does it have meaning at the length scale of atoms, but a power-law correlation over some finite range represents an approximate fractal. When the value of parameter D is less than the Euclidean dimension (in 3D space, $D_E = 3$), the value D represents the mass fractal dimension, and the autocorrelation $B_n(r)$ is proportional to r^{D-3} for values of r much less than L_n. As r approaches L_n, the function

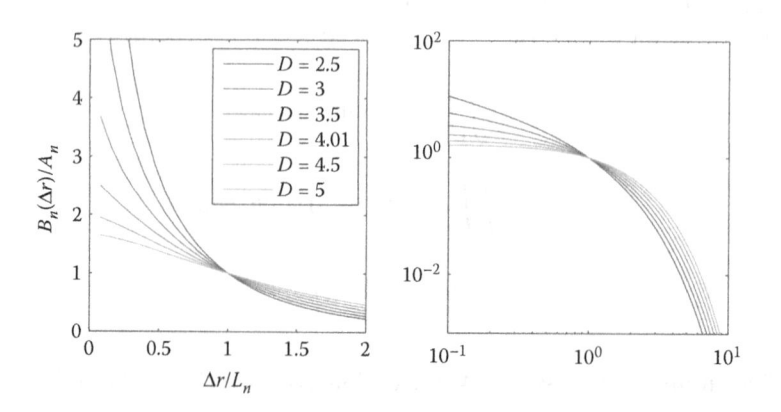

FIGURE 5.1
Whittle–Matérn correlation functions plotted for different values of parameter D.

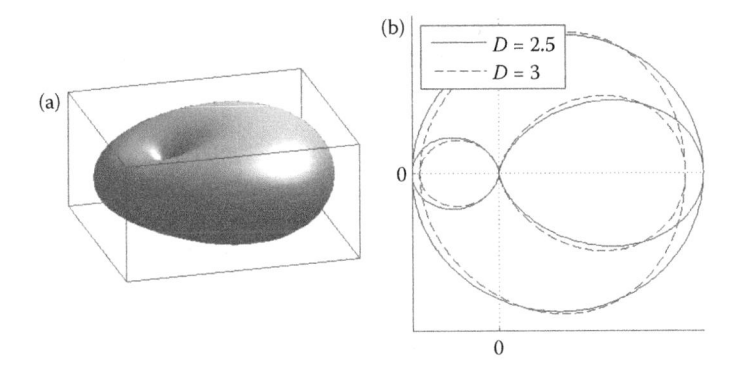

FIGURE 5.2
(a) Differential scattering cross-section for parameters $L_n 2\pi/\lambda = 1/2$ and $D = 2.5$. The incident plane wave is polarized in the vertical direction. (b) Comparison of the shape for two values of D.

falls off more quickly than the power law, and L_n can be considered the outer scale of the fractal (Figure 5.2).

One of the conveniences of using the Whittle–Matérn correlation family to describe the refractive index correlation function is that the Fourier transform has an analytical form, and the dependence of scattering properties on parameter D can be evaluated [5,6]. The resulting analytic equation for differential scattering cross-section

$$\sigma(\theta,\varphi) = A_n 2^{(D-3)/2} \Gamma(D/2)/\left(\pi^{1/2} K_{(D-3)/2}(1)\right)$$
$$\left(1 - \cos^2(\varphi)\sin^2(\theta)\right) k^4 L_n^3 / \left(1 - 2k^2 L_n^2 (1 - \cos(\theta))\right)^{D/2} \tag{5.9}$$

can be integrated as in Equations 5.4 through 5.6 to obtain expressions for the scattering properties.

Of key importance is the wavelength dependence of the experimentally measurable scattering properties such as μ_s'. As can be seen from Figure 5.3, the spectral dependence of μ_s' follows a power law that depends on D when $\lambda = L_n$. In this case, μ_s' is proportional

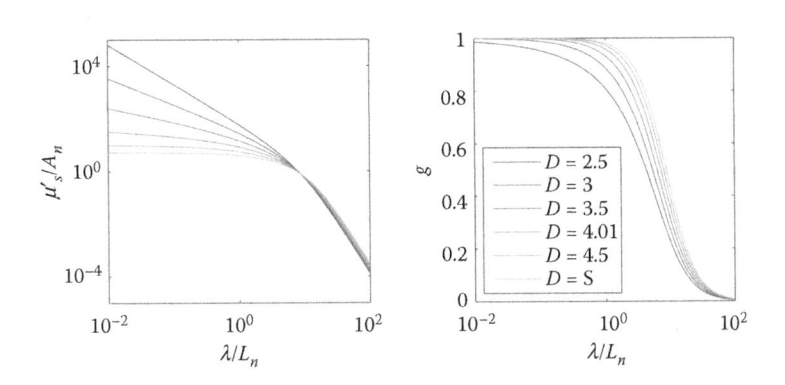

FIGURE 5.3
Dependence of reduced scattering coefficient and anisotropy coefficient on wavelength.

to λ^{D-4} when $D < 4$. Fortunately, the fact that g for tissue is close to 1 indicates that tissue falls in this range, and this in turn provides a means to measure D for tissue.

Range of Validity

It is important to examine the range of validity for applying the Born approximation. In random media, the scattering coefficient is defined as total scattering cross-section per unit volume. The first Born approximation requires "weak scattering" or that the incident field be a constant plane wave over the volume dV. This is sometimes called the single-scattering approximation, since within dV, second- or higher-order scattering events are assumed to be negligible compared to the incident wave. But how can the validity of this approximation be assessed?

As discussed at the beginning of this section, the volume dV is introduced as a conceptual tool to formulate the problem, but the scattering properties are defined per unit volume and should not depend on the volume considered. The volume must be large enough to contain the correlation function, or, equivalently, the statistics of the medium must be well represented within the volume. If the volume is too large, multiple scattering will be significant, and the incident wave will begin to lose power and change by the time it passes to the far side of the volume. In other words, the medium must be weakly scattering so that the length over which the medium is correlated is much less than the mean free path ($l_s = 1/\mu_s$). When this is not satisfied, the very definition of scattering cross-section per unit volume is invalid, since the scattering depends on the volume considered. The condition for validity can be written as [7]

$$\sigma_n^2 (kL_n)^2 \ll 1 \tag{5.10}$$

Finally, it is worth noting that while the Whittle–Matérn model is useful, it is by no means the only possible form of the correlation function. If the refractive index correlation function can be determined, either through measurement or by inference, the known function should be compared to the model, and the merit of this model should be assessed. The Whittle–Matérn model does not itself provide a way to invert the scattering process and determine the correlation function of the medium. The power-law dependence of scattering on wavelength only provides some insight into the relative contribution to scattering from length scales larger than or smaller than the wavelength. Since scattering properties in tissue are observed to be smooth functions without periodic oscillations, one can be reasonably sure that a range or distribution of length scales is present. Lacking any additional information, the relative contribution from large- and small-length scales can be quantified through the model by evaluating parameter D. Other model correlation functions could be devised to account for the scattering equally well. The choice of the Whittle–Matérn function family to model refractive index correlation function in tissue is one of convenience and plausibility but should not be taken as unique.

Phase Function–Corrected Diffusion Approximation

Introduction

The diffusion approximation is a first-order approximation to the equation of radiative transfer in the sense that scattering in the diffusion approximation is assumed to be either isotropic or near isotropic. However, in most of biomedical light scattering applications, the phase function, which describes the angular distribution for a single scattering event, is predominantly forward directed. Nevertheless, the transport of light at large distances is still accurately described by the diffusion approximation because the effects of anisotropic scattering become averaged out over many scattering events. This makes the diffusion approximation a powerful tool for many biomedical applications. However, many biomedical applications require the prediction of radiance near the source or the point of entry (POE). Here, a more rigorous solution of the radiative transfer equation is needed, either computational (e.g., the Monte Carlo simulation) or analytical. In recent years, several types of perturbation analysis have been introduced in an attempt to improve the performance of the diffusion approximation for biological media near the POE. Each of these methods includes higher-order terms of the scattering-phase function in addition to the terms that are normally included in the diffusion approximation. In this section, we will describe a new method that has been shown to have significantly improved accuracy over both the diffusion approximation and the other analytical techniques. As required by many biomedical applications, we will assume a narrow-collimated beam is normally incident on a semi-infinite random medium.

Transport Equation for the Phase Function–Corrected Radiance

The phase function–corrected diffusion approximation [8] can be obtained from the equation of radiative transfer by dividing an arbitrary phase function $p(\mathbf{s} \cdot \mathbf{s}')$ with anisotropy factor $g = \int p(\mathbf{s} \cdot \mathbf{s}')\,(\mathbf{s} \cdot \mathbf{s}')d\mathbf{s}'$ into the sum of the delta-isotropic phase function $p_{DI}(\mathbf{s} \cdot \mathbf{s}')$ with the same anisotropy factor as $p(\mathbf{s} \cdot \mathbf{s}')$, and a deviation term $\Delta p(\mathbf{s} \cdot \mathbf{s}')$

$$p(\mathbf{s} \cdot \mathbf{s}') = p_{DI}(\mathbf{s} \cdot \mathbf{s}') + \Delta p(\mathbf{s} \cdot \mathbf{s}'), \tag{5.11}$$

where $\mathbf{s}'$ is the incident direction and $\mathbf{s}$ is the emerging direction of the scattering event. The delta-isotropic phase function can be written using the Dirac delta function δ as [9]

$$p_{DI}(\mathbf{s} \cdot \mathbf{s}') = [1 - g + 2g\delta(1 - \mathbf{s} \cdot \mathbf{s}')] / 4\pi, \tag{5.12}$$

The delta-isotropic phase function which is uniform in all directions except the forward direction, where $\mathbf{s} \cdot \mathbf{s}' = 1$, is perhaps the simplest anisotropic phase function one can select. We should note here that the delta-isotropic phase function [10] should not be confused with an often-used delta-Eddington phase function, which in addition to the isotropic and delta function terms has a term depending on the cosine of the scattering angle [11].

The deviation term in Equation 5.11 is then $\Delta p(\mathbf{s} \cdot \mathbf{s}') = p(\mathbf{s} \cdot \mathbf{s}') - p_{DI}(\mathbf{s} \cdot \mathbf{s}')$. By substituting expression (5.11) for the phase function into the equation of radiative transfer

$$\mathbf{s} \cdot \nabla L(\mathbf{r}, \mathbf{s}) = -(\mu_a + \mu_s)L(\mathbf{r}, \mathbf{s}) + \mu_s \iint p(\mathbf{s}, \mathbf{s}')L(\mathbf{r}, \mathbf{s}')d\mathbf{s}', \tag{5.13}$$

where $L(\mathbf{r}, \mathbf{s})$ is the radiance of light at position $\mathbf{r}$ traveling in a direction of the vector $\mathbf{s}$, and μ_a and μ_s are the absorption and scattering coefficients of the medium, respectively, we can rewrite it as

$$\mathbf{s} \cdot \nabla L(\mathbf{r},\mathbf{s}) = -(\mu_a + \mu_s)L(\mathbf{r},\mathbf{s}) + \frac{\mu_s}{4\pi} \iint [1 - g + 2g\delta(1 - \mathbf{s} \cdot \mathbf{s}')]L(\mathbf{r},\mathbf{s}')d\mathbf{s}'$$
$$+ \mu_s \iint \Delta p(\mathbf{s} \cdot \mathbf{s}')L(\mathbf{r},\mathbf{s}')d\mathbf{s}'. \tag{5.14}$$

Since $\mu_s / 4\pi \int\int 2g\delta(1 - \mathbf{s} \cdot \mathbf{s}')L(\mathbf{r},\mathbf{s}')d\mathbf{s}' = g\mu_s L(\mathbf{r},\mathbf{s})$, this term can be combined with the first right-hand-side term in Equation (5.14), which will become $-(\mu_a + \mu_s(1 - g))L(\mathbf{r},\mathbf{s})$. Using the definition for the reduced scattering coefficient, $\mu_s' = \mu_s(1 - g)$, we can now rewrite the radiative transfer equation in the following form

$$\mathbf{s} \cdot \nabla L(\mathbf{r},\mathbf{s}) = -(\mu_a + \mu_s')L(\mathbf{r},\mathbf{s}) + \frac{\mu_s'}{4\pi} \iint L(\mathbf{r},\mathbf{s}')d\mathbf{s}' + \mu_s \iint \Delta p(\mathbf{s} \cdot \mathbf{s}')L(\mathbf{r},\mathbf{s}')d\mathbf{s}'. \tag{5.15}$$

Without the second integral term on the right-hand side, which depends on the deviation term $\Delta p(\mathbf{s} \cdot \mathbf{s}')$, Equation 5.15 would be the equation of the radiative transfer for the case of an isotropically scattering medium, which, as mentioned previously, can be described using the diffusion approximation.

Let us solve the transfer Equation 5.15 for the radiance in a semi-infinite turbid medium with no internal sources illuminated with a narrow-collimated beam (see Figure 5.4), an important, general problem often treated using the diffusion approximation.

To obtain a solution in this geometry, we must first take the boundary conditions into account. For a semi-infinite medium, we require that no light on the boundary that is directed outward ($\mathbf{s} \cdot \hat{\mathbf{z}} > 0$) can re-enter the medium. We also require the presence of an infinitely narrow-collimated beam positioned at $\rho = 0$ and $z = 0$. Thus, the boundary condition can be written as

$$L|_{z=0,\,\mathbf{s}\hat{\mathbf{z}}>0} = \frac{P_0}{(2\pi)^2} \frac{\delta(\rho)}{\rho} \delta(1 - \mathbf{s} \cdot \hat{\mathbf{z}}), \tag{5.16}$$

where P_0 is the power of the incoming beam, ρ and z are the cylindrical coordinates, and $\hat{\mathbf{z}}$ is a unit vector along the z axis.

Since the boundary condition features the infinitely narrow-collimated beam, while the first and second right-hand-side terms of the radiative transfer equation, when it is written

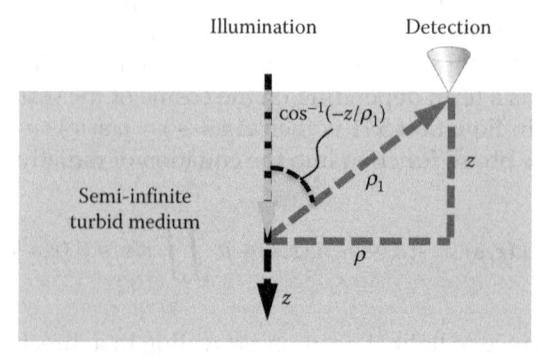

FIGURE 5.4
Semi-infinite turbid medium geometry.

in form (5.15), are characteristic of the diffusion approximation, it appears logical to present the radiance $L(\mathbf{r},\mathbf{s})$ as the sum of an on-axis part due to the narrow-collimated beam $L_c(\mathbf{r},\mathbf{s})$, a predominantly isotropic diffuse part $L_d(\mathbf{r},\mathbf{s})$, and the remaining phase function dependent part due to the deviation term in the phase function $L_p(\mathbf{r},\mathbf{s})$

$$L(\mathbf{r},\mathbf{s}) = L_c(\mathbf{r},\mathbf{s}) + L_d(\mathbf{r},\mathbf{s}) + L_p(\mathbf{r},\mathbf{s}), \tag{5.17}$$

By substituting Equation 5.17 into Equation 5.15 and collecting $L_d(\mathbf{r},\mathbf{s})$ and $L_p(\mathbf{r},\mathbf{s})$ terms, we can obtain a system of three interconnected partial differential equations

$$\mathbf{s}\cdot\nabla L_c(\mathbf{r},\mathbf{s}) = -(\mu_a + \mu_s')L_c(\mathbf{r},\mathbf{s}), \tag{5.18}$$

$$\mathbf{s}\cdot\nabla L_d(\mathbf{r},\mathbf{s}) = -(\mu_a + \mu_s')L_d(\mathbf{r},\mathbf{s}) + \frac{\mu_s'}{4\pi}\iint L_d(\mathbf{r},\mathbf{s}')d\mathbf{s}' + \mu_s\iint \Delta p(\mathbf{s}\cdot\mathbf{s}')L_d(\mathbf{r},\mathbf{s}')d\mathbf{s}'$$
$$+ \frac{\mu_s'}{4\pi}\iint L_c(\mathbf{r},\mathbf{s}')d\mathbf{s}', \tag{5.19}$$

$$\mathbf{s}\cdot\nabla L_p(\mathbf{r},\mathbf{s}) = -(\mu_a + \mu_s')L_p(\mathbf{r},\mathbf{s}) + \frac{\mu_s'}{4\pi}\iint L_p(\mathbf{r},\mathbf{s}')d\mathbf{s}' + \mu_s\iint \Delta p(\mathbf{s}\cdot\mathbf{s}')L_p(\mathbf{r},\mathbf{s}')d\mathbf{s}'$$
$$+ \mu_s\iint \Delta p(\mathbf{s}\cdot\mathbf{s}')L_c(\mathbf{r},\mathbf{s}')d\mathbf{s}' \tag{5.20}$$

In dividing the terms between these three equations, we have intentionally included the scattering integral contributions related to the narrow-collimated beam in their respective equations for the diffuse radiance $L_d(\mathbf{r},\mathbf{s})$ and phase function correction to the radiance $L_p(\mathbf{r},\mathbf{s})$. These terms will serve as the effective sources. Equation 5.18 can be easily solved to obtain $L_c(\mathbf{r},\mathbf{s})$ using the boundary conditions described by Equation 5.16

$$L_c(\mathbf{r},\mathbf{s}) = \frac{P_0}{(2\pi)^2}\frac{\delta(\rho)}{\rho}\delta(1 - \mathbf{s}\cdot\hat{\mathbf{z}})\exp[-(\mu_a + \mu_s')z]. \tag{5.21}$$

Equation 5.21 describes a collimated source within the scattering medium that is being attenuated at a rate of $\exp[-(\mu_a + \mu_s')z]$. In assuming that the other terms in the system of partial differential equations do not contribute to the solution of $L_c(\mathbf{r},\mathbf{s})$, we have implicitly neglected the spread of the incident beam due to single scattering. In other words, the unscattered collimated beam attenuates at an exponential rate of μ_s, but when we include low-order scattering events in order to include the z axis and its immediate surroundings, we approximate it as a μ_s' dependence. This approximation is often used in diffusion theory.

As we mentioned previously, if not for the $\mu_s\iint \Delta p(\mathbf{s}\cdot\mathbf{s}')L_d(\mathbf{r},\mathbf{s}')d\mathbf{s}'$ term, Equation 5.19 would be the equation of the radiative transfer for the case of an isotropically scattering medium, which can be accurately described using the diffusion approximation. Let us evaluate this term by employing the standard diffusion approximation expression for $L_d(\mathbf{r},\mathbf{s})$ [11]

$$L_d(\mathbf{r}, \mathbf{s}') = \frac{1}{4\pi}[\phi_d(\mathbf{r}) - 3\mathbf{s} \cdot \mathbf{j}_d(\mathbf{r})], \tag{5.22}$$

where $\phi_d(\mathbf{r})$ is the diffuse fluence rate, and $\mathbf{j}_d(\mathbf{r}) = -\left[3(\mu_a + \mu_s')\right]^{-1}\nabla\phi_d(\mathbf{r})$ is the diffuse current density. We then get

$$\mu_s \iint \Delta p(\mathbf{s}\cdot\mathbf{s}')L_d(\mathbf{r},\mathbf{s}')d\mathbf{s}' = \frac{\mu_s}{4\pi}\iint \Delta p(\mathbf{s}\cdot\mathbf{s}')\phi_d(\mathbf{r})d\mathbf{s}' - \frac{3\mu_s}{4\pi}\iint \Delta p(\mathbf{s}\cdot\mathbf{s}')(\mathbf{s}'\cdot\mathbf{j}_d(\mathbf{r}))d\mathbf{s}'. \tag{5.23}$$

The first integral here

$$\frac{\mu_s}{4\pi}\iint \Delta p(\mathbf{s}\cdot\mathbf{s}')\phi_d(\mathbf{r})d\mathbf{s}' = \frac{\mu_s\phi_d(\mathbf{r})}{4\pi}\iint\left[p(\mathbf{s}\cdot\mathbf{s}') - p_{DI}(\mathbf{s}\cdot\mathbf{s}')\right]d\mathbf{s}' = 0$$

since

$$\iint p(\mathbf{s}\cdot\mathbf{s}')d\mathbf{s}' = \iint p_{DI}(\mathbf{s}\cdot\mathbf{s}')d\mathbf{s}' = 4\pi.$$

Let us now calculate the integral term $\iint p(\mathbf{s},\mathbf{s}')(\mathbf{s}'\cdot\mathbf{j}_d(\mathbf{r}))d\mathbf{s}'$ by expressing it in a spherical coordinate system (Figure 5.5).

The integral becomes

$$|\mathbf{j}_d(\mathbf{r})|\iint p(\cos\theta)(\sin\theta_j\cdot\sin\varphi\cdot\sin\theta + \cos\theta\cdot\cos\theta_j)\sin\theta\,d\theta\,d\varphi,$$

where $\hat{\mathbf{j}}$ is the unit vector in the direction of vector $\mathbf{j}_d(\mathbf{r})$, $\theta = \cos^{-1}(\mathbf{s}\cdot\mathbf{s}')$, and $\theta_j = \cos^{-1}(\hat{\mathbf{j}}_d\cdot\mathbf{s})$. The part proportional to the $\sin\theta_j\sin\varphi\sin\theta$ is zero since $\int_0^{2\pi}\sin\varphi\,d\varphi = 0$. The part proportional to $\cos\theta\cdot\cos\theta_j$ gives $|\mathbf{j}_d(\mathbf{r})|\cos\theta_j\int\int p(\cos\theta)\cos\theta\sin\theta\,d\theta\,d\varphi$. Here

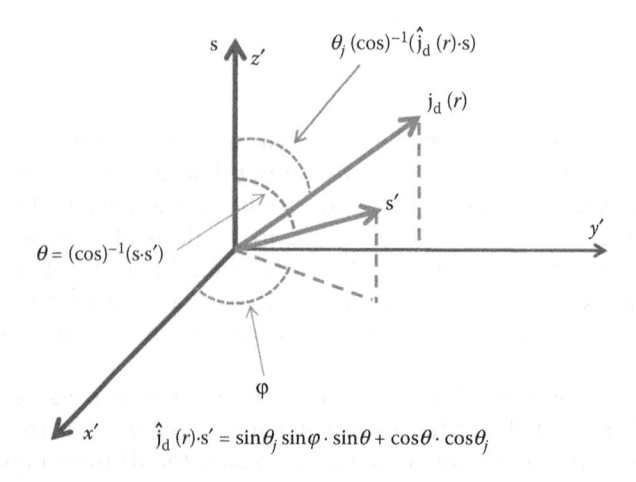

FIGURE 5.5
Angles between vectors $\mathbf{s}$, $\mathbf{s}'$, and $\mathbf{j}_d(\mathbf{r})$. The vector $\mathbf{s}$ defines the z'. The vectors $\mathbf{s}$ and $\mathbf{j}_d(\mathbf{r})$ define the (y', z') plane.

$|\mathbf{j}_d(\mathbf{r})|\cos\theta_0 = \mathbf{s}\cdot\mathbf{j}_d(\mathbf{r})$ and, by definition, $\iint p(\cos\theta)\cos\theta\sin\theta d\theta d\varphi = g$. Thus, we see that $\iint p(\mathbf{s},\mathbf{s}')(\mathbf{s}'\cdot\mathbf{j}_d(\mathbf{r}))ds' = g(\mathbf{s}\cdot\mathbf{j}_d(\mathbf{r}))$. By substituting this into the remaining portion of the integral $(3\mu_s/4\pi)\iint \Delta p(\mathbf{s}\cdot\mathbf{s}')(\mathbf{s}'\cdot\mathbf{j}_d(\mathbf{r}))ds'$, we get

$$\frac{3\mu_s}{4\pi}\iint \left[p(\mathbf{s}\cdot\mathbf{s}') - p_{DI}(\mathbf{s}\cdot\mathbf{s}')\right](\mathbf{s}'\cdot\mathbf{j}_d(\mathbf{r}))ds' = \frac{3\mu_s}{4\pi}\left[g(\mathbf{s}\cdot\mathbf{j}_d(\mathbf{r})) - g(\mathbf{s}\cdot\mathbf{j}_d(\mathbf{r}))\right] = 0, \quad (5.24)$$

as both phase functions $p(\mathbf{s}\cdot\mathbf{s}')$ and $p_{DI}(\mathbf{s}\cdot\mathbf{s}')$ have the same anisotropy factor g. Thus,

$$\mu_s \iint \left[p(\mathbf{s}\cdot\mathbf{s}') - p_{DI}(\mathbf{s}\cdot\mathbf{s}')\right]L_d(\mathbf{r},\mathbf{s}')ds' = 0. \quad (5.25)$$

Equation 5.20 can also be simplified. First, we can substitute the definition of $\Delta p(\mathbf{s}\cdot\mathbf{s}')$ into the second-to-last term to obtain a cancellation of the two $(\mu_s'/4\pi)\iint L_p(\mathbf{r},\mathbf{s}')ds'$ terms

$$\mathbf{s}\cdot\nabla L_p(\mathbf{r},\mathbf{s})$$
$$= -(\mu_a + \mu_s')L_p(\mathbf{r},\mathbf{s}) + \mu_s \iint \left[p(\mathbf{s}\cdot\mathbf{s}') - \frac{g\delta(1 - \mathbf{s}\cdot\mathbf{s}')}{2\pi}\right]L_p(\mathbf{r},\mathbf{s}')ds'$$
$$+ \mu_s \iint \Delta p(\mathbf{s}\cdot\mathbf{s}')L_c(\mathbf{r},\mathbf{s}')ds'. \quad (5.26)$$

We also assume that the contribution of the second term on the right-hand side is negligible relative to the source term. The validity of this assumption is verified later.

Thus, we have now approximated the radiative transfer equation with the system of two equations, one for predominantly isotropic diffuse part, $L_d(\mathbf{r},\mathbf{s})$, and the other for the phase function–dependent part, $L_p(\mathbf{r},\mathbf{s})$

$$\mathbf{s}\cdot\nabla L_d(\mathbf{r},\mathbf{s}) = -(\mu_a + \mu_s')L_d(\mathbf{r},\mathbf{s}) + \frac{\mu_s'}{4\pi}\iint L_d(\mathbf{r},\mathbf{s}')ds' + \frac{\mu_s'}{4\pi}\iint L_c(\mathbf{r},\mathbf{s}')ds', \quad (5.27)$$

$$\mathbf{s}\cdot\nabla L_p(\mathbf{r},\mathbf{s}) = -(\mu_a + \mu_s')L_p(\mathbf{r},\mathbf{s}) + \mu_s \iint \Delta p(\mathbf{s}\cdot\mathbf{s}')L_c(\mathbf{r},\mathbf{s}')ds', \quad (5.28)$$

with $L_c(\mathbf{r},\mathbf{s})$ being defined in Equation 5.21. The integration of $L_c(\mathbf{r},\mathbf{s})$ over angle is trivial due to the presence of an angular Dirac delta function

$$\mathbf{s}\cdot\nabla L_d(\mathbf{r},\mathbf{s}) = -(\mu_a + \mu_s')L_d(\mathbf{r},\mathbf{s}) + \frac{\mu_s'}{4\pi}\iint L_d(\mathbf{r},\mathbf{s}')ds' + P_0 \frac{\mu_s'}{8\pi^2}\frac{\delta(\rho)}{\rho}\exp\left[-(\mu_a + \mu_s')z\right], \quad (5.29)$$

$$\mathbf{s}\cdot\nabla L_p(\mathbf{r},\mathbf{s}) = -(\mu_a + \mu_s')L_p(\mathbf{r},\mathbf{s}) + P_0 \frac{\mu_s}{2\pi}\frac{\delta(\rho)}{\rho}\left[p(\mathbf{s}\cdot\hat{\mathbf{z}}) - \frac{1-g}{4\pi}\right]\exp\left[-(\mu_a + \mu_s')z\right]. \quad (5.30)$$

Here, Equation 5.29 exactly matches the equation of radiative transfer for a turbid medium with an isotropic phase function, illuminated with a narrow-collimated beam. It can be solved by employing the standard diffusion approximation expression (5.22) for $L_d(\mathbf{r,s})$. The solution is described elsewhere (e.g., in References 12 and 13) and is not the subject of this section.

Transport Equation 5.30 provides the phase function–corrected radiance; its solution is described in the next section.

Derivation of the Phase Function–Corrected Radiance, Fluence Rate, and Reflectance

By multiplying both sides of Equation 5.30 by an unknown function $u(\mathbf{r,s})$, we obtain

$$u(\mathbf{r,s})[\mathbf{s}\cdot\nabla L_p(\mathbf{r,s})] = -\left(\mu_a + \mu_s'\right)u(\mathbf{r,s})L_p(\mathbf{r,s}) + u(\mathbf{r,s})Q(\mathbf{r,s}), \qquad (5.31)$$

where $Q(\mathbf{r,s})$ is the source term in Equation 5.30

$$Q(\mathbf{r, s}) = P_0 \frac{\mu_s}{2\pi} \frac{\delta(\rho)}{\rho}\left[p(\mathbf{s}\cdot\hat{\mathbf{z}}) - \frac{1-g}{4\pi}\right]\exp\left[-\left(\mu_a + \mu_s'\right)z\right], \qquad (5.32)$$

Since $\mathbf{s}\cdot\nabla[u(\mathbf{r,s})L_p(\mathbf{r,s})] = u(\mathbf{r,s})[\mathbf{s}\cdot\nabla L_p(\mathbf{r,s})] + L_p(\mathbf{r,s})[\mathbf{s}\cdot\nabla u(\mathbf{r,s})]$, we see that if $\mathbf{s}\cdot\nabla u(\mathbf{r,s}) = \left(\mu_a + \mu_s'\right)u(\mathbf{r,s})$, or, in other words $u(\mathbf{r,s}) = \exp\left[\left(\mu_a + \mu_s'\right)(\mathbf{r}\cdot\mathbf{s})\right]$, Equation 5.32 becomes

$$\mathbf{s}\cdot\nabla(\exp[\left(\mu_a + \mu_s'\right)(\mathbf{r}\cdot\mathbf{s})]L_p(\mathbf{r,s})) = \exp\left[\left(\mu_a + \mu_s'\right)(\mathbf{r}\cdot\mathbf{s})\right]Q(\mathbf{r,s}), \qquad (5.33)$$

By introducing vector $\mathbf{F(r,s)} = \mathbf{s}\,L_p(\mathbf{r,s})\exp\left[\left(\mu_a + \mu_s'\right)(\mathbf{r}\cdot\mathbf{s})\right]$ related to the energy flux, Equation 5.33 can be written as

$$\operatorname{div}\mathbf{F(r,s)} = \exp[(\mu_a + \mu_s')(\mathbf{r}\cdot\mathbf{s})]Q(\mathbf{r,s}), \qquad (5.34)$$

and integrated using the divergence theorem, $\int\int\int_V \operatorname{div}\mathbf{F(r,s)}dV = \oiint_S (\mathbf{F(r,s)}\cdot\mathbf{n})\,dS$. It is convenient to choose the integration volume V and boundary S as depicted in Figure 5.6

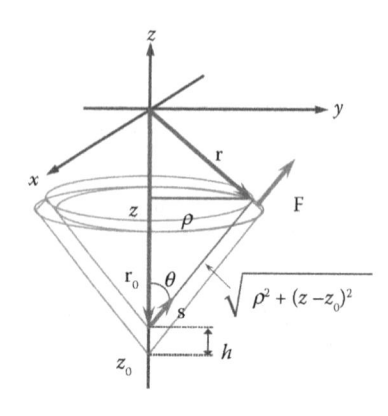

FIGURE 5.6
Integration volume for Equation 5.34. The volume is confined by the boundaries depicted with red lines and consists of two circular conical surfaces with the apex located at point z_0 on the collimated beam.

with red lines, so that there is no flux of vector $\mathbf{F}(\mathbf{r},\mathbf{s})$ through the conical parts of the boundary, and only the narrow tilted ring at the top of the boundary next to green arrow $\mathbf{F}(\mathbf{r},\mathbf{s})$ has a nonzero contribution.

Assuming h to be infinitesimally small, the area of the ring contributing to the right-hand side of the divergence integral is $2\pi\rho h \sin\theta$. At the same time, due to the presence of the radial delta function in the source term (5.32), the volume integral

$$\iiint_V \exp\left[(\mu_a + \mu_s')(\mathbf{r}\cdot\mathbf{s})\right]Q(\mathbf{r},\mathbf{s})dV = hP_0\mu_s\left[p(\mathbf{s}\cdot\hat{\mathbf{z}}) - \frac{1-g}{4\pi}\right]\exp\left[-(\mu_a + \mu_s')(z_0 - (\mathbf{r}_0\cdot\mathbf{s}))\right].$$

After integrating Equation 5.34 over the volume and using the divergence theorem, we get

$$\exp\left[(\mu_a + \mu_s')(\mathbf{r}\cdot\mathbf{s})\right]L_p(\mathbf{r},\mathbf{s})2\pi\rho h \sin\theta = hP_0\mu_s\left[p(\mathbf{s}\cdot\hat{\mathbf{z}}) - \frac{1-g}{4\pi}\right]\exp\left[-(\mu_a + \mu_s')(z_0 - (\mathbf{r}_0\cdot\mathbf{s}))\right]$$

or

$$L_p(\mathbf{r},\mathbf{s}) = \frac{P_0\mu_s}{2\pi\rho\sin\theta}\left[p(\mathbf{s}\cdot\hat{\mathbf{z}}) - \frac{1-g}{4\pi}\right]\exp\left[-(\mu_a + \mu_s')(z_0 + [(\mathbf{r}-\mathbf{r}_0\cdot\mathbf{s})])\right]. \tag{5.35}$$

Taking into account that $(\mathbf{r}-\mathbf{r}_0)\cdot\mathbf{s} = \sqrt{\rho^2 + (z-z_0)^2}$ and $\mathbf{s}\cdot\hat{\mathbf{z}} = \cos\theta$, we finally get the following expression for the phase function correction to the radiance

$$L_p(\rho,z,\theta) = \frac{P_0\mu_s}{2\pi\rho\sin\theta}\left[p(\cos\theta) - \frac{1-g}{4\pi}\right]\exp\left[-(\mu_a + \mu_s')\left(z_0 + \sqrt{\rho^2 + (z-z_0)^2}\right)\right]. \tag{5.36}$$

The PFC correction to the fluence rate is obtained by integrating L_p over the entire solid angle $\Phi_p = \int_{4\pi} L_p(\mathbf{r},\mathbf{s})d\Omega$, where $\underline{d\Omega = \sin\theta\,d\theta\,d\varphi}$. From straightforward geometrical considerations, $\cos\theta = (z - z_0/\sqrt{\rho^2 + (z-z_0)^2})$; thus, $d\theta = (\rho/\rho^2 + (z-z_0)^2)dz_0$. By using these expressions for $\cos\theta$ and $d\theta$, and expression for $L_p(\rho,z,\theta)$ from Equation 5.36, we obtain the fluence rate correction

$$\Phi_p(\rho,z) = P_0\mu_s\int_0^\infty\left[p\left(\frac{z-z_0}{\sqrt{\rho^2 + (z-z_0)^2}}\right) - \frac{1-g}{4\pi}\right]\exp\left[-(\mu_a + \mu_s')\left(z_0 + \sqrt{\rho^2 + (z-z_0)^2}\right)\right]$$

$$\frac{dz_0}{\rho^2 + (z-z_0)^2}, \tag{5.37}$$

We can also obtain the phase function correction to the reflectance at the surface by integrating $L_p(\rho,z,\theta)|_{z=0}$ that emerges per unit area of the surface

$$R_p = \frac{1}{P_0}\int_{\mathbf{s}\cdot\hat{\mathbf{z}}<0}(\hat{\mathbf{z}}\cdot\mathbf{s})\,L_p(\mathbf{r},\mathbf{s})|_{z=0}\,d\Omega,$$

where $\hat{\mathbf{z}}$ is a unit vector normal to the surface of the medium pointing outside (−z direction). Using the same procedure as for the integration of the fluence rate, we obtain phase function–corrected reflectance term

$$R_p(\rho) = \mu_s \int_0^\infty \left[p\left(\frac{-z_0}{\sqrt{\rho^2 + z_0^2}}\right) - \frac{1-g}{4\pi} \right] \exp\left[-(\mu_a + \mu_s')\left(z_0 + \sqrt{\rho^2 + z_0^2}\right) \right] \frac{z_0}{\left[\rho^2 + z_0^2\right]^{3/2}} \, dz_0. \quad (5.38)$$

To get the total reflectance, $R_p(\rho)$ from Equation 5.81 should be added to the diffuse reflectance term $R_d(\rho)$, calculated following the standard diffusion approximation procedure [11] from Equation 5.29.

We can also now verify that the contribution of the integral term $\mu_s \int \int [p(\mathbf{s}\cdot\mathbf{s}') - (2\pi)^{-1} g \cdot \delta(1 - \mathbf{s}\cdot\mathbf{s}')] L_p(\mathbf{r}, \mathbf{s}') d\mathbf{s}'$ neglected in Equation 5.26 is indeed insignificant.

First, let us calculate the phase function correction to the radially dependent reflectance with and without that integral. We then can plot the difference $\Delta R_p(\rho)$ and see it is just several percent of $R_p(\rho)$ for small $\rho\mu_s'$ (see Figure 5.7). We also include $R_p(\rho)$ to show that at greater $\rho\mu_s'$, both $\Delta R_p(\rho)$ and $R_p(\rho)$ corrections can be neglected. Thus, we conclude that neglecting the above integral is indeed justified.

Accuracy of the Phase Function–Corrected Diffusion Approximation

To evaluate the accuracy of PFC diffusion approximation, we compare the reflectance calculated by Monte Carlo simulation to the reflectance predicted by the PFC theory. We include results for two phase functions commonly used in turbid media: the **Henyey–Greenstein phase function** [14] $p_{HG}(\mathbf{s}\cdot\mathbf{s}') = [(1 - g^2)/4\pi]/[1 + g^2 - 2g(\mathbf{s}\cdot\mathbf{s}')]^{3/2}$ and the **"Mie" phase function,** which can be described using Mie theory [15].

Examples of the PFC diffusion theory performance are presented in Figure 5.8. While the diffusion approximation cannot predict the reflectance in the vicinity of the source ($\rho\mu_s' < 0.1$), PFC theory shows excellent agreement for both the Henyey–Greenstein and Mie phase functions. The presented comparison is for an anisotropy factor $g = 0.95$, although for other g, the agreement is equally good. We note that both the Henyey–Greenstein and Mie phase functions used in these examples have identical anisotropy factors yet show marked differences at small source–detector separations. This illustrates how the reflectance near the POE is sensitive to subtle alterations in the phase function and how the PFC theory is capable of capturing these subtle phase function–dependent effects.

We can also evaluate the accuracy of the correction at predicting the fluence within the medium. The approach presented previously employs the standard diffusion approximation to provide the isotropic solution, which is then corrected with the PFC. However, the standard diffusion approximation is not accurate in predicting the distribution of light near the POE within the medium, even for $g = 0$, due to the approximate treatment of the boundary conditions. We therefore employ Monte Carlo simulation for the isotropic scattering as a more accurate replacement for the diffusion approximation. The phase function correction to the fluence is then compared with Monte Carlo calculated correction, as obtained by subtracting the fluence in an anisotropic scattering simulation from the fluence in an isotropic scattering simulation. Figure 5.9 presents the comparison for the Henyey–Greenstein phase function and $g = 0.9$.

Physical Interpretation of the Phase Function–Corrected Radiance

The physical nature of the phase function correction (5.81) can be qualitatively understood if we notice that $z_0 + \sqrt{\rho^2 + z_0^2}$ is the total distance through the turbid medium, first from the surface along the collimated beam to depth z_0 and then to the detector (see Figure 5.4). However, note that the attenuation (the exponential in Equation 5.81) has a characteristic inverse scattering length of μ_s'. The reduced scattering coefficient μ_s' characterizes the attenuation of light due to diffuse scattering; that is, only diffusely scattered photons are removed from the $z_0 + \sqrt{\rho^2 + z_0^2}$ path; unscattered and low-angle scattered photons remain. If we were considering only unscattered photons, the attenuation factor would have the scattering coefficient μ_s in the exponent instead of the reduced scattering coefficient μ_s'. The $z_0 + \sqrt{\rho^2 + z_0^2}$ path is not the actual path of real photons but rather the virtual path of quasi-ballistic photons that are described by Equation 5.81. The contribution of the term describing this behavior can be positive or negative since it depends on the difference of the contributions of the actual phase function and the delta-isotropic phase function. Physically, the PFC term accounts for the contribution of these photons, which have undergone multiple low-angle scattering events plus a single large-angle scattering event along the $z_0 + \sqrt{\rho^2 + z_0^2}$ path. That single large-angle turn is described by the phase function, thus providing the phase function correction.

Summary and Concluding Remarks

We have described an analytical formulation for an accurate correction to the diffusion approximation. This approximation allows for the inversion of parameters of the phase function by providing a simple analytical forward model of reflectance near the POE.

One requirement for applying the PFC theory is the availability of an accurate isotropic solution ($g = 0$). Though this solution is available for diffuse reflectance, an accurate solution for the fluence rate near the POE is not commonly available. One approach for obtaining the fluence rate within the medium would be to use Monte Carlo simulation for isotropic scattering and combine it with PFC to obtain the fluence for arbitrary phase functions. This is the approach we used for obtaining the results in Figure 5.7. Thus, the result of a single Monte Carlo simulation can be used to obtain solutions for a large array

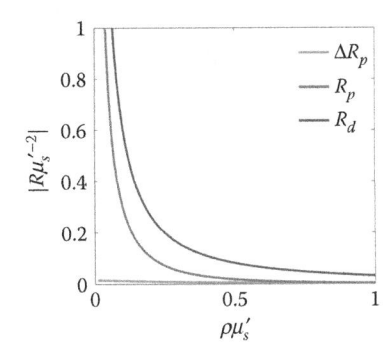

FIGURE 5.7

Evaluation of the contribution of the neglected integral term. The calculations are performed for Henyey–Greenstein phase function, $g = 0.9$ and $\mu_a / \mu_s' = 0.01$. Black line—radially dependent diffuse reflectance $R_d(\rho)$, blue line—contribution of the PFC term $R_p(\rho)$, red line—contribution of the neglected integral term $\Delta R_p(\rho)$. The contribution of the neglected integral term $\Delta R_p(\rho)$ is less than 4% of $R_p(\rho)$ for $\rho\mu_s' = 0.1$. The contribution of $\Delta R_p(\rho)$ is less than 7% of $R_d(\rho)$ for $\rho\mu_s' = 1$.

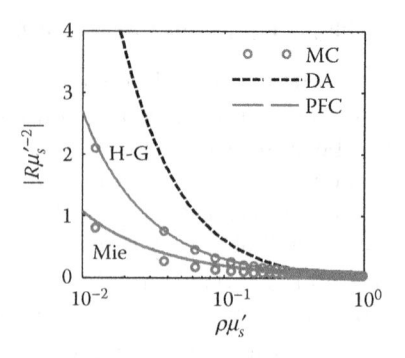

FIGURE 5.8

Dimensionless reflectance for Henyey–Greenstein (H-G) and Mie phase functions with $g = 0.95$ and $\mu_a/\mu_s' = 0.01$. The blue lines and circles represent the Henyey–Greenstein phase function, while the red lines and circles represent the Mie phase function. The lines are for PFC theory and circles are for Monte Carlo simulations (MC). The black dashed line is the standard diffusion approximation (DA).

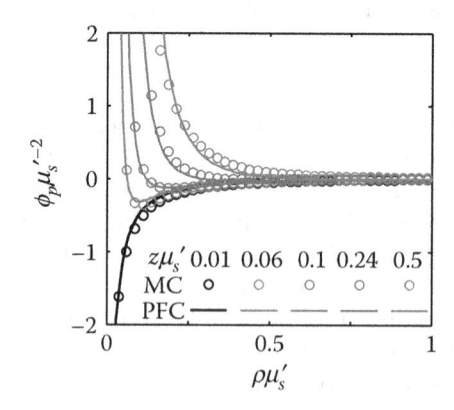

FIGURE 5.9

Comparison of the PFC for internal fluence rate with Monte Carlo simulations. The phase function correction to the fluence rate within the medium (solid lines) is calculated for the Henyey–Greenstein phase function with $g = 0.9$. The Monte Carlo data (circles) are obtained taking the difference in the fluence rate from a simulation with isotropic scattering ($g = 0$) and a simulation with $g = 0.9$. In all cases, $\mu_a/\mu_s' = 0.001$. The depth values $z\mu_s'$ are 0.01, 0.06, 0.1, 0.24, and 0.5.

of phase functions and anisotropy parameters. It is also likely that PFC could be used to describe angular dependences in a similar fashion.

In conclusion, PFC, the phase function–corrected diffusion theory for light transport in turbid media, addresses the deficiencies of the standard diffusion theory near the POE by accounting for the specific form of the phase function. The technique accurately predicts the correct scattering behavior for two frequently used but very different types of phase functions, the Henyey–Greenstein and Mie theory phase functions (Figure 5.5). The PFC approach results in a substantial improvement over the predictions of other existing approximations that seek to improve the performance of the diffusion approximation near the POE and will likely see many applications for the characterization of biomedical samples in the near future.

Laboratory 1: Building a Spectrometer

It should not come as a surprise that a ubiquitous piece of equipment that underlines essentially all modalities of tissue spectroscopy is a spectrometer. Therefore, it is only fitting to start exploration of this field with understanding how spectrometers work. This is precisely the goal of this laboratory: to learn the basics of a typical diffraction grating-based spectrometer. Students will learn how to design a spectrometer according to desired specifications such as the central wavelength, the spectral resolution, and bandwidth.

Different spectroscopic applications require spectrometers with different technical specifications. The three most relevant parameters that describe the optical performance of a spectrometer include the center wavelength, the bandwidth, and the spectral resolution. Typically, either a two-dimensional (2-D) or line scan camera is equipped to record the spectrum dispersed by the grating. Since the sensor chip size and the pixel number of the camera are always limited, selecting suitable components (grating and lens) for a spectrometer is essential.

In optical spectroscopy, it is desired to separate a multiwavelength light beam into individual wavelengths for spectral analysis. One of the means is by using dispersive material, an optical component such as a glass prism that has slight refractive index differences in different wavelengths, often called wavelength dispersion. The dispersion then provides angular separations for light in different wavelengths and creates a rainbow spectrum when a light is obliquely incident on the prism. The angular separation depends on the dispersion: The greater the dispersion, the better the material can separate the different wavelengths. However, it is very difficult to have a highly dispersive material yield a highly spread spectrum. The other means is by using a diffraction grating, an optical component with a periodic structure that can disperse wavelengths in much greater angular separation than a conventional dispersive material.

The dispersion of the light off a diffraction grating is governed by the grating equation

$$\lambda f = \sin\theta_i + \sin\theta_d, \tag{5.39}$$

where θ_i and θ_d are the incident and diffraction angles with respect to the grating surface normal. λ is the wavelength of the light, and f is the frequency of the grating periodic structures in lines/mm. The equation states that the diffraction angle is determined by the incident angle, the wavelength, and the frequency of the grating. Note that λf in general should be written as $m\lambda f$, where m is the order of the diffraction. Since in most cases the first-order diffraction is desired, the above equation omits m.

The other important parameter for a given grating is dispersion, defined as the rate of change of the angle of diffraction with wavelength for a fixed angle of incidence, or $d\theta_d/d\lambda$

$$\frac{d\theta_d}{d\lambda} = \frac{f}{\cos\theta_d},$$

Therefore, the ability to separate different wavelengths into angular distribution is determined by the grating frequency and the cosine of the diffraction angle.

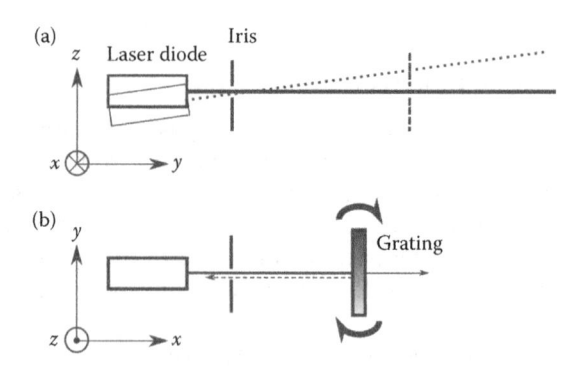

FIGURE 5.10
Schematic of laser alignments. The x–y plane is parallel to a table surface, and y is perpendicular to the table surface. (a) The laser beam is first aligned to be parallel to the table surface, and then (b) the grating surface is aligned to be perpendicular to the beam.

Suggested Lab Components

A transmission diffraction grating (Thorlabs GT25-12), laser diode module with collimated beam output (Thorlabs, CPS180, CPS532) or wavelength-known laser pointers, a rotational mechanical mount (Thorlabs, RP01), and a positive lens with 200-mm focal length.

Laboratory Protocol

1. Align laser beam to be normal to the grating surface. Secure the laser diode module kit on an optical table, and align the collimated beam to pass the aperture of an iris diaphragm as Figure 5.10 shows. Move the iris diaphragm farther away from the light source, and adjust the beam path through the iris. Iterate this procedure so that the collimated beam is aligned parallel to the table surface. The height of the iris may need to be adjusted during the alignment. Place the grating on the beam path with a rotation stage and a grating mount. Adjust the rotation stage and the grating mount so that the reflected beam off the grating surface can pass the same iris that passes the incident beam. Secure the grating position.

2. Change incident angle to observe the diffraction equation. Place a screen adhered by a piece of paper on the other side of the grating. Measure the distance of the screen to the grating, and depict the position of the transmitted beam on the paper. Record the angle reading of the rotation stage, and rotate the grating by 5° as in Figure 5.11. The diffracted beams start to deviate from the original transmitted beam on the screen. Depict the position of the first-order diffracted beam. Repeat

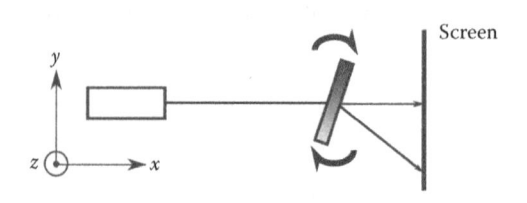

FIGURE 5.11
Rotation of the grating changes the incident angle, and the diffracted beam on the other side is projected to the screen.

this procedure, and have a series of marked beam positions after every 5° increment. Calculate the diffracted angle, and calculate λf according to above grating diffraction equation. Compare to the theoretical value of λf from the wavelength of laser diode and the grating frequency.

3. Change incident laser wavelength to observe dispersion equation. Keep the rotation stage 45° from the original position, and calculate the diffraction angle. Calculate the dispersion according to the right-hand side of Equation 5.39. Replace the light source by another laser diode, and calculate the corresponding diffraction angle similarly. Calculate the dispersion according to the left side of Equation 5.39 given the measured diffraction angles and two wavelengths from two laser sources. Compare the result from two sides of the equation. Rotate the stage 10° from the original position, and repeat the previous procedures.

4. Align a lens to focus on the diffracted beam, and measure separation of two different wavelengths (Figure 5.12). Keep the rotation stage 45° from the original position. Place the iris after the grating to pass diffracted beam through the aperture. Place the lens on the beam path between grating and the iris, and keep the distance between the lens and the iris about one focal length. Adjust the lens to be perpendicular to the beam. Adjust the position of the lens so that the focused beam is centered at the aperture as well. Replace the iris with a screen, and depict the focused beam spot on the screen. Switch the laser diode and depict the beam spot, too. Measure the displacement of two beam spots, and calculate the theoretical displacement given the diffraction angles and the focal length of the lens.

Analysis

1. Measurement of the diffraction angle. In Exercise 2, after measuring the distance between the screen and the grating, the diffraction angle can be calculated by the trigonometric relationship

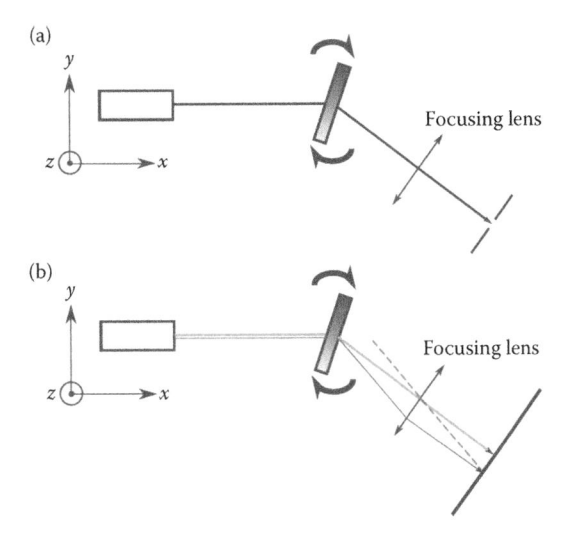

FIGURE 5.12
Schematic of aligning the lens. (a) The axis of the lens is aligned to overlap the diffracted beam path, and (b) a second laser diode with a different wavelength is used to calculate the dispersion.

$$\theta_d = a\tan\frac{d}{L}$$

where L is the distance between the screen and the grating, and d is the displacement of the beam spot on the screen.

2. Calculate the theoretical displacement of the focused beam spots. A similar trigonometric relationship can be applied

$$\Delta d = f\tan\Delta\theta_d$$

where Δd is the displacement of the focused beam spots, f is the focal length of the lens, and $\Delta\theta_d$ is the dispersion difference of the two laser beams.

Discussion

1. The incident angle can be arbitrary to disperse the spectrum. However, a grating is usually designed to yield the best efficiency by equalizing the incident and the diffraction angle. Unless there is specific requirement on the incident, one can then simplify Equation 1 by the same values of θ_i and θ_d.

2. Given the size of the camera chip, the bandwidth coverage is determined by the grating dispersion and the focal length of the focusing lens.

3. Spectral resolution is determined by the magnification of a two-lens system. For example, if a single-mode fiber is used as an input to a spectrometer, the first lens with focal length $f1$ is used to collimate the beam for grating, and a second lens with focal length $f2$ is used in front of the camera. The spectral resolution is equal to the beam size of the focused spot with a single-wavelength incidence. The lens forms an imaging system and projects the single-mode fiber port on the camera. The spectral resolution is then equal to $(f2/f1)d$, in which d is the mode FWHM of the single-mode fiber.

Questions

1. For the same coverage, what is the difference of the following two designs: using a low-dispersion grating and a long focal length lens or using a high-dispersion grating and a short focal length lens?

2. In Exercise 3, should the measured dispersion be equal to theoretical dispersion and why?

Laboratory 2: Diffuse Reflectance Spectroscopy

Measurement of Optical Properties

This lab will allow students to measure the optical properties of a tissue model—tissue models are conventionally referred to as "phantoms"—using diffuse reflectance

TABLE 5.1

Common Constituents of Tissue-Simulating Phantoms

Matrix Element	Scattering Agent	Absorption Agent
Water suspension	Lipids	Whole blood/hemoglobin
Gelatin/agar	Polymer microspheres	Ink
Polyacrylamide gel	TiO_2/Al_2O_3 powders	Molecular dyes
Polyester/epoxy resin	Quartz glass microspheres	Fluorophores
Polyurethane resin		
RTV silicone		

spectroscopy. Phantoms are artificial constructs that mimic the scattering and absorption properties of tissue. They have been used in the biomedical optics community to test system designs, to perform quality control, and to compare results between different systems. Phantoms generally consist of three elements: a matrix that forms the structural bulk of the phantom, a scattering agent, and an absorption agent. In some cases, a fluorophore can also be added to a phantom. Table 5.1 lists some common matrix, scattering, and absorption materials used in the biomedical optics community [16].

In this lab, we will create phantoms using water as the matrix element, lipids as the scattering agent, and a molecular dye (methylene blue) as the absorbing agent. The goal of the lab will be to measure the optical properties of the phantoms using optical spectroscopy based on multiple elastic scattering. The technique has acquired several names, including elastic scattering spectroscopy and diffuse reflectance spectroscopy. The latter terminology is, strictly speaking, not always accurate. This is because the term "diffusion" implies a specific kind of light transport process. The terminology "diffused reflectance," however, is widely used with the understanding that multiply scattered light is being detected regardless of whether the requirements of diffusion are satisfied. In this section, we will continue using this term in the same context.

Materials

1. Diffuse reflectance probe (Ocean Optics QR200-7-VIS-NIR). Consists of a tight bundle of seven optical fibers in a stainless-steel ferrule—six illumination fibers around one read fiber. The fibers are 200 µm in diameter.
2. LS-1 tungsten halogen light source (Ocean Optics).
3. USB-2000+VIS-NIR Ocean Optics fiber spectrometer.
4. Liposyn 10% (Hospira, Inc.): Intralipid scatterer.
5. Methylene blue absorber.
6. Computer with SpectraSuite software.
7. Diffuse reflectance standard (Ocean Optics).

Lab Protocol

1. Connect the illumination leg of the probe to the light source and the collection leg of the probe to the spectrometer.
2. Connect the spectrometer to a computer, and start the SpectraSuite software.

3. Turn on the light source, and take a measurement from the diffuse reflectance standard with the probe perpendicular to the surface. Select an integration time such that the detector does not saturate. Make sure that the ambient light does not interfere with the measurement.

4. Place the probe in a dark area, and take a measurement at the same integration time as the reflectance standard measurement. This will serve as a background reading. Save both the standard and background measurements for later analysis.

5. Phantom preparation and measurements

   a. The reduced scattering coefficient for 10% stock solution of Intralipid is given as $\mu_s'(\lambda) = 9.3\lambda^{-4} - 1.6\lambda^{-2.4}$ in units of mm^{-1}. Use the titration relation $\mu_s'^{,before} \times V^{before} + \mu_s'^{,Intralipid} \times V^{Intralipid} = \mu_s'^{,after} \times V^{after}$ to make a 15-mL mixture of deionized water and 10% Intralipid having a $\mu_s' = 0.6\,mm^{-1}$ at $\lambda = 600$ nm. Note that in this case $\mu_s'^{,before}$ and V^{before} are the reduced scattering coefficient and volume of water before Intralipid is added. Set the μ_s' of water to be zero.

   b. Take five measurements with the probe slightly immersed in the phantom and away from any edges of the container or the bottom surface.

   c. Use the titration relation again to create phantoms having $\mu_s' = [0.8, 1, 1.2, 1.4, 1.6, 1.8]\,mm^{-1}$. This can be done by successive additions of the appropriate volume of Intralipid. Take measurements as in (b). Clean the probe tip between measurements.

   d. Add an appropriate volume of methylene blue to the last phantom from (c) to obtain a μ_a of 1 mm^{-1} at 600 nm. The absorption spectra for methylene blue can be found at http://omlc.ogi.edu/spectra/mb/index.html. Take measurements on this phantom.

   e. Take several measurements with the probe from your skin (e.g., forearm).

   f. Place the probe in a dark area, and take a background reading at the same integration time as the phantom measurements.

Analysis

1. Normalize the raw measurements taken in the phantom according to the following equation

$$I(\lambda) = \frac{I_{raw}(\lambda) - I_{bg}^{phantom}(\lambda)}{I_{DS}(\lambda) - I_{bg}^{DS}(\lambda)}$$

where I_{DS} is the measurement from the diffuse reflectance standard and I_{bg} is the background measurement. If the phantom measurement has a different integration time than the reflectance standard, multiply its intensity by the ratio of the standard to phantom integration time to rescale it.

2. Plot the normalized reflectance at $\lambda = 600$ nm from the nonabsorbing phantoms as a function of μ_s'.

3. To model the diffuse reflectance spectra from the phantoms, we will use the diffusion approximation expression given by Zonios et al. [17].

$$I(\lambda) = \frac{I_0}{2} \frac{\mu_s'}{\mu_s' + \mu_a} \left\{ \exp(-\mu z_0) + \exp\left(-(1 + \frac{4}{3}A)\mu z_0\right) - z_0 \frac{\exp(-\mu r_1')}{r_1'} \right.$$
$$\left. - \left(1 + \frac{4}{3}A\right) z_0 \frac{\exp(-\mu r_2')}{r_2'} \right\}$$
$$(5.40)$$

with,

$$r_1' = \left(z_0^2 + r_c^2\right)^{1/2}, \quad r_2' = \left(z_0^2\left(1 + \frac{4}{3}A\right)^2 + r_c^2\right)^{1/2},$$
$$\mu = \left(3\mu_a(\mu_a + \mu_s')\right)^{1/2}, \quad z_0 = \frac{1}{\mu_a + \mu_s'}.$$

This expression was obtained by integrating the solution of the diffusion expression up to an effective probe radius r_c that is dependent on probe geometry. The parameter A depends on the refractive index mismatch between the sample and surrounding medium, while I_0 is an overall intensity calibration factor. The values of r_c, A, and I_0 need to be determined by a one-time calibration on a phantom. For the probe used in this lab, $r_c = 1.3$ mm and $A = 1.065$. Determine the value of I_0 by choosing one of the phantoms and fitting Equation 5.40 to the value of reflectance that was measured at $\lambda = 600$ nm.

4. Knowing the wavelength dependence of μ_s' of Intralipid and the absorption spectra of methylene blue, fit Equation 5.40 to the diffuse reflectance data over a wavelength range of 500–700 nm. Determine the measured values of μ_s' and μ_a of each phantom at 600 nm. Compare these values with those determined from the titration relation.

5. The reduced scattering coefficient of skin has a linear relationship with wavelength, and oxyhemoglobin/deoxyhemoglobin, melanin, and water are the main absorbers in skin. Use the absorption spectra given at http://omlc.ogi.edu/spectra/index.html to extract the reduced scattering coefficient and the concentrations of hemoglobin, melanin, and water from your measured reflectance measurements.

Questions

1. Discuss reasons why normalization is necessary to analyze the diffuse reflectance signal.

2. How does the diffuse reflectance behave as a function of the reduced scattering coefficient? Does this agree or disagree with Equation 5.40?

3. Based on the phantom data, with what accuracy would you expect to measure μ_s' and μ_a from tissue?

4. If you were to use a different probe geometry (e.g., larger fibers or interfiber spacing), how would that impact the terms in Equation 5.40? What steps would you take before using this probe to measure optical properties?

5. The μ_a term in Equation 5.40 assumes that the absorption is distributed homogeneously throughout the sample. Is this a valid assumption for the phantom? Is this a valid assumption for tissue?

6. Do a literature search and see if the optical properties of skin measurements agree with what has been previously reported. Discuss reasons for disagreement if there are any.

Learning Objectives

1. How to construct a simple tissue-simulating optical phantom.

2. How to take probe-based diffuse reflectance measurements.

3. How to use the diffusion approximation to extract optical properties of a sample.

4. Apply diffuse reflectance spectroscopy to biological tissue (skin).

Laboratory 3: Measurement of Optical Properties of Tissue Using Spectroscopic Optical Coherence Tomography

In the preceding laboratory, we learned how to measure the optical properties of tissue by means of elastic scattering spectroscopy based on multiple scattering (also referred to as diffused reflectance spectroscopy). Although these measurements can be made depth selective in the sense that optical properties of tissue up to or around a particular depth are assessed, diffuse light techniques are not able to provide imaging information. In this laboratory, we will learn how three-dimensional (3-D) tissue imaging is possible by using optical imaging techniques such as optical coherence tomography. The goal of this laboratory is to learn how OCT works and how it can be used to measure the optical properties of tissue.

Background

Optical coherence tomography has been developed as an important 3-D imaging technique with micron-level resolutions, analogous to that from histological imaging [18]. Its noninvasive nature promises that OCT will be an excellent platform for *in vivo* applications in clinics, spanning ophthalmology, cardiology, dermatology, gastroenterology, and oncology. In addition to providing a morphological image in 3-D, tissue optical properties can be also quantified using OCT. The backscattering coefficient μ_b has been used to quantify by the amplitude of the OCT image intensity, while the scattering coefficient μ_s can be calculated by Beer–Lambert's law from the image intensity decay rate along the penetration depth [19,20]. However, due to the backscattering scheme of the detection, it has been a great challenge to quantify higher-order optical properties such as the anisotropic factor g in conventional OCT.

Inverse scattering optical coherence tomography (ISOCT) is a methodology proposed to quantify a complete set of optical properties, including μ_b, μ_s, and g, and to quantify the physical properties of the tissue, the complete mass density correlation functional form. Compared to conventional OCT, ISOCT inverts the physical process of light scattering to quantify the tissue physical properties at a nanoscopic scale for each microscopic voxel of the 3-D tissue image.

Biological tissue is composed of complex macromolecules organized in a sophisticated manner. The mass density of those macromolecules is linearly proportional to the refractive index (RI) such that a denser structure gives a higher RI. [21]. Since RI heterogeneity induces light scattering, an optical measurement can be used to interpret structural properties. Furthermore, the complex interconnected structure of tissue, even at nanometer scale, leads one to consider tissue a continuously varying RI medium [22,23,24,25]. Given the random and continuous nature of RI distribution, tissue structures can be quantified by the statistical autocorrelation function of the RI spatial fluctuation (RI correlation function). Although the spatial resolution in conventional 3-D OCT is on the order of several microns in both the lateral and axial directions due to the diffraction limit, at a fundamental level, the OCT spectra extracted within a diffraction-limited volume are determined by the subdiffractional RI correlation function in the following way

$$\Pi^2(k, z') \approx \left(\frac{a\,|u_0|^2}{\pi f} \right)^2 k^4 \mathcal{F}_z\{C_n(\rho)\}, \tag{5.41}$$

where $\Pi(k, z')$ is the OCT spectrum extracted from a confined microscopic volume at specific depth z' by performing a short-frequency Fourier transform (SFFT): $k = 2\pi/\lambda$ is the wave number, a is the reflectance from the OCT reference arm, u_0 is the amplitude of the illumination, f is the focal length of the imaging lens, $C_n(\rho)$ denotes the 3-D RI correlation function with the spatial distance displacement ρ, and $F_z\{\cdot\}$ denotes the Fourier transform in the backscattering direction.

The heterogeneity of the RI distribution gives rise to the OCT spectrum. After applying the Wiener–Khinchin theorem, the square intensity of the OCT spectrum is thus proportional to the Fourier transform of the 3-D RI correlation function if an isotropic medium is assumed (Equation 5.41). The physical meaning of the right-hand side of Equation 5.41 is the backward power spectral density of the 3-D RI correlation function.

To quantify the tissue RI correlation function, we next introduce the Whittle–Matérn family of correlation functions [26], which we parameterize using three properties: the normalization factor of the correlation function $\mathcal{N}_c$, the length scale l_c, and a unitless deterministic factor D defining the function type

$$C_n(\rho) = \mathcal{N}_c 2^{\frac{5-D}{2}} \left(\frac{\rho}{l_c} \right)^{(D-3)/2} K_{(D-3)/2} \left(\frac{\rho}{l_c} \right), \tag{5.42}$$

in which $K\nu(\cdot)$ denotes the modified Bessel function of the second kind, and ρ is the distance displacement (Figure 5.13). The meaning of D and l_c is described in a previous section. Briefly, when $0 < D < 3$, the tissue is organized in a fractal manner, and the correlation function has the form of a power law. D defines the mass fractal dimension ranging from 0 to 3 in Euclidean space. When $3 < D < 4$, the correlation function forms a stretched exponential. When $D = 4$, the correlation function evolves into an exponential and approaches a Gaussian function when D approaches infinity. Qualitatively speaking, a smaller D value indicates a more drastic change of RI in a smaller length scale, or in other words, a smaller length scale of tissue heterogeneity. l_c determines the physical

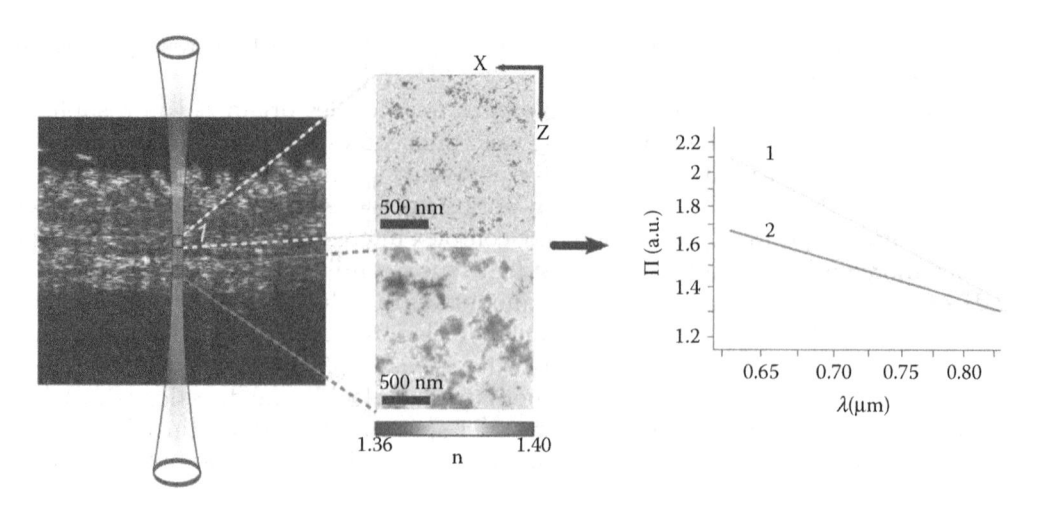

FIGURE 5.13
Illustration of the principle of ISOCT. The diffraction-limited voxel in an OCT image contains finer RI fluctuations with nanometer-length scales (top middle), which can alter OCT spectrum (top right) in the far field due to the different RI heterogeneities. This submicron RI correlation function thus can be inversely deduced from the OCT spectrum. The exploded-view voxels 1 and 2 illustrate RI distribution defined by two RI correlation functions (above: power law, below: stretched exponential).

length scale of the correlation function. When in the fractal regime, $0 < D < 3$, l_c defines the outer length scale beyond which the correlation function is no longer a power law and decays to 0 quickly. When $D > 3$, l_c determines how fast the function decays to 0. An extreme case is that when l_c approaches 0, the correlation function is compressed into a delta function acting as a dipole scatterer. The scattering will be fairly isotropic. On the other hand, biological tissue usually satisfies $kl_c \gg 1$, so that the scattering is highly anisotropic and forward directed. The shape of the Whittle–Matérn function can be fully described by D and l_c. $\mathcal{N}_c$ is simply a scaling factor of the correlation function quantifying the RI fluctuation amplitude. If we define dn^2 as the physical RI variance, we can specify $\mathcal{N}_c = dn^2 / |\Gamma((D-3)/2)|$ when $D > 3$ so that $C_n(0) = dn^2$. This relationship loses accuracy when $0 < D < 3$ because the function goes to infinity when ρ approaches 0. A round-off of the function at a very small length scale ρ_{min} can be performed to introduce a finite normalization with negligible error so that $C_n(0)$ is bounded, and an explicit relationship can be drawn to dn^2. Therefore, let $C_n(\rho < \rho_{min}) = C_n(\rho_{min}) = dn^2$ so that the value of the correlation function is approximated to be dn^2.

Thus, three physical parameters: dn^2, l_c, and D, are intended to fully describe the correlation functional form, for which we measure three optical quantities: the OCT spectrum, μ_s, and μ_b. If we plug in $C_n(\rho)$ defined in Equation 5.42 into Equation 5.41, when $kl_c \gg 1$

$$\Pi^2(k,z') \approx R\left(\frac{a\,|u_0|^2}{\pi f}\right)^2 \mathcal{N}_c (kl_c)^{4-D}, \tag{5.43}$$

in which $R = \Gamma(D/2)\pi^{-3/2}$, and Γ is the gamma function. The square intensity of the OCT spectrum then follows a power law of the wave number k, and the exponent (often referred to as scattering power) is equal to $4-D$, so that D can be calculated from the OCT spectrum.

The μ_b and μ_s are calculated based on the absolute OCT intensity magnitude and the intensity decay rate along the depth. The ratio of μ_b and μ_s is used to calculate l_c, deduced with known D according to the following equation

$$\frac{\mu_b}{\mu_s} = \frac{16(kl_c)^6 \Gamma\left(\dfrac{D}{2}\right)}{\Gamma\left(\dfrac{D}{2}-3\right)}\left[1+4k^2l_c^2\right]^{-\frac{D}{2}} \times \left[1 + \left(2k^2l_c^2\left(\frac{D}{2}-1\right)-1\right)\left(2k^2l_c^2\left(\frac{D}{2}-3\right)\right)\right.$$

$$\left. - \left(1+2k^2l_c^2\left(\frac{D}{2}+1\right)+4k^4l_c^4\left(4-\frac{3D}{2}+\frac{D^2}{4}\right)(1+4k^2l_c^2)^{1-D/2}\right)\right]^{-1}. \tag{5.44}$$

The next step is to plug in μ_b, l_c, and D values for $\mathcal{N}_c$

$$\mathcal{N}_c = \frac{\mu_b(1+4k^2l_c^2)^{D/2}}{8\sqrt{\pi}k^4l_c^3\Gamma(D/2)}. \tag{5.45}$$

The final step is to obtain dn^2 by a renormalization to $\mathcal{N}_c$ according to the above-calculated D and l_c

$$\begin{cases} dn^2 = \mathcal{N}_c \times \Gamma\left(\dfrac{D-3}{2}\right) & D > 3 \\[2ex] dn^2 = \mathcal{N}_c \times 2^{\frac{5-D}{2}}\left(\dfrac{\rho_{min}}{l_c}\right)^{(D-3)/2}K_{(D-3)/2}\left(\dfrac{\rho_{min}}{l_c}\right) & 0 < D \leq 3 \end{cases}, \tag{5.46}$$

where the value of ρ_{min} is specified to be an inner length scale below which the tiny structural features have very little scattering power to which a finite bandwidth system cannot be sensitive.

The anisotropic factor is determined by D and l_c according to the following equation

$$g = \left[\left[(1+4k^2l_c^2)^{\frac{D}{2}}\left(3+2k^2l_c^2\left(\frac{D}{2}-4\right)\right)\left(-3-4k^2l_c^2\left(k^2l_c^2\left(\frac{D}{2}-2\right)-1\right)\left(\frac{D}{2}-3\right)\right)\right]^{-1}\right.$$

$$-\left.(1+4k^2l_c^2)\left(3+6k^2l_c^2\left(\frac{D}{2}+2\right)+4k^6l_c^6D\left(10+\left(\frac{D}{2}-5\right)\frac{D}{2}\right)+8k^4l_c^4\left(6+\frac{D}{2}\left(\frac{D}{2}-1\right)\right)\right)\right]$$

$$\times\left[2k^2l_c^2\left(\frac{D}{2}-4\right)\times\left((1+4k^2l_c^2)^{\frac{D}{2}}\left(-1-2k^2l_c^2\times\left(2k^2l_c^2\left(\frac{D}{2}-2\right)-1\right)\left(\frac{D}{2}-3\right)\right)\right.\right.$$

$$\left.\left. +(1+4k^2l_c^2)\left(1+2k^2l_c^2\left(\frac{D}{2}+1\right)+4k^4l_c^4\left(4+\frac{D^2}{4}-\frac{3D}{2}\right)\right)\right)\right]^{-1}, \tag{5.47}$$

which can be further simplified under several limiting cases

$$g = \begin{cases} \dfrac{2}{5}D(k^2 l_c^2) & \text{if } kl_c \ll 1 \\[2ex] 1 - 0.45(kl_c)^{-\frac{1}{4}} & \text{if } kl_c \gg 1 \text{ and } D = 2 \\[2ex] 1 - 0.36(kl_c)^{-\frac{7}{4}} & \text{if } kl_c \gg 1 \text{ and } D = 4 \\[2ex] 1 - \left[\dfrac{-1 + (kl_c)^2(-4+D) + (1+4(kl_c)^2)^{1-\frac{D}{2}}(1+D(kl_c)^2)}{(kl_c)^2(D-4)(1-(1+4(kl_c)^2)^{1-\frac{D}{2}})} \right] & \text{if } kl_c \gg 1 \text{ and } D = elsewhere \end{cases}$$

$$(5.48)$$

Laboratory Protocol

The purpose of this lab is to demonstrate the ISOCT methodology of quantifying the complete set and the physical mass density correlation function of tissue. Students should learn how to rigorously process and normalize the data and accurately recover the different measured properties, including D, μ_s, and μ_b.

Suggested Lab Components

Fourier domain OCT system, MATLAB, polystyrene microspheres (80 nm, 0.82 μm in diameter).

Exercises

1. *Obtain one A-scan Fourier domain OCT raw spectrum and process time–frequency analysis to extract a depth-resolved spectral profile.* A simplified equation for the OCT signal can be written as

$$I = I_r + I_s + 2\sqrt{I_r I_s}\, \cos(2kz), \tag{5.49}$$

where I_r and I_s are the intensity from the reference and sample arm, k is the wave number, and z is the distance between the sample and the reference.

The first step is to normalize the raw spectrum with the reference spectrum so that I_r becomes a DC component in the spectral domain. The next step is to substrate the normalized spectrum by the averaged value of the whole spectrum to eliminate the DC components. Before performing the inverse Fourier transform (IFT), the spectrum should be resampled equally in k space. One A-scan depth profile can then be obtained by taking IFT on the spectrum and a series of continuous A-scan profiles to create a B-scan 2-D OCT image.

To obtain the depth-resolved spectral profile, a short-frequency Fourier transform (SFFT) is realized on the spectrum as illustrated in Figure 5.14. Create a Gaussian spectral wavelet window (equivalent FWHM = 30 nm), and multiply the window on the interferometric spectrum. After taking the IFT on the filtered spectrum, generate a 2-D map with two dimensions on depth and wavelength by scanning the window throughout the entire spectrum range. Each wavelength

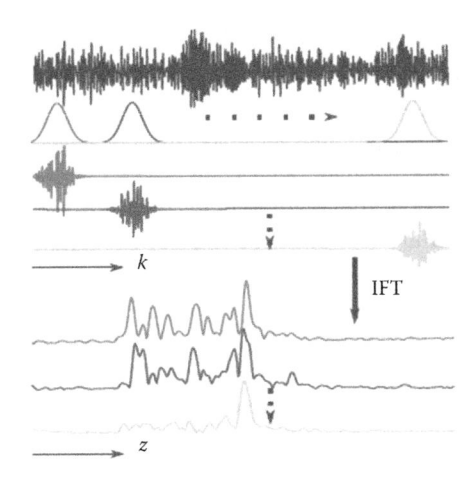

FIGURE 5.14
Schematic of short-frequency Fourier transform. A filtering wavelet filters the k spectrum before taking the inverse Fourier transform to obtain the depth profile contributed from the filtered bandwidth. Then, each point of the depth profile contains a spectrum as the center of the wavelet scans through the whole spectrum.

represents the center of the Gaussian spectral window. Then, the wavelength-dependent depth profile can be obtained.

2. *Use dipole scatterers to create the spectral normalization map.* Prepare a spectral normalization phantom with 2% 80-nm polystyrene sphere solution. The rationale for this phantom is that the scattering coefficient μ_b can be controlled at the same level of tissue, ~0.8 mm^{-1} at 700-nm wavelength, and the scattering power is well known to be k^4. Obtain a B-scan image of the solution, and extract the spectrum from superficial 50 μm by SFFT as the spectral point spread function (PSF) at the depth of the phantom surface. During the measurement, maintain a flat surface of the solution on the image so that the spectral PSF can be considered from an identical depth when the B-scan image is averaged along lateral direction.

3. *Evaluate the spectral PSF at different depths.* Repeat Exercise 2, placing the solution surface at different depths. Compare the spectral PSF and the total intensity from different depths.

4. *Calibrate μ_s and μ_b* Prepare another phantom with 10% w/w 0.82-micron polystyrene microspheres. Dilute with 1% boiled agarose solution with a volume ratio of 1:30. Agarose powder becomes soluble in water to form a gel solution that solidifies when cooled down to room temperature. Use a microwave to boil the solution. The values of μ_s and μ_b are calculated to be 8.8 and 0.4 mm^{-1} at 700-nm wavelength based on Mie theory. Obtain the B-scan OCT image of the phantom. Calculate the intensity decay rate along the depth from the A-scan depth profile for μ_s by fitting an exponential curve on the decayed signal, and calculate the absolute intensity magnitude at $z = 0$ for μ_b based on Beer–Lambert's law

$$I(z) = a\mu_b e^{-b\mu_s z},$$ (5.50)

where a and b are calibration constants. Plug the phantom-designed μ_s and μ_b into the equation, and deduce a and b values based on the calculated decay rate and the absolute intensity magnitude.

5. *Measure optical properties from chicken liver.* Take a B-scan OCT measurement of chicken liver. Calculate μ_s and μ_b from each A-scan signal according to the equation in Exercise 4, and take the average over all the A-scan measurements. The μ_s and μ_b need to be calculated from the surface of the tissue.

Use SFFT to extract the spectrum from a superficial 30 microns from each A-scan signal. Normalize the OCT spectrum by the spectral PSF obtained in Exercise 3 at the depth of the tissue surface. Calculate the exponent of the spectrum and deduce the value of D according to Equation 5.43. Note: The exponent should be calculated to the wave number k.

Deduce l_c, N_c, and dn^2 according to Equations 5.44 and 5.46. Deduce the anisotropic factor g according to Equation 5.47. Search literatures and find the optical properties of chicken liver or similar organs measured by other optical techniques. Compare your results with existing data.

Discussion

1. Ideally, the spectral PSF is independent of depth, while in reality, it is usually not. The depth-dependent spectral PSF is mainly subject to the fact that the objectives used for imaging have chromatic aberrations, which cause different focal positions for different wavelengths. If a well-corrected objective over the probing wavelength range is used, this aberration may be largely reduced. Otherwise, careful depth-dependent normalization will be required for an accurate D value measurement. The purpose of Exercise 2 is to provide an approach to the depth-dependent normalization map. With the well-known k^4 dipole scattering spectral dependent as a control, the student should be able to calculate the scattering power from biological tissue.

2. The value of factors a and b in Equation 5.50 can be arbitrary depending on the systematic setup, for example, the numerical aperture of the objectives. Careful calibration using microsphere phantoms is used to obtain these values further used in tissue measurement.

Questions

1. What would be axial resolution of measuring D value? It is the same as axial resolution in conventional OCT images?

2. When you calculate the D value and μ_b in deeper tissue, how can you compensate for the error induced by the wavelength-dependent attenuation by the scattering coefficient?

3. What is the effect of blood absorption? In what conditions can the effect of blood absorption be neglected? What effects can blood absorption have on the measurement of D, μ_s, and μ_b?

Laboratory 4: Depth-Selective Tissue Spectroscopy: Measuring Reflectance Impulse Response Function by Using Enhanced Backscattering

As we have learned in this chapter, reflectance impulse response function $p(r)$ (i.e., the radial intensity distribution of multiply scattered light at the sample surface) is a crucial property of light transport. In Laboratory 2, we learned how reflectance spectroscopy could be used to measure optical properties of tissue, which can be further used in multiple tissue diagnostics applications. In Laboratory 5, we will see how fiber-optic probes can be designed to measure tissue optical properties at different tissue depths. In all these applications, being able to measure or model $p(r)$ is critical. In this laboratory exercise, we will learn how this function can be experimentally measured. There are several approaches to measuring $p(r)$. The most straightforward approach would involve illuminating tissue by a pencil beam and then either acquiring an image of the tissue surface by a CCD or scanning the surface with the use of a fiber-optic probe. Although feasible and conceptually simple, in practice these approaches are not so easy to implement. In this exercise, we will measure $p(r)$ by taking advantage of the physical phenomenon known as enhanced backscattering. First, we review the principles of enhanced backscattering.

Enhanced Backscattering: An Introduction

Enhanced backscattering (EBS), also known as coherent backscattering, is a phenomenon in which rays traveling time-reversed paths within a scattering medium constructively interfere, resulting in an angular intensity peak centered in the exact backscattering direction [27,28,29]. As we will see shortly, the shape of this peak is dictated by the radial intensity distribution of multiply scattered light at the sample surface $p(r)$ through a Fourier transform relationship. As such, the EBS peak is sensitive to the optical scattering, absorption, and polarization properties that alter the radial intensity distribution. Using this sensitivity, EBS has been studied in many objects, including fractals [30,31,32], amplifying random media [33], cold atoms [34], liquid crystals [35,36], and biological tissue [37,38,39,40,41].

For biomedical applications, EBS offers an extremely sensitive tool to noninvasively interrogate the optical properties of different tissue specimens at subdiffusion length scales (i.e., $r <$ transport mean free path l_s') where information about the scattering phase function is preserved. Using this sensitivity, EBS has the ability to quantify the absorption coefficient μ_a, the scattering coefficient μ_s, the anisotropy factor g, and a second shape parameter of the phase function m with a single spectral measurement [41,42,43]. Combining these sensitivities with the ability to selectively interrogate different layers of tissue through implementation of a partial spatial coherence source, EBS has shown to be a promising technique for the characterization and detection of cancers, including colorectal, pancreatic, and lung [44,45,46].

The Scalar Theory of EBS

The origin of the EBS intensity peak is the presence of time-reversed path pairs. A time-reversed path pair consists of two rays that travel through the same sequence of scattering events but in the opposite direction from each other. Additionally, both rays enter the medium with the same angle θ_i and exit the medium with the same angle θ_s. The appearance of time-reversed path pairs is guaranteed by the reciprocity theorem (and Maxwell's equations). In everyday experience, the reciprocity theorem is at work when

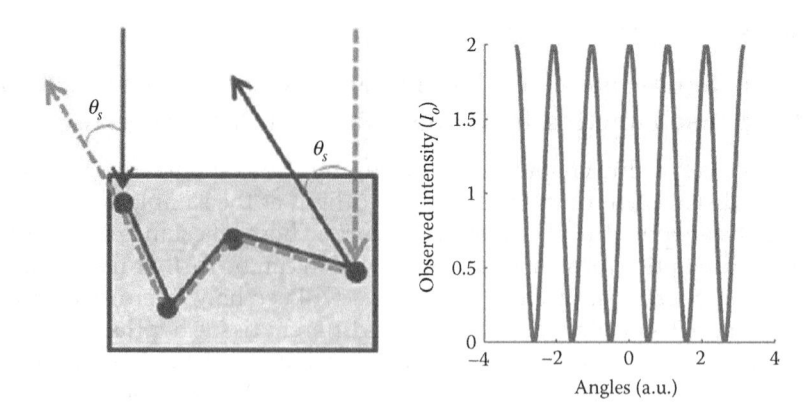

FIGURE 5.15
Illustration of a single time-reversed path pair (left) and the observed angular intensity pattern it creates (right).

two people look at each other through reflections in a mirror ("If I can see you, you can see me") [47].

The left panel of Figure 5.15 shows a single time-reversed path pair for a scattering medium under plane wave illumination with normal incidence. When $\theta_s = 0$, the forward and reverse propagating rays will have traveled through the same path length and will, therefore, exit with the same accumulated phase. As a result, in the exact backward direction, the rays will constructively interfere, and the observed intensity will be 2 times the incident intensity I_o. As θ_s is increased, a path length difference between forward and reverse propagating will occur, and the angular intensity pattern will alternate between complete constructive and destructive interference, forming a cosine squared pattern. Using this as a starting point, we can draw a direct analogy with Young's double-slit experiment by viewing the first and last scattering events in the time-reversed path pair as two centers of spherical waves (i.e., double pinholes). With this understanding, the observed angular intensity pattern can be thought of as the diffraction pattern from two pinholes separated by the lateral distance between the first and last scattering event. In other words, the angular intensity pattern is the Fourier transform of the spatial distribution of first and last scatters.

In a semi-infinite scattering medium, an infinite number of time-reversed path pairs with different spatial separations will combine to form the EBS peak. Therefore, if we neglect polarization effects as a first approximation, the EBS peak is the summed diffraction pattern from all possible sets of time-reversed path pairs. In other words, the EBS peak is the Fourier transform of the spatial reflectance profile of light in a random scattering medium illuminated by an infinitely narrow pencil beam $I_{ms}(x,y)$. In this case, $I_{ms}(x,y)$ represents the intensity of *multiply* scattered light that exits the medium at a position (x,y) away from where it entered and in a direction that is antiparallel to the incident beam. Since rays undergoing single scattering (I_{ss}) do not have separated time-reversed paths, they do not contribute to the EBS interference signal. If we normalize by the incoherent baseline intensity that contains both single and multiple scattering contributions, the resulting EBS peak can be found as

$$I_{EBS}(\theta_x,\theta_y) = \frac{\displaystyle\iint_{-\infty}^{\infty} I_{ms}(x,y)e^{-jk(\theta_x x + \theta_y y)}dx\,dy}{\displaystyle\iint_{-\infty}^{\infty} [I_{ms}(x,y) + I_{ss}(x,y)]dx\,dy} \qquad (5.51)$$

or more simply

$$I_{EBS}(\theta_x, \theta_y) = FT[P(x, y)] \tag{5.52}$$

where the appropriately normalized spatial intensity distribution is written as $P(x,y)$, and the Fourier transform operation can be denoted with *FT*. In either case, it should be noted that these equations give the purely coherent portion of the angular backscattering with values between 0 and 1 (where 1 signifies constructive interference and 0 signifies no interference).

In the following discussion, when it is useful to observe the radial intensity distribution, we use the rotational sum of $P(x,y)$, which we denote as $P(r)$. This represents the summed intensity that exits the scattering medium in radial annuli that is located at a distance r away from its entrance point

$$P(r) = \int_{0}^{2\pi} P(x = r\cos\varphi, y = r\sin\varphi) \mathrm{d}\phi \tag{5.53}$$

Figure 5.16 shows the Fourier pair of $P(r)$ and $I_{EBS}(\theta)$. The left panel shows a Monte Carlo simulation of $P(r)$ using the Henyey–Greenstein phase function with $g = 0$ and $l_s^* = 200\ \mu m$. This plot is normalized such that $\int P(r)dr = 1$. The right panel shows the EBS peak calculated according to Equation 5.51. In the exact backscattering direction, there is full constructive interference and $I_{EBS}(\theta_s) = 1$. As θ_s, the interference fringes from different path pairs cancels out, and the interference signal is decreased.

Under a scalar diffusion approximation, Akkermans et al. derived an equation that governs the shape of the EBS peak [48]

$$I_{EBS}(\theta) = \frac{3}{8\pi} \left[1 + \frac{2z_0}{l_s^*} + \frac{1}{(1 + k\theta l_s^*)^2} \cdot \left(1 + \frac{1 - \exp(-2k\theta z_0)}{k\theta l_s^*} \right) \right] \tag{5.54}$$

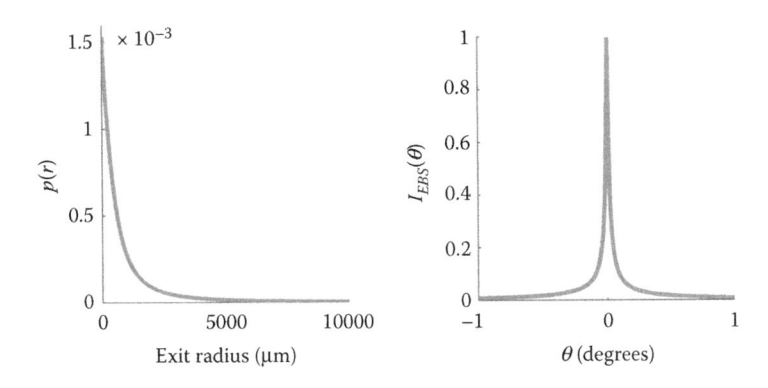

FIGURE 5.16
Illustration of $p(r)$ for a sample with $g = 0$ and $l_s^* = 200\ \mu m$ (left panel). Corresponding EBS peak after Fourier transformation (right panel).

where $k = 2\pi/\lambda$ is the wave number and z_0 is the location of the trapping plane (can be assumed to be equal to 0.7 l_s^*).

Vector Theory of EBS

For the previous calculations, we have neglected the effects of polarization as a first approximation. Under this scalar theory, each path pair can completely interfere, and the predicted peak height (enhancement factor) reaches a value that is twice the incoherent baseline. However, since EBS is a coherent phenomenon, both the phase and the polarization of the detected wave must be the same for constructive interference to occur. As we will see, the effects of polarization can have a huge impact on the shape and height of the EBS peak.

One effect of polarization on the EBS peak is a simple reduction in enhancement factor due to the contribution from single scattering. Since rays that undergo single scattering do not possess time-reversed partners, only the fraction of light that is multiply scattered can contribute to the EBS peak. As a demonstration, we implement polarized light Monte Carlo simulations (PMC) to measure the multiple scattering ratio ($MSR = I_{ms}/(I_{ms} + I_{ms})$) for suspensions of microspheres. Figure 5.17 shows the MSR for the different polarization channels as a function of the anisotropy factor (g).

For the helicity-preserving (++) and linear cross (xy) channels, MSR is exactly equal to 1 for all values of g, since these channels completely reject single scattered light. However, for other channels, MSR is dependent on the shape of the scattering phase function. For Rayleigh scatterers (i.e., *radius* $\ll \lambda$) with $g = 0$, MSR is 0.8357, 0.7514, and 0.7342 for the unpolarized, linear co (xx), and opposite helicity (+−) channels, respectively. As g increases, the phase function becomes more forward directed, and the MSR approaches 1 for all channels. Because of the single scattering contribution, the enhancement factor will never be greater than 1 + MSR.

In addition to the effect of single scattering, the peak is further reduced in the orthogonal polarization channels due to scattering rotations that are irreversible for these channels. In the strictest sense, the reciprocity theorem is only fully satisfied if the illumination

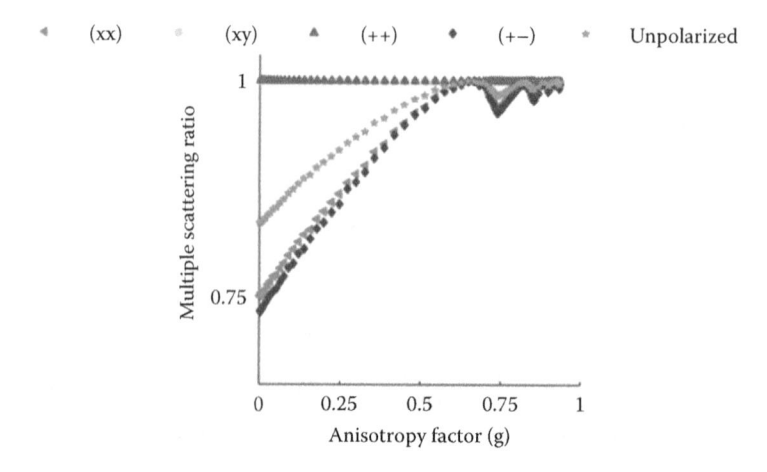

FIGURE 5.17

Multiple scattering ratio as a function of anisotropy factor (g). Polarizations: linear co-polarized (xx), linear cross-polarized (xy), helicity preserving (++), and opposite helicity (+−).

and collection polarization states are exactly the same. Thus, to a first approximation, an EBS peak is expected to appear only in the polarization-preserving channels (i.e., linear co-polarized [*xx*] and helicity preserving [++]). In these channels, each multiply scattered ray is guaranteed to possess a time-reversed partner that exits with the same accumulated phase. As a result, path pairs exiting at any (*x*,*y*) separation have perfectly correlated phases and can fully interfere. However, even in the orthogonal polarization channels (i.e., linear cross-polarized [*xy*] and opposite helicity [+−]), an EBS peak can still be observed. In this case, the reciprocity theorem no longer guarantees that each sequence of scattering events is fully reversible. As a result, light rays that are scattered into the orthogonal polarized channel may or may not have a time-reversed partner with which to interfere, and the height of the EBS peak is greatly reduced.

A more mathematically rigorous way to describe the effect of polarization on the EBS peak is using linear matrix operations (Jones calculus formalism). Any linear transformation of light can be calculated as

$$\begin{pmatrix} E'_\parallel \\ E'_\perp \end{pmatrix} = \begin{bmatrix} J_{11} & J_{12} \\ J_{21} & J_{22} \end{bmatrix} \begin{pmatrix} E_\parallel \\ E_\perp \end{pmatrix} \tag{5.55}$$

$$\widetilde{E'} = J\tilde{E} \tag{5.56}$$

where $\tilde{E}$ and $\widetilde{E'}$ are the incoming and outgoing complex electric field vectors, respectively, and J is a complex linear operation that represents either an optical element (i.e., linear polarizer, quarter wave plate) or a scattering event. A more complete description of Jones calculus along with the Jones matrices for various different optical elements can be found in the textbook by Hecht [49].

For a single scattering event, the electric field undergoes three transformations [41,50]

$$\widetilde{E'} = R(\zeta)S(\theta)R(\varphi)\tilde{E} \tag{5.57}$$

where $R(\varphi)$ is a transformation from the meridian plane into the scattering plane, $S(\theta)$ is the amplitude scattering matrix, and $R(\zeta)$ is a rotation back into the meridian plane. These matrices can be expanded as [51,52]

$$R(x) = \begin{vmatrix} \cos(x) & \sin(x) \\ \sin(x) & \cos(x) \end{vmatrix} \tag{5.58}$$

$$S(\theta) = \begin{vmatrix} S_2(\theta) & S_3(\theta) \\ S_4(\theta) & S_1(\theta) \end{vmatrix} \tag{5.59}$$

The calculation of the various scattering amplitudes is discussed by van de Hulst [52] and is not important for the current discussion.

In a multiple scattering medium, we continue transforming the electric field with these operations until the light exits the medium. In this case, the full equation can be written as

$$\widetilde{E'} = R_n(\zeta)S_n(\theta)R_n(\varphi)\cdots R_2(\zeta)S_2(\theta)R_2(\varphi)R_1(\zeta)S_1(\theta)R_1(\varphi)\tilde{E} \tag{5.60}$$

$$\widetilde{E}' = M\widetilde{E} \tag{5.61}$$

where M represents the complex effective scattering matrix for a multiply scattered ray, and the subscript indicates the number of scattering events. The matrix elements of M can take an infinite number of different values depending on the composition of the sample, incident polarization, and path with which the light travels. While these values are not trivial to calculate for a ray undergoing thousands of scattering events, they are fairly straightforward to calculate using Monte Carlo simulations. Still, to gain a general understanding of the effects of polarization on EBS, we can generalize M for a forward propagating path as

$$M_{forward} = \begin{bmatrix} a_r + ia_i & b_r + ib_i \\ c_r + ic_i & d_r + id_i \end{bmatrix} \tag{5.62}$$

where the subscripts indicate the coefficients for the real and imaginary parts, and $i = \sqrt{-1}$.

In a mathematical sense, the reciprocity theorem states that scattering matrix of the reverse propagating path is simply the "antitransposition" of the scattering matrix for the forward propagating path [47,53]. This can be expressed as

$$M_{reverse} = \begin{bmatrix} a_r + ia_i & -c_r - ic_i \\ -b_r - ib_i & d_r + id_i \end{bmatrix} \tag{5.63}$$

Using Equations 5.62 and 5.63, we can calculate the path coherence for any arbitrary time-reversed path. The path coherence represents the ability for the forward and reverse paths to interfere with each other and is calculated as the coherent summation divided by the incoherent summation.

$$coherence = \frac{2 \cdot \mathrm{Re}\left[\widetilde{E}_{forward} \cdot \widetilde{E}_{reverse}^*\right]}{\left|\widetilde{E}_{forward}\right|^2 + \left|\widetilde{E}_{reverse}\right|^2} \tag{5.64}$$

Taking incident linear polarization with $\begin{pmatrix} E_\parallel \\ E_\perp \end{pmatrix} = \begin{pmatrix} 1 \\ 0 \end{pmatrix}$ and performing the appropriate matrix operations, the *coherence* can be found as

$$\begin{pmatrix} coherence\, E_\parallel \\ coherence\, E_\perp \end{pmatrix} = \begin{pmatrix} \dfrac{2 \cdot (a_r^{\,2} + a_i^{\,2})}{(a_r^{\,2} + a_i^{\,2}) + (a_r^{\,2} + a_i^{\,2})} \\ \dfrac{-2 \cdot (b_r c_r + b_i c_i)}{(b_r^{\,2} + b_i^{\,2}) + (c_r^{\,2} + c_i^{\,2})} \end{pmatrix} \tag{5.65}$$

From Equation 5.65, we can see that the co-polarized channel (i.e., $E_\parallel$) simplifies to 1 regardless of the values of $a, b, c,$ or d. In other words, for co-polarized channels, the forward and reversed paths **must** interfere independently of the specific path or optical properties of the sample. On the other hand, in the cross-polarized channel (i.e., $E_\perp$), the *coherence* is a mix between b and c and can take any value between -1 (destructive interference) and 1 (constructive interference). In other words, for cross-polarized channels, the forward and reverse paths do not necessarily exit the medium with the same phase and can, therefore, interfere to different extents.

For a multiple scattering medium, the reflectance at any spatial position on the sample will contain the contribution from an infinite number of different time-reversed path pairs. Therefore, in order to describe the ability of rays arriving at a particular separation to interfere, we introduce the degree of phase correlation function $PC(r)$, which modulates the shape of $P(x,y)$ for the orthogonal polarization channels. When $PC = 1$, the entire portion of intensity arriving at particular separation can interfere, and when $PC = 0$, none of the intensity can interfere. Modifying the calculations of Lenke and Maret [54] for no external magnetic field, PC can be calculated from the degree of linear ($dlp = ((xx)-(xy))/((xx)+(xy))$) and circular ($dcp = ((++)-(+-))/((++)+(+-))$) polarization

$$PC_{xy}(r) = \frac{dlp(r) + dcp(r)}{1 - dlp(r)}$$

$$PC_{+-}(r) = \frac{2 \cdot dlp(r)}{1 - dcp(r)} \tag{5.66}$$

To generate the distributions in Equation 5.66, dlp and dcp can easily be calculated using polarized light Monte Carlo simulations [41]. Figure 5.18 shows the Monte Carlo simulated PC functions in the four polarization channels for an aqueous suspension of latex microspheres. As described previously, function PC is identically equal to 1 at all exit radii for the polarization-preserving channels. On the other hand, for the cross-polarized channels only, light rays exiting at very short radii remain fully reversible, and PC is nearly 1. As r increases, a larger proportion of rays travels through irreversible sequences of scattering events, and $PC \to 0$ as $r \to \infty$.

Figure 5.19 combines the effects of single scattering and irreversible rotations of light in the orthogonal polarization channels to demonstrate the maximum theoretically

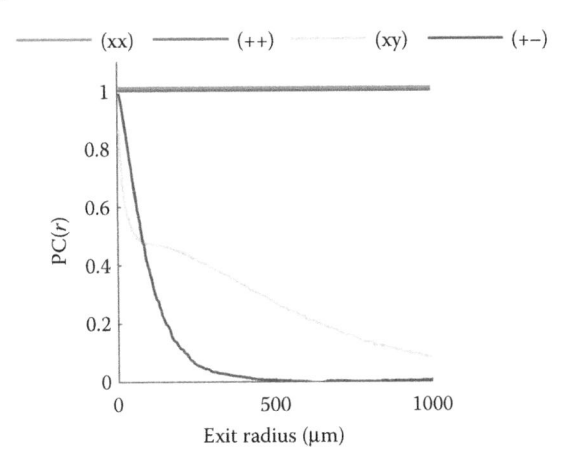

FIGURE 5.18
Phase correlation function (PC) for the four polarization channels calculated using polarized light Monte Carlo. The simulated sample was an aqueous suspension of latex microspheres with 0.65-μm-diameter spheres with transport mean free path of 205 μm with illumination at 633 nm.

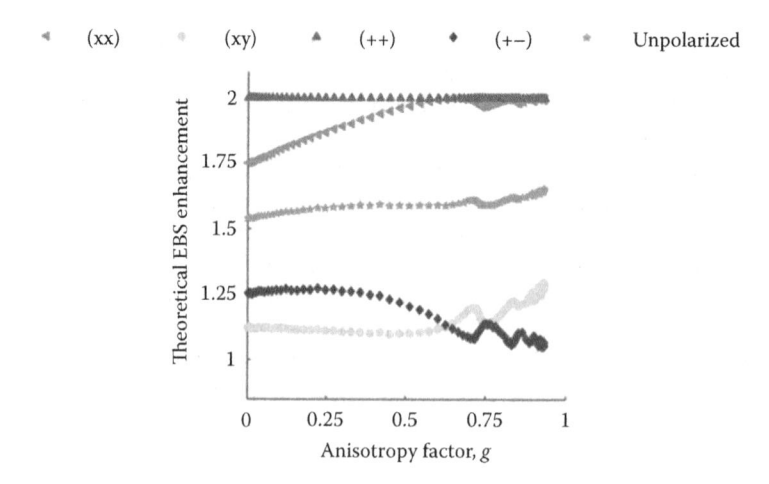

FIGURE 5.19

Theoretical EBS enhancement factor for the four polarization channels calculated using polarized light Monte Carlo. The simulated sample was an aqueous suspension of latex microspheres.

achievable enhancement factor for spherical particles with different values of *g*. For the (++) channel, single scattering is completely suppressed, and each time-reversed pair is fully reversible according to the reciprocity theorem. As such, the theoretical enhancement factor in this channel is identically equal to 2 for all *g*. For the (*xx*) channel, the reduction in enhancement factor is solely due to single scattering and therefore follows the shape of the MSR shown in Figure 5.19. The strong reduction in the enhancement factor for the (*xy*) and (+−) channels occurs because only a small percentage of the initial polarized intensity is transferred into the orthogonal channels through a fully reversible path.

In addition to the alteration in enhancement value due to polarization effects, the azimuthal pattern is also greatly dependent on the illumination and collection polarization states. As a starting point to understand the origin of the different patterns, we consider the shapes of the phase function for scattering media illuminated with different polarizations. Under circularly polarized illumination, the phase function (shown in Figure 5.20a) is rotationally symmetric about the *x–y* plane. This means that in the (+o) polarization channel (i.e., circular illumination with unpolarized collection), *P(x,y)* for a homogeneous scattering sample (shown in Figure 5.20b) must also be rotationally symmetric. Under linear polarized illumination, the phase function (shown in Figure 5.20c) has a reduced probability in the direction of the incident polarization vector due to the dipole radiation pattern; therefore, less light will be scattered in that direction. This means that in the (*x*o) polarization channel (i.e., linear illumination with unpolarized collection), *P(x,y)* will be elongated in the direction orthogonal to the polarization vector (shown in Figure 5.20d). After taking the Fourier transform of *P(x,y)*, the resulting EBS peak will therefore be elongated in the direction of the polarization vector.

By further decomposing the patterns in Figure 5.20b, d, we obtain the more commonly measured (*xx*), (*xy*), (++), and (+−) polarization channels shown in Figure 5.21. The first two rows of this figure show the (++) and (+−) channels obtained by decomposing the (+o) channel. Since there is no breaking of symmetry, these channels are also rotationally symmetric. As shown in the left column, *P(x,y)* for the (++) channel is broader than for the (+−) channel. After Fourier transformation, this relationship is flipped, and the (++) channel is narrower than the (+−) channel, as shown in the right column. The final two rows

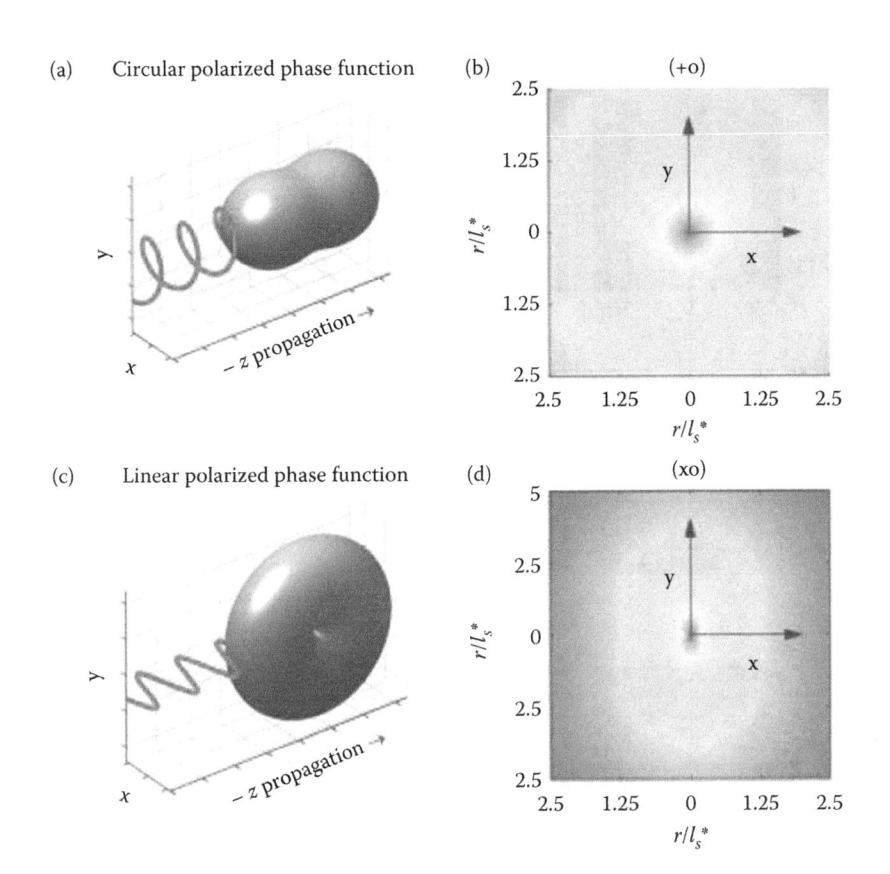

FIGURE 5.20
Phase function and Monte Carlo simulated $P(x,y)$ for a dipole (Rayleigh) scattering sample under linear and circular polarization illumination. (a) Phase function for circular polarized illumination and the corresponding $P(x,y)$ (b) for unpolarized collection. (c) Phase function for linear polarized illumination and the corresponding $P(x,y)$ (d) for unpolarized collection.

of Figure 5.21 show the complex azimuthal pattern for the (xx) and (xy) channels obtained by decomposing the (xo) channel. Since light is transferred into the cross-polarized channel most efficiently at 45° with respect to the x and y axes, the (xy) component exhibits a "four-leaf clover" or "X" pattern [47]. The remaining rays that are not rotated into the cross channel then form the (xx) component. As a result, the (xx) peak is in general elongated in the direction of the polarization but also shows decreased intensity in the diagonal directions due to depolarization of light into the cross channel.

Effects of Partial Spatial Coherence Illumination and Finite Illumination Spot Size on EBS

For many light-scattering applications, it is advantageous to isolate the scattering signal generated by rays exiting at short radial exit positions (also known as source–detector separations). The reasons for this are that short radial exit positions (1) preserve details about the shape of the phase function and (2) allow for the interrogation of superficial tissues. Two ways to achieve this selectivity are through use of illumination with partial spatial coherence and finite spot size.

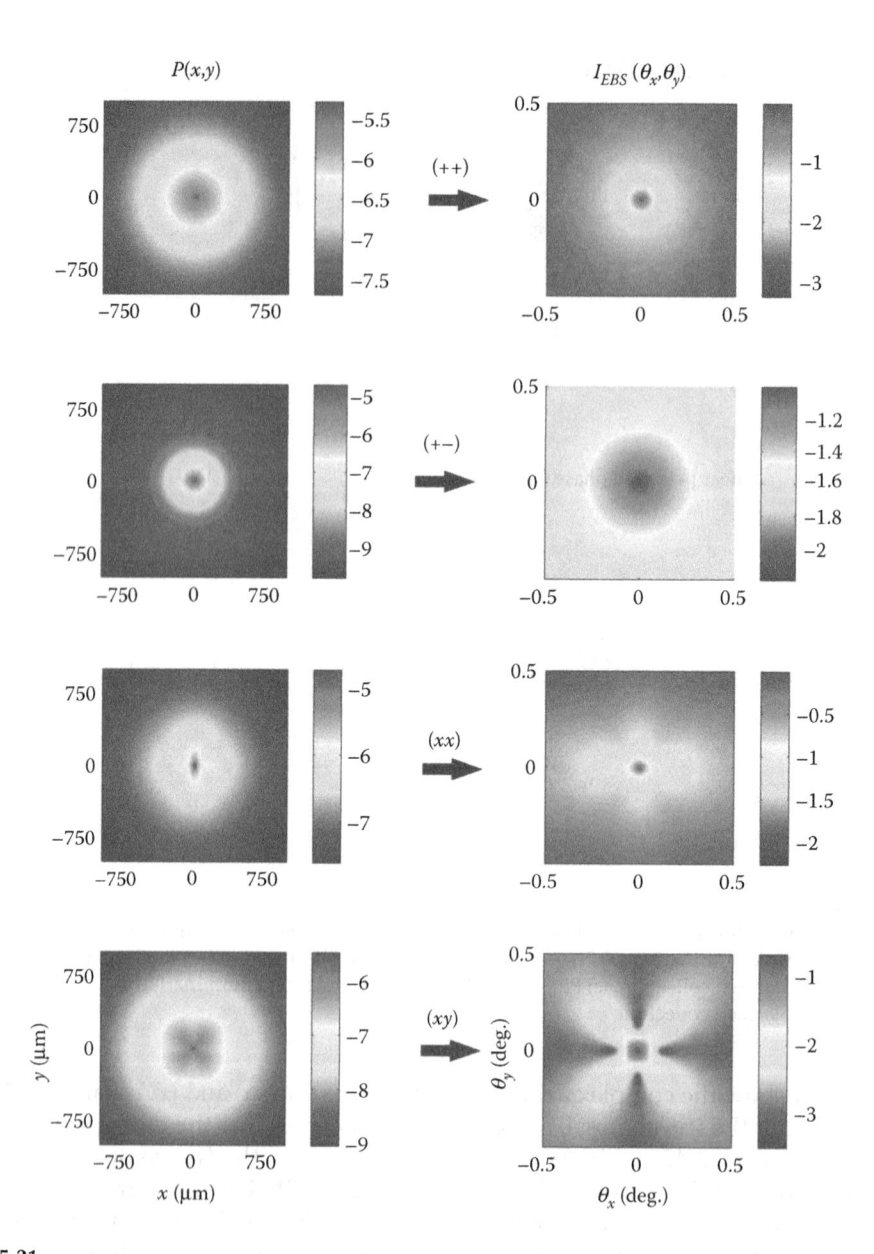

FIGURE 5.21

Monte Carlo simulated $P(x,y)$ (left column) and $I_{ebs}(\theta_x,\theta_y)$ (right column) for an aqueous suspensions of polystyrene microspheres (0.65-μm diameter, transport mean free path of 205 μm, 633-nm illumination) for the four polarization channels. First row: helicity preserving (++); Second row: opposite helicity (+−); Third row: linear co-polarized; Fourth row: linear cross-polarized.

Spatial coherence describes the ability of light waves at different spatial separations to interfere. Waves that can fully interfere at any separation are said to be spatially coherent, while waves that lose their ability to interfere for larger separations are said to be partially spatially coherent. In an experimental setup, partial spatial coherence occurs when an incoherent source of finite size is collimated with a lens. This is typically done in microscopy setups where incoherent illumination from a xenon lamp is first focused onto a circular aperture, which acts as a secondary source and is collimated with a lens. For this configuration, the light is partially spatially coherent because the independent oscillations from different points on the source are combined in a particular angular direction after collimation. According to the van Cittert–Zernike theorem [55], the spatial coherence function $C(x,y)$ is the Fourier transform of the angular intensity distribution of the source $S(\theta_x,\theta_y)$.

$$
C(x,y) = \frac{\displaystyle\iint_{-\infty}^{\infty} S(\theta_x,\theta_y)e^{-jk(\theta_x x+\theta_y y)}d\theta_x\,d\theta_y}{\displaystyle\iint_{-\infty}^{\infty} S(\theta_x,\theta_y)d\theta_x\,d\theta_y}
$$

$$
= FT[S(\theta_x,\theta_y)] \tag{5.67}
$$

In the case where a circular aperture is used as the secondary source, the Fourier transform can be calculated as

$$
C(x,y) = \frac{2J_1\left(\dfrac{\sqrt{x^2+y^2}}{L_{sc}}\right)}{\dfrac{\sqrt{x^2+y^2}}{L_{sc}}}; \quad L_{sc} = \frac{\lambda f}{\pi d} \tag{5.68}
$$

where J_1 is the first-order Bessel function of the first kind, L_{sc} is the spatial coherence length, λ is the illumination wavelength, f is the focal length of the collimating lens, and d is the diameter of the circular aperture. L_{sc} is defined as the radial distance at which $C(x,y)$ reaches a value of 0.88. Figure 5.22 shows the functional form of Equation 5.68 for the unitless parameter distance/L_{sc}.

Since EBS is a coherence phenomenon formed by time-reverse path pairs exiting at different spatial separations, $C(x,y)$ acts as a spatial filter that limits the interferences from waves exiting at large separations. In this case, the phenomenon is known as low-coherence

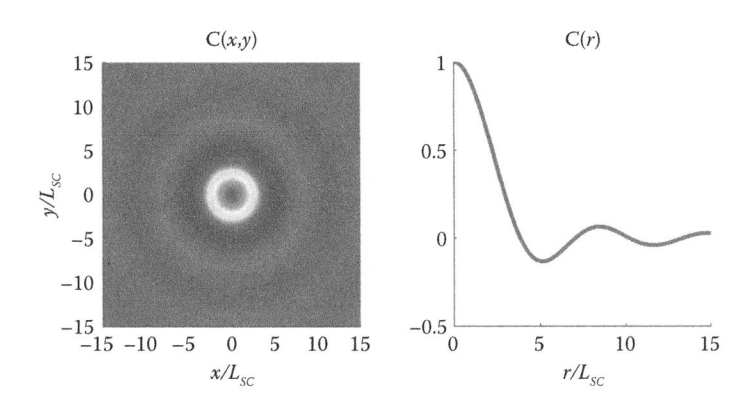

FIGURE 5.22
Coherence function for a circular aperture secondary source. (left) Two-dimensional shape for $C(x,y)$. (right) Cross-section through the center of function $C(x,y)$.

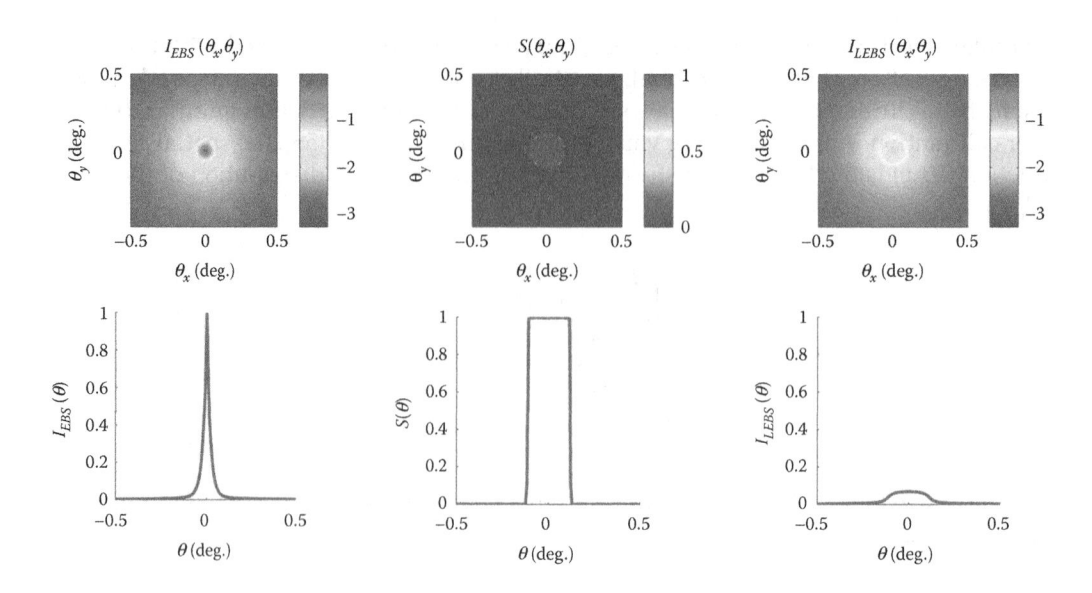

FIGURE 5.23
Demonstration of the formation of the LEBS peak under partial spatial coherence illumination using Monte Carlo simulation. Simulation parameters: Aqueous suspensions of polystyrene microspheres with 0.65-μm diameter, transport mean free path of 205 μm, 633-nm illumination, and $Lsc = 50$ μm. The top row shows the 2-D distribution for each term in Equation 5.65. The bottom row shows a 1-D cross-section through the center of the 2-D peak.

enhanced backscattering (LEBS). The LEBS peak will, therefore, be the Fourier transform of $P(x,y) \cdot C(x,y)$

$$I_{LEBS}(\theta_x, \theta_y) = FT[P(x,y) \cdot C(x,y)] \tag{5.69}$$

As a useful alternative for understanding the shape of the LEBS peak, we employ the convolution theorem to represent EBS as [56]

$$I_{LEBS}(\theta_x, \theta_y) = I_{EBS}(\theta_x, \theta_y) \circledast S(\theta_x, \theta_y) \tag{5.70}$$

where $\circledast$ denotes the convolution operation and $S(\theta_x, \theta_y)$ is the angular intensity distribution or source function. As a result, LEBS can be conceptualized as either a filtering operation in spatial coordinates or a convolution operation in angular coordinates. Figure 5.23 demonstrates each term in Equation 5.70 using the (++) polarization channel. After convolving with the source function, the LEBS enhancement value is greatly reduced, and the peak is broadened.

For samples such as biological tissue in which the spot size is not significantly larger than l_s^*, it is necessary to find the effective spatial intensity distribution of light that remains inside the illumination spot size. This is because rays that exit outside the illumination spot do not have a time-reversed partner and therefore will not contribute to the EBS interference signal. Typically, the effective intensity distribution within an illumination spot is found by convolving $P(x,y)$ with the illumination spot size [57]. However, the EBS interference signal is dependent on the relative entrance and exit position of path pairs as opposed to their absolute position within the illumination spot. As a result, the effective

EBS intensity distribution $P_{eff}(x,y)$ should be calculated by averaging truncated versions of $P(x,y)$ from every position within the illumination spot

$$P_{eff}(x,y) = \frac{\iint^{\Omega} P(x,y) \cdot A(x-\alpha, y-\beta) d\alpha \, d\beta}{\iint^{\Omega} A(\alpha, \beta) d\alpha \, d\beta} \tag{5.71}$$

where A is a function that represents the illumination spot intensity distribution, and the area of integration Ω is over the area of the illumination spot size. Note that the numerator in Equation 5.71 is different than either a convolution or cross-correlation. This can be simplified as

$$P_{eff}(x,y) = P(x,y) \frac{\iint_{-\infty}^{\infty} A(\alpha, \beta) A(x-\alpha, y-\beta) d\alpha \, d\beta}{\iint_{-\infty}^{-\infty} A(\alpha, \beta) d\alpha \, d\beta}$$

$$= P(x,y) \cdot ACF[A] \tag{5.72}$$

where ACF is the normalized autocorrelation function. Assuming the illumination intensity is flat across the entire spot size and the spot is formed by a circular aperture, the function A can be taken as a top-hat function (i.e., $A = 1$ within the spot and 0 elsewhere). In this case, an analytical solution exits for $ACF(A)$

$$S(r) \equiv ACF[A(r)] = \frac{D^2}{2} \cos^{-1}\left(\frac{r}{D}\right) - 0.5\sqrt{r^2 + D^2} \tag{5.73}$$

where D is the beam diameter, and we define $S(r)$ as a function that represents the illumination spot size. Figure 5.24 shows Equation 5.73 for the unitless parameter r/D. $ACF[A]$ begins at a value of 1 and decreases to 0 when $r - D$. Rays exiting with $r > D$ will fall outside the illumination spot and therefore have no time-reversed partner.

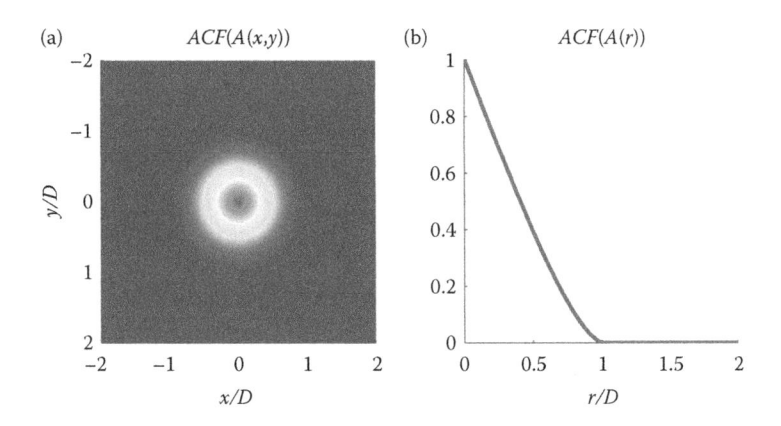

FIGURE 5.24
Effective illumination spot size function for a circular beam. (a) Two-dimensional shape for $S(x,y)$. (b) Cross-section through the center of function $S(x,y)$.

Partial spatial coherence illumination and finite illumination beam spot size both offer methods to selectively interrogate the shape of $P(r)$ at short length scales. For partial spatial coherence illumination, we can choose how broad or narrow $C(r)$ is by selecting the size of the secondary source and/or the focal length of the collimating lens. For a finite illumination beam spot, we can choose how broad or narrow $S(r)$ is by selecting the size of the illumination spot.

Full EBS Equation

As a final consideration for the realistically measurable EBS peak, we must incorporate the effect of the optical system's modulation transfer function, or MTF. For any imaging system, the MTF describes the ability to capture information from different spatial frequencies. As such, the measurement of $P(r)$ will also be modulated by the system's $MTF(r)$. Further discussion of the measurement of the MTF can be found in the exercises section.

Combining the contributions from the spatial backscattering impulse–response $P(r)$, the phase correlation function $PC(r)$ (Equation 5.66), the spatial coherence function $C(r)$ (Equation 5.68), the illumination spot size function $S(r)$ (Equation 5.73), and the modulation transfer function $MTF(r)$, the effective equation for the measurement of EBS can be found as

$$I_{EBS}(\theta_x,\theta_y) = FT[P(x,y) \cdot PC(x,y) \cdot C(x,y) \cdot S(x,y) \cdot MTF(x,y)] \qquad (5.74)$$

Penetration Depth and Measurement of Optical Properties Using Low-Coherence EBS

As discussed previously, a benefit of using LEBS is that $C(x,y)$ acts as a spatial filter that limits the contribution from light rays exiting at larger exit positions. Since rays that exit the medium at larger separations tend on average to penetrate deeper within the sample, the use of partial spatial coherence provides a method to selectively restrict the depth from which the optical signal is measured [42].

Using Monte Carlo simulations for media composed of a continuous distribution of refractive index fluctuations (using the Whittle–Matérn phase function), an equation for the average penetration depth of LEBS under a scalar approximation can be obtained for different scattering properties. Under the scalar approximation, the unity-normalized Whittle–Matérn phase function $F_{wm}(\theta)$ can be written as [58]

$$F_{wm}(\theta) = \frac{2(kl_c)^2(m-1)}{1-[1+2(kl_c)^2]^{1-m}} \cdot \frac{1}{\left[1 + \left(2kl_c\sin\left(\frac{\theta}{2}\right)\right)^2\right]^m} \qquad (5.75)$$

where θ is the polar scattering angle, k is the wave number, and l_c and m are two parameters of the Whittle–Matérn function, as discussed in Chapter 6. The anisotropy factor of $F_{wm}(\theta)$ can be found as $g = 1-[1+4(kl_c)^2 - 1/2(kl_c)^2]$.

After performing the simulations for a range of expected tissue-relevant optical scattering properties, a saturation curve can be constructed for the LEBS enhancement factor by varying the thickness of the sample. The normalized derivative of this curve is the depth probability function $p(z)$. The function $p(z)$ describes the probability that the photons that contribute to the LEBS enhancement factor have maximum depth of z. The average penetration depth D_p can then be calculated as the first moment of this distribution

$$D_p = \int_0^\infty z\, p(z)\, dz \tag{5.76}$$

Finally, an expression for the average penetration depth for different values of L_{sc} and l_s' is obtained empirically by fitting with Monte Carlo simulations according to

$$D_p = a(Lsc)^{1-b}\left(l_s^*\right)^b \tag{5.77}$$

where the coefficients a and b are given as

$$\begin{aligned} a &= a_1(1-g)^{a_2} \\ b &= b_0 + b_1(1-g)^{b_2} \end{aligned} \tag{5.78}$$

For the expected range of optical properties, these coefficients are $a_1 = 1.071$, $a_2 = -0.121$, $b_0 = 0.300$, $b_1 = 0.510$, and $b_2 = 0.459$. The comparison between this model and MC simulations is shown in Figure 5.25. In Figure 5.25a, it is shown that the penetration depth increases monotonically with increasing l_s^*, as predicted by our model. Conceptually, this makes sense, since with increasing l_s^*, photons tend to travel longer paths before returning to the sample surface. An important point to note is that the penetration depth is essentially independent of the parameter m, as shown in Figure 5.25a and b. This means that the penetration depth is dominated by g and l_s^*, as shown in Figure 5.25c. With increasing values of g, the penetration depth is decreased for a constant value of l_v^*. The rationale behind this is that for increasing values of g, the value

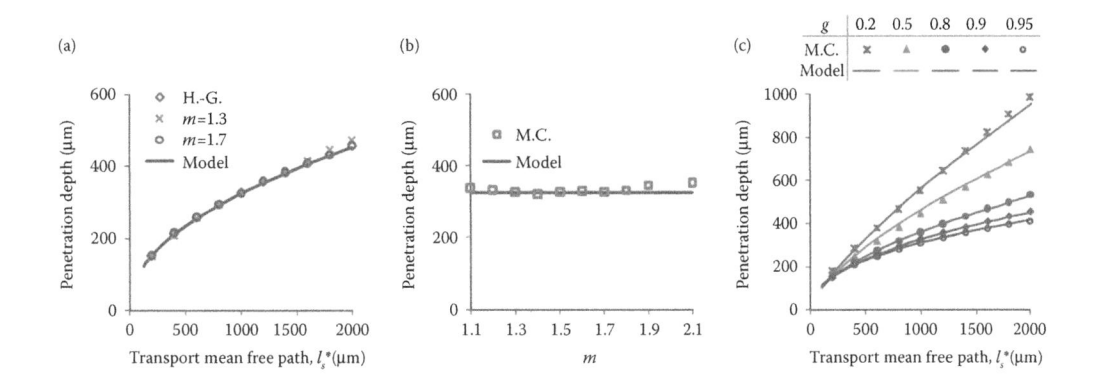

FIGURE 5.25

Penetration depth model. (a) Penetration depth as a function of l_s' for a range of different values of m (dots) along with the model presented in Equation 5.77 (solid line). (b) Demonstration of the independence of the penetration depth on the parameter m. (c) Penetration depth as a function of l_s' for a range of different values of g. For each case, the simulation (dots) agree well with the model (solid lines).

of the mean free path $l_s = l_s^*(1-g)$ is decreased. As a result, photons tend to penetrate more superficially into the sample.

While the depth from which a signal originates within a tissue specimen is important for most biomedical applications, it is also very desirable to quantify that sample's optical properties. Doing so helps to gain a more physical understanding of how the sample is composed and, for the case of cancer biology, to understand how the tissue changes as it progresses to a more cancerous state. The quantification of optical properties under the scalar Whittle–Matérn approximation was first presented by Turzhitsky et al. [59].

Within the low coherence regime (i.e., $L_{sc}/l_s^* = L_{sc}\mu_s^* < 1$), the LEBS enhancement factor E can be approximated as being linear with a constant $C_E \sim 0.2$

$$E = C_E L_{sc}\mu_s^* \approx 0.2 L_{sc}\mu_s^* \tag{5.79}$$

As a result, if we experimentally measure a value for E for a given L_{sc} within the low-coherence regimes, we can approximate the value of μ_s^*. The linear range of E vs. $L_{sc}\mu_s^*$ can be seen for a wide range of g and m, as shown in Figure 5.26.

In biological tissue with $kl_c \gg 1$, the values of μ_s^* are given as [60]

$$\mu_s \propto \begin{cases} \Delta n^2 k (kl_c)^{3-2m} & m < 1 \\ \Delta n^2 k^2 l_c & m > 1 \end{cases} \tag{5.80}$$

Additionally,

$$1 - g \propto \begin{cases} (kl_c)^0 & m < 1 \\ (kl_c)^{2-2m} & 1 < m < 2 \\ (kl_c)^{-2} & m > 2 \end{cases} \tag{5.81}$$

If we calculate $\mu_s^* = \mu_s(1-g)$ from the above equation, we find the spectral relationship

$$\mu_s^* \propto \lambda^{2m-4} \tag{5.82}$$

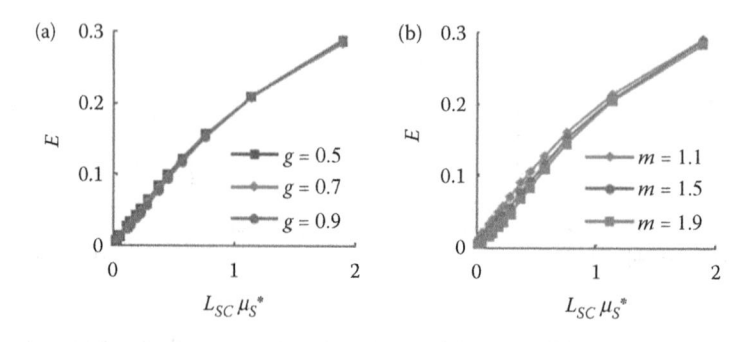

FIGURE 5.26
Enhancement factor E versus dimensionless parameter $L_{sc}\mu_s^*$. (a) E for a range of different g values with a constant $m = 1.5$. (b) E for a range of different m values for $g = 0.9$.

Applying Equation 5.82 into Equation 5.79, we can find a simple relationship between the values of E and m

$$E(\lambda) \propto L_{sc}\mu_s^* = C_1\lambda^{2m-3} \tag{5.83}$$

where C_1 is an unimportant constant. In order to extract m, we can measure the slope of the log–log plot of $E(\lambda)$. Alternatively, this can be expressed as

$$m = \frac{1}{2}\left\{\frac{d[E(\lambda)]}{d\lambda}\frac{\lambda}{E} + 3\right\} \tag{5.84}$$

where λ is the central wavelength of the range being used and E is the average enhancement factor over the entire range. Finally, once the values of μ_s^* and m are known, the value of g can be obtained by comparing the shape of the LEBS peak at different angles with the theoretically expected values. In Reference 43, this was achieved by creating a lookup table for the LEBS peak width. Alternatively, the entire shape of the peak can be fit to reduce the variability due to noise.

Laboratory

Materials

- Xenon lamp
- Aperture wheel (AW)
- Apertures of varying diameter ($\sim$100s to 1000s of microns)
- HeNe laser
- Lenses of varying size (L_1, L_2, L_3, L_4)
- Two linear polarizers (P and A)
- One quarter wave plate (QWP)
- One mirror (M)
- One beamsplitter (B)
- One bandpass filter at He-Ne laser line
- One camera (CCD or CMOS)
- Suspensions of latex microspheres (10% solids by weight)
- Chicken thigh
- Dental floss

The full schematic for the EBS instrument is shown in Figures 5.27 and 5.28. To obtain spatial coherence and partial spatial coherence, two different light sources are needed. These two configurations are shown in Figure 5.27. For partial spatial coherence illumination, a spatially incoherent xenon arc lamp source is needed. This source is first collimated by lens C and focused onto aperture wheel AW with lens L_1. Essentially, this re-images the xenon

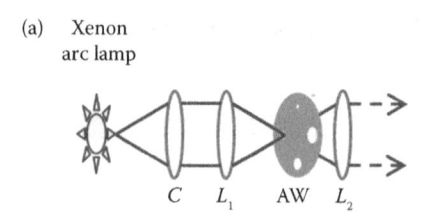

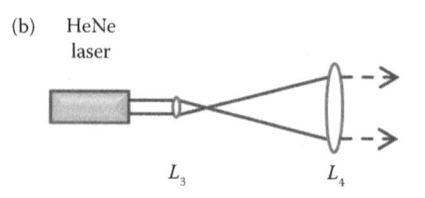

FIGURE 5.27
Illumination sources for EBS instrumentations. (a) Incoherent broadband xenon arc lamp source for partial spatial coherence illumination. Condenser (C), focusing lens (L_1), aperture wheel (AW), and collimating lens (L_2). (b) Helium-neon laser for essentially spatially coherent illumination. Telescope system with lenses L_3 and L_4.

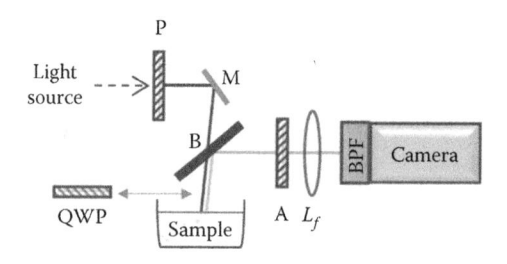

FIGURE 5.28
Main EBS instrument schematic. Linear polarizer (P), mirror (M), beamsplitter (B), analyzer (A), Fourier Lens (L_f), band pass filter (BPF), and quarter-wave plate (QWP).

arc into the plane of AW. The aperture within AW then acts as an incoherent secondary source that gains spatial coherence as it propagates through space until being collimated by lens L_s. The shape of the coherence function can be changed by selecting secondary source apertures of different sizes or L_s with a different focal length according to Equation 5.68. The sizes of the apertures in AW are suggested to be on the order of 100s to 1000s of microns, and L_s is suggested to be 200 mm. The values of C and L_1 should be chosen such that the intensity falling on the secondary source is both maximized and homogeneous over the entire aperture. The intensity distribution within the aperture can be measured by simply imaging it onto a camera.

For spatially coherent illumination, a HeNe laser can be used. Note that although any realistic illumination source will have a finite L_{sc}, we can assume the HeNe laser to be spatially coherent, since its L_{sc} is typically much greater than the beam spot size and/or extent of $P(r)$. Typically, a HeNe laser with have a spot size of < 2 mm, while a spot size of ~ 10 mm is desirable for experiments. Because of this, a telescope system can be created using lenses L_3 and L_4 to expand the beam size.

After the initial light source conditioning, the beam is polarized by linear polarizer P before being directed onto the scattering sample via a mirror M and nonpolarizing 50/50 beamsplitter B. The backscattered light will then be collected by B and pass through a linear analyzer A before being focused onto a charge-coupled device (CCD) camera with a Fourier lens L_f. A bandpass filter (BPF) attached to the CCD separates the backscattered light into its component wavelengths. The BPF wavelength should be chosen so that it is at the laser line of the HeNe laser being used. Different polarization combinations can be measured by rotating P and/or inserting a quarter-wave plate (QWP) between B and the sample. The suggested focal length of L_f is 100 mm with camera pixel resolution of $\sim$20 μm or smaller.

Experimental data collection consists of four CCD camera measurements: (1) scattering sample, (2) ambient background, (3) reflectance standard (Spectralon >98% reflectance, Ocean Optics), and (4) flat field. The flat field is collected by illuminating the sample with essentially spatially incoherent light ($L_{sc} < 10$ μm) directly from the xenon lamp and collecting the linear cross-polarized channel. This measurement configuration ensures that the flat field contains no coherent portion and thus provides a measurement of the isolated incoherent baseline shape.

To process the data, the scattering sample and reflectance standard are first background subtracted. The scattering sample is then normalized by the total unpolarized incoherent intensity measured from the reflectance standard and divided by the flat field that corrects both aberrations in the camera as well as vignetting in the incoherent baseline, which occurs due to rays that exit the scattering medium outside the illumination spot. The EBS peak is obtained by subtracting the incoherent baseline with a plane fit using data from an annular ring that is 3° away from the maximum peak intensity.

The unpolarized incoherent intensity with which we normalize the scattering sample is the incoherent backscattered intensity that would be measured if the illumination and collection were completely unpolarized. The incoherent intensity for a particular polarization channel is measured by sampling an angular region on the CCD that is >3° away from the maximum peak intensity of the Spectralon reflectance standard. The unpolarized incoherent intensity arriving at the sample can then be found based on *a priori* knowledge of the depolarization characteristics of the reflectance standard, as well as the amount of intensity lost due to specular reflection at the sample surface.

Conventionally, the EBS peak for a particular sample is normalized with respect to its own incoherent baseline. However, we choose to normalize by a reflectance standard because it makes measurements from different polarization channels and from samples with different absorption levels and thicknesses directly comparable. In addition, it avoids the common practice of placing an aperture at the sample surface to reject light that has traveled outside of the illumination spot [47].

Exercises

1. Using the Akkermans equation (Equation 5.54), explore the shape of the EBS under the scalar approximation.

   a. What happens to the enhancement factor as l_s^* is increased?

   b. What happens to the width of the EBS peak as l_s^* is increased?

   c. Calculate the angular resolution for a typical experimental setup with $L_f = 100$ mm and a camera pixel size of 20 μm. What potential problems may arise when trying to measure biological tissue with l_s^* on the order of 1 mm?

   d. Use Equation 5.54 to calculate the 2-D EBS peak, and then find $P(x,y)$ according to Equation 5.52. What happens to the shape of $P(x,y)$ as l_s^* is increased? Explain how this agrees with the observations from Part b.

e. Use Equation 5.68 to calculate $C(r)$ for different L_{sc}, and compute the LEBS peak according to Equations 5.69 and 5.70. What happens to the enhancement factor value as a function of L_{sc}? What happens to the enhancement factor value as a function of l_s^*? Can you reconcile these two trends into a single plot? Hint: Try plotting the enhancement factor as a function of some unitless parameter.

f. Use Equation 5.73 to calculate $S(r)$ for different illumination spot sizes, and compute the EBS peak according to Equation 5.74 (with $C(r) = 0$ and $MTF(r) = 0$). What happens to the enhancement factor value as a function of beam diameter? What happens to the enhancement factor value as a function of l_s^*? Can you reconcile these two trends into a single plot?

2. Using a HeNe laser, assemble and align the EBS instrument according to Figures 5.27 and 5.28 for linear co-polarized illumination and detection. Using the stock suspension of latex microspheres at 10% solids by weight, verify that the general shape of the peak agrees with the Akkermans equation (Equation 5.54). Optimize the position of the camera by adjusting it along the optical axis of L_f until the enhancement factor is maximized. Prepare a set of microsphere dilutions with values of l_s^* between 100 and 1500 μm in 200-μm steps.

a. Take measurements on the prepared set of microspheres using a 5-mm spot size, and plot the enhancement factor and width versus l_s^*. Compare these results with the ones calculated from the Akkermans equation, and explain any discrepancies.

b. Take a measurement from the muscular section of a chicken thigh. Using the measurements from Exercise (a) as a calibration curve, estimate the value of l_s^*.

c. Using a suspension with $l_s^* \sim 500$ μm, measure the EBS peak for spot size between 1 and 10 mm. Compare these results with those derived from the Akkermans equation, and explain any discrepancies.

3. Using a solution of polystyrene microspheres with $l_s^* \sim 200$ μm, measure the shape of the EBS peak in the (xx), (xy), $(++)$, and $(+-)$ polarization channels. Explain the measured values for enhancement factor and the azimuthal intensity distribution for each of these channels.

4. Measure the stationary Spectralon reflectance standard, and describe the appearance of the peak. You should observe a pattern called "speckle," which occurs when measuring stationary samples.

a. Does the speckle pattern occur in the suspension of microspheres you previously measured? Why or why not?

b. Now gently shake the reflectance standard, and describe how the appearance of the speckle pattern changes. Building on this observation, qualitatively explain the origin of the speckle pattern.

5. Optimize the system for the $(++)$ polarization channel.

a. Measure a single piece of dental floss, and explain the pattern. Which term in Equation 5.74 is causing the peak to behave in this way?

b. Wrap the dental floss around your finger to a thickness of ~ 1 mm. Take measurements and explain the observed pattern.

c. Randomly jumble all the dental floss from your finger into a single ball. Take measurements and explain the observed pattern.

    d. Crack the bone from the chicken thigh, and measure the inside surface. You should view some degree of anisotropy in the peak. Explain this pattern and describe the underlying structures that cause it.

6. Implement the xenon light source setup in Figure 5.27 with a minimum of four different aperture sizes.

    a. Use Equation 5.68 to calculate the shape of $C(r)$ for the four apertures. What happens to $C(r)$ as the aperture size is increased? What happens to $C(r)$ as the aperture size is decreased?

    b. Place a mirror in the sample plane, and adjust the reflected light so it is in the exact backscattering direction. This can be done by making sure the reflected beam follows along the direction of the incident beam. After this is done, the aperture in AW can be imaged directly onto the camera. (Be careful not to saturate the camera!) Measure the angular intensity distributions from your four apertures, and calculate $C(r)$ according to Equation 5.67. Verify these agree with the calculations in part (a).

    c. Measure the stationary Spectralon reflectance standard for the different values of L_{sc}. What trends do you observe in the speckle pattern as you change L_{sc}? What benefits can you see in using LEBS for characterization of biological tissue? Hint: The extreme case of the spatially coherent HeNe laser speckle pattern in Exercise 4 may provide additional help.

    d. Using a solution of polystyrene microspheres with $l_s^* \sim 200$ μm, plot the observed LEBS enhancement factor versus l_s^*, and compare with the Akkermans equation.

Laboratory 5: Depth-Selective Tissue Spectroscopy: Polarization-Gated Spectroscopy

Diffuse light typically travels several millimeters deep into tissue before being detected. As a result, the diffuse signal is determined by the sample properties averaged over the several-millimeter penetration depth. Many applications, however, require light to selectively probe a sample's superficial (e.g., located under 1 mm below tissue surface) structures. For example, early carcinomas and precancerous (dysplastic) lesions originate in the mucosae (which can be on order of a few hundred microns thick) of various organs such as the lung, breast, prostate, and colon. To achieve depth selectivity, several techniques have been developed that take advantage of the fact that deeper-traveling photons reemerge from tissue at farther radial distances from their point of entry [61]. In other words, at each radial distance r, the radial impulse response function (i.e., the radial intensity distribution of multiply scattered light emerging at the surface of a scattering medium) $p(r)$ samples a distribution of light paths within the medium, the penetration depth of these paths increasing with r. Therefore, a fiber-optic probe designed in such a way that most of the recorded light corresponds to larger r will, for the same tissue, collect light that propagated deeper into tissue than a probe that measures its signal corresponding to smaller r.

    How can a probe be designed to record signals from different r values? Recall that r is distance between a pencil source and a point detector, and there are multiple means to control for the source–detector separation r. Mechanically, fiber-optic probes with small source–detector separations have been designed that collect light emerging only at small radial distances and therefore smaller penetration depths [62,63,64]. Another approach involves an angulated illumination and collection fibers, which also restricts the path lengths of recorded light. Other methods, such as low-coherence enhanced backscattering and polarization gating, enhance sensitivity to photons emerging at small distances by modulating the properties of the incident itself. LEBS limits the spatial coherence of the incident light to control the radial extent of light contributing to the enhanced backscattering peak [65,66]. Polarization gating, on the other hand, takes advantage of the polarization properties of the incident light.

Polarization gating is based on the observation that scattering depolarizes the incident light. On average, the deeper light penetrates into tissue, the more depolarizing scattering events it will undergo. In contrast, light that only penetrates superficially into tissue before being collected will likely have undergone only a few scattering events and will maintain most of its initial polarization state. Consequently, if a sample is illuminated with linearly polarized light, the reflected intensities polarized parallel ($I_{\parallel}$) and perpendicular ($I_{\perp}$) to the incident polarization will sample different penetration depths, with the $I_{\perp}$ signal emanating from a deeper depth than the $I_{\parallel}$ signal. It is also possible to further limit the penetration depth of polarized light by analyzing the differential signal

$$\Delta I = I_{\parallel} - I_{\perp}.$$

This signal probes an even shallower depth than the $I_{\parallel}$ signal. To understand why, consider the two-layer model consisting of a top layer A and a bottom layer B depicted in Figure 5.28. Assume that layer A is thin enough such that any light reflected from it maintains its initial polarization $\left(I_{\parallel}^{A} \gg I_{\perp}^{A}\right)$. Light reflected from layer B will be depolarized due to the number of scattering events it takes to reach it and travel back to the top surface $\left(I_{\parallel}^{B} \approx I_{\perp}^{B}\right)$. The differential signal can then be simplified to $I_{\parallel}^{A} - I_{\perp}^{A}$ and is determined solely by the properties of the top layer A, whereas both $I_{\parallel}$ and $I_{\perp}$ have contributions from layer B. The unique penetration-depth properties of the differential signal can also be understood in terms of the radial intensity distribution of the co-polarized and cross-polarized light reflectance intensities [67]. Figure 5.29 illustrates the radial spread of the ΔI, $I_{\parallel}$, and $I_{\perp}$ intensities. For small radial distances, $I_{\parallel} \gg I_{\perp}$, while for large radial distances, $I_{\parallel} \sim I_{\perp}$. The differential signal ΔI is therefore peaked at small radial distances and approaches zero at large ones. By effectively limiting the radial collection area, polarization gating selects

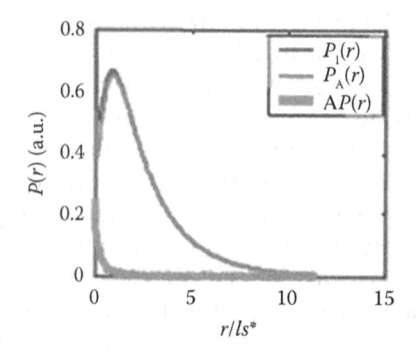

FIGURE 5.29
Radial intensity distributions of the polarization-gated signals.

TABLE 5.2

Functional Forms and Fitting Coefficients of Penetration-Depth Equation Parameters

Parameter	Functional Form	x_1	x_2	x_3	x_4	x_5	x_6
a	$(x_1 + x_2g + x_3\theta_c^2)^{-1}$ $\exp\left(-\left(\dfrac{\mu_a}{\mu_s}\right)^{0.75}\left(x_4 + x_5\sqrt{g} + x_6\theta_c^3\right)\right)$	1.875	−1.881	2.19E-05	4.009	−3.932	8.26E−08
b	$\exp(x_1 + x_2g^3 + x_3\theta_c^{1.5})$ $\exp\left(-\left(\dfrac{\mu_a}{\mu_s}\right)^{0.75}\exp\left(x_4 + \dfrac{x_5}{\ln(g)} + \dfrac{x_6}{\ln(\theta_c)}\right)\right)$	0.602	−2.082	−0.00121	0.281	0.169	−2.89
c	$\left((x_1 + x_2g^2 + x_3\ln\theta_c)\right)$ $\exp\left(-\left(\dfrac{\mu_a}{\mu_s}\right)^{0.75}\exp\left(x_4 + x_5g^2 + x_6\theta_c^3\right)\right)$	2.521	−0.728	−0.195	1.108	−0.000573	9.96E−06

photons that have traversed short penetration depths by the radial extent/penetration depth relationship discussed in [61]. Polarization gating therefore allows depth-selective interrogation of tissue with the signals ΔI, $I_{||}$, and $I_\perp$ probing progressively deeper penetration depths. Experimental studies have demonstrated that the average penetration depth of these signals can be on the order of 100, 150, and 300 μm, respectively, for the three signals [68].

An equation for the average penetration depth of the ΔI signal is

$$\tau = a\left(1 - \exp\left(-b(R\mu_t)^c\right)\right), \tag{5.85}$$

where τ is the product of the physical penetration depth and the total scattering coefficient μ_t, R is the size of the illumination–collection area (assumed to be overlapping), and [a, b, c] are fitting parameters. The functional dependence of [a,b,c] on the optical properties of the medium and on the collection angle (θ_c) is shown in Table 5.2. Equation 5.85 allows one to investigate the sensitivity of the polarization-gated penetration depth to sample optical properties, and it gives insight into how the penetration depth may be tuned by altering the illumination–collection geometry (R, θ_c).

Polarization-gated spectroscopy has found several clinical applications, especially since it can be readily incorporated into a fiber-optic probe for *in vivo* measurements. Figure 5.30 shows an example schematic of a polarization-gated fiber-optic probe. It consists of an illumination fiber and two collection fibers, one of which collects light polarized parallel to the incident light and another that collects light polarized perpendicular to the incident light. Polarization gating has been used to measure the size and the refractive index distribution of epithelial cells with the goal of detecting dysplasia [69–73]. The wavelength dependence of the ΔI signal also can reveal information about the size distribution of submicron intraepithelial structures (macromolecules and organelles) [74]. In particular, the intensity of the polarization-gated signals can be written as

$$I(\lambda) = I_s(\lambda)\exp\left(\mu_a(\lambda)\overline{\langle L\rangle}\right), \tag{5.86}$$

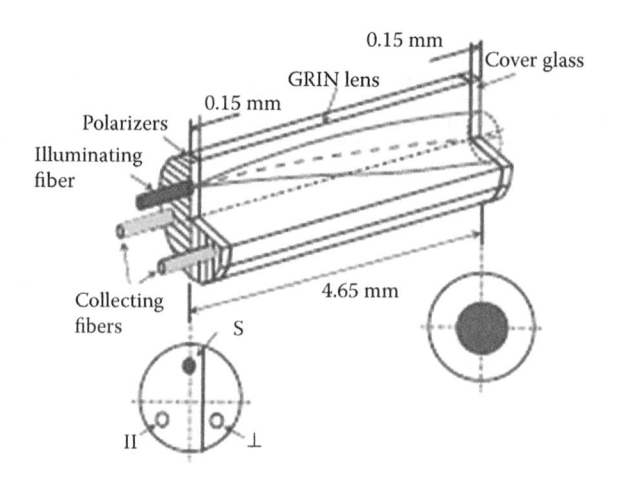

FIGURE 5.30
Example schematic of polarization-gated probe.

where $\overline{\langle L \rangle}$ is the effective pathlength and I_s is the intensity of the medium if it were to be devoid of absorbers. ΔI, $I_{||}$, and $I_\perp$ can be substituted for I in Equation 5.86. I_s is often a function of μ_s' that is proportional to λ^{2m-4}, where m is a parameter related to the shape of the refractive index correlation function [75]. By fitting Equation 5.86 to spectroscopic data from polarization gating, the values of m as well as absorber concentration can be quantified depth-selectively. For, example, polarization gating has been used to selectively probe microvascular blood supply in early carcinogenesis [76–79]. Polarization gating has both allowed for a better understanding of where biological changes occur in disease and helped researchers target their signals to diagnostically relevant depths.

Laboratory

Materials

1. Light source: 75 W Xenon lamp (Newport).
2. Three lenses (Melles Griot) (see Figure 5.31): L1 with focal length = 160 mm and diameter 40 mm, L2 with focal length = 300 mm and diameter = 50 mm, L3 = 31 mm, diameter = 17.5 mm.
3. Two field diaphragms.
4. Mirror (Thorlabs BB1-E02P).
5. Broadband nonpolarizing beamsplitter (Thorlabs BS007).
6. Two dichroic sheet polarizers (Melles Griot).
7. Spectrometer with CCD camera (Acton/Pixis).
8. 100-μm-thick quartz coverslips.
9. 150-μm-thick plastic or metallic shims (McMaster-Carr).

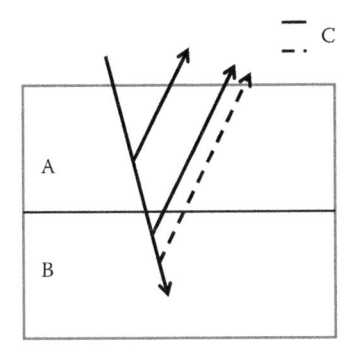

FIGURE 5.31
Two-layer model for light depolarization.

Protocol

1. Two-layer phantom
   a. Make a $\mu_s' = 1.8\,\text{mm}^{-1}$ mixture of Intralipid and deionized water as outlined in the diffuse reflectance section in the lab topic.
   b. Make a 150-μm-thick layer of this phantom by placing it between two 100-μm-thick quartz coverslips separated vertically by 150-μm spacers (plastic or metallic shims) as shown in Figure 5.32a.
   c. Make another >1-mm-thick phantom having $\mu_s' = 1.8\,\text{mm}^{-1}$ and $\mu_a = 1\,\text{mm}^{-1}$ with methylene blue as the absorber. Place this phantom layer below the top layer made in (b) as in Figure 5.32b. These two layers should be separated by a single 150-μm quartz coverslip.

2. Construct the optical setup shown in Figure 5.33. Set the apertures to have 1-mm diameters. Have the sample beam intercept the sample at a 15° angle to avoid the collection of specular reflection.

3. Orient the polarizer in front of the detector (P2) to have the same polarization direction as the incident light polarization (P1). Take a measurement on the two-layer phantom. Rotate the polarizer 90° and take another measurement.

Analysis

1. Plot the reflectance values ΔI, $I_{||}$, and $I_\perp$ as a function of wavelength from 500 to 700 nm.

2. Methylene blue absorbs considerably more at 600 nm versus 500 nm. Calculate the ratio $I_{500}\,\text{nm}/I_{600}\,\text{nm}$ for the ΔI, $I_{||}$, and $I_\perp$ signals.

3. Plot the spectrum of Intralipid's reduced scattering coefficient, and compare the shape with the polarization-gated signals from the two-layer phantom.

(a) (b)

FIGURE 5.32
Two-layer phantom design.

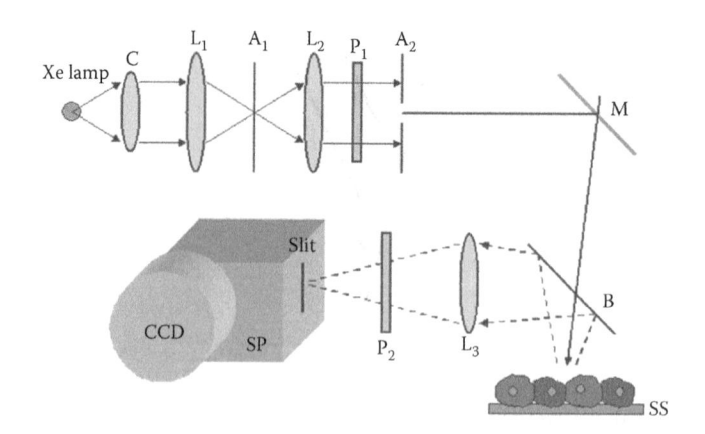

FIGURE 5.33
Polarization-gating setup. (Adapted from V. Turzhitsky et al., *Dis. Markers.* 2008;25:313–321.)

Questions

1. Which polarization-gated signal most closely matches the spectrum of Intralipid? Why would you expect this *a priori*?

2. How does the intensity ratio change between the different polarization-gated signals?

3. Explain how the answers to 1 and 2 relate to the sampling depth of each signal.

Learning Objectives

1. Design and build a two-layer phantom.

2. Learn how to take polarization-gated measurements.

3. Understand how depth selectivity of polarization gating allows one to probe different layers of a sample.

References

1. E. Vitkin, V. Turzhitsky, L. Qiu, L. Guo, I. Itzkan, E. B. Hanlon, and LT. Perelman, Photon diffusion near the point-of-entry in anisotropically scattering turbid media. *Nature Commun.* 2011;2:587.
2. P. Dirac, *Principles of Quantum Mechanics.* Oxford, U.K.: Clarendon Press, 1958.
3. R. A. J. Groenhuis, H. A. Ferwerda, and J. J. Tenbosch, Scattering and absorption of turbid materials determined from reflection measurements. 1: Theory. *Appl. Opt.* 1983;22:2456.
4. J. H. Joseph, W. J. Wiscombe, and J. A. Weinman, The delta-Eddington approximation for radiative flux transfer. *J. Atmos. Sci.* 1976;33:2452.
5. L. V. Wang and H. I. Wu, *Biomedical Optics: Principles and Imaging.* New Jersey: Wiley, 2007.
6. T. J. Farrell, M. S. Patterson, and B. Wilson, A diffusion theory model of spatially resolved, steady-state diffuse reflectance for the noninvasive determination of tissue optical properties *in vivo. Med. Phys.* 1992;19:879.
7. L. G. Henyey and J. L. Greenstein, Diffuse radiation in the galaxy. *Astrophys. J.* 1941;93:70.
8. D. Toublanc, Henyey-Greenstein and Mie phase functions in Monte Carlo radiative transfer computations. *Appl. Opt.* 1996;35:3270.

9. A. Ishimaru, *Wave Propagation and Scattering in Random Media*. IEEE Press, 1997.

10. M. Born and E. Wolf, *Principles of Optics*, 7th ed. Cambridge U. Press, 1999.

11. P. Guttorp and T. Gneiting, *National Research Center for Statistics and the Environment*, Technical Report Series, 2005.

12. B. Mandelbrot, *The Fractal Geometry of Nature*. W.H. Freeman and Company, 1982.

13. C. J. R. Sheppard, Effect of an annular pupil on confocal imaging through highly scattering media. *Opt. Lett.* 1996;21:1310.

14. J. D. Rogers, I. R. Capoglu, and V. Backman, Nonscalar elastic light scattering from continuous random media in the Born approximation. *Opt. Lett.* 2009;34:1891.

15. I. R. Capoglu, J. D. Rogers, A. Taflove, and V. Backman, Accuracy of the Born approximation in calculating the scattering coefficient of biological continuous random media. *Opt. Lett.* 2009;34:2679.

16. B. W. Pogue and M. S. Patterson, Review of tissue simulating phantoms for optical spectroscopy, imaging and dosimetry. *J. Biomed. Opt.* 2006;11:041102.

17. G. Zonios, L. T. Perelman, V. Backman, R. Manoharan, M. Fitzmaurice, J. Van Dam, and M. S. Feld, Diffuse reflectance spectroscopy of human adenomatous colon polyps *in vivo*. *Appl. Opt.* 1999;38:6628–6637.

18. D. Huang, E. A. Swanson, C. P. Lin, J. S. Schuman, W. G. Stinson, W. Chang et al. Optical coherence tomography. *Science* 1991;254(5035):1178–1181.

19. J. M. Schmitt, A. Knuttel, M. Yadlowsky, and M. A. Eckhaus, Optical-coherence tomography of a dense tissue: Statistics of attenuation and backscattering. *Phys. Med. Biol.* 1994;39:1705.

20. N. Bosschaart, D. J. Faber, T. G. van Leeuwen, and M. C. G. Aalders, Measurements of wavelength dependent scattering and backscattering coefficients by low-coherence spectroscopy. *J. Biomed. Opt.* 2011;16:030503.

21. R. Barer and S. Tkaczyk. Refractive index of concentrated protein solutions. *Nature*. 1954;173:821–822.

22. J. M. Schmitt and G. Kumar, Turbulent nature of refractive-index variations in biological tissue. *Opt. Lett.* 1996;21:1310–1312.

23. M. Moscoso, J. B. Keller, and G. Papanicolaou, Depolarization and blurring of optical images by biological tissue. *J. Opt. Soc. Am.* 2001;A18:948–960.

24. M. Xu and R. R. Alfano, Fractal mechanisms of light scattering in biological tissue and cells. *Opt. Lett.* 2005;30:3051–3053.

25. C. J. R. Sheppard, Fractal model of light scattering in biological tissue and cells. *Opt. Lett.* 2007;32:142–144.

26. P. Guttorp and T. Gneiting, On the Whittle-Matérn correlation family. Technical Report no. NRCSE-TRS No. 080. 2005.

27. Y. Kuga and A. Ishimaru, Retroreflectance from a dense distribution of spherical-particles. *J. Opt. Soc. Am. A* 1984;1:831–835.

28. M. P. Van Albada and A. Lagendijk, Observation of weak localization of light in a random medium. *Phys. Rev. Lett.* 1985;55:2692–2695.

29. P. E. Wolf and G. Maret, Weak localization and coherent backscattering of photons in disordered media. *Phys. Rev. Lett.* 1985;55:2696–2699.

30. E. Akkermans, P. E. Wolf, R. Maynard, and G. Maret, Theoretical-study of the coherent backscattering of light by disordered media, *J Phys.* 1988;49:77–98.

31. K. Ishii, T. Iwai, and T. Asakura, Polarization properties of the enhanced backscattering of light from the fractal aggregate of particles. *Opt. Rev.* 1997;4:643–647.

32. A. Dogariu, J. Uozumi, and T. Asakura, Enhancement factor in the light backscattered by fractal aggregated media. *Opt. Rev.* 1996;3:71–82.

33. D. S. Wiersma, M. P. Van Albada, and A. Lagendijk, Coherent backscattering of light from amplifying random-media. *Phys. Rev. Lett.* 1995;75:1739–1742.

34. G. Labeyrie, F. de Tomasi, J. C. Bernard, C. A. Muller, C. Miniatura, and R. Kaiser, Coherent backscattering of light by cold atoms. *Phys. Rev. Lett.* 1999;83:5266–5269.

35. H. K. M. Vithana, L. Asfaw, and D. L. Johnson, Coherent backscattering of light in a nematic liquid-crystal., *Phys. Rev. Lett.* 1993;70:3561–3564.
36. R. Sapienza, S. Mujumdar, C. Cheung, A. G. Yodh, and D. Wiersma, Anisotropic weak localization of light. *Phys. Rev. Lett.* 2004;92(3):033903.
37. K. M. Yoo, G. C. Tang, and R. R. Alfano, Coherent backscattering of light from biological tissues. *Appl. Opt.* 1990;29:3237–3239.
38. M. H. Eddowes, T. N. Mills, and D. T. Delpy, Monte Carlo simulations of coherent backscatter for identification of the optical coefficients of biological tissues *in-vivo*. *Appl. Opt.* 1995;34:2261–2267.
39. G. Yoon, D. N. G. Roy, and R. C. Straight, Coherent backscattering in biological media—Measurement and estimation of optical-properties. *Appl. Opt.* 1993;32:580–585.
40. K. M. Yoo, F. Liu, and R. R. Alfano, Biological-materials probed by the temporal and angular profiles of the backscattered ultrafast laser-pulses. *J. Opt. Soc. Am. B* 1990;7:1685–1693.
41. A. J. Radosevich, J. D. Rogers, V. M. Turzhitsky, N. N. Mutyal, J. Yi, H. K. Roy, and V. Backman, Polarized enhanced backscattering spectroscopy for characterization of biological tissues at subdiffusion length-scales. *IEEE J. Sel. Top. Quantum Electron.* 2012;18(4):1313–1325.
42. A. J. Radosevich, V. M. Turzhitsky, N. N. Mutyal, J. D. Rogers, V. Stoyneva, A. K. Tiwari, M. De La Cruz, D. P. Kunte, R. K. Wali, H. K. Roy, and V. Backman, Depth-resolved measurement of mucosal microvascular blood content using low-coherence enhanced backscattering spectroscopy. *Biomed. Opt. Express* 2010;1:1196–1208.
43. V. Turzhitsky, A. J. Radosevich, J. D. Rogers, N. N. Mutyal, and V. Backman, Measurement of optical scattering properties with low-coherence enhanced backscattering spectroscopy. *J. Biomed. Opt.* 2011;16(6):067007.
44. H. K. Roy, V. Turzhitsky, Y. Kim, M. J. Goldberg, P. Watson, J. D. Rogers, A. J. Gomes et al. Association between rectal optical signatures and colonic neoplasia: Potential applications for screening. *Cancer Res.* 2009;69:4476–4483.
45. Y. Liu, R. E. Brand, V. Turzhitsky, Y. L. Kim, H. K. Roy, N. Hasabou, C. Sturgis, D. Shah, C. Hall, and V. Backman, Optical markers in duodenal mucosa predict the presence of pancreatic cancer. *Clin. Cancer Res.* 2007;13:4392–4399.
46. V. Turzhitsky, Y. Liu, N. Hasabou, M. Goldberg, H. K. Roy, V. Backman, and R. Brand, Investigating population risk factors of pancreatic cancer by evaluation of optical markers in the duodenal mucosa. *Dis. Markers* 2008;25:313–321.
47. R. Lenke and G. Maret, Multiple scattering of light: Coherent backscattering and transmission. In W. Brown and K. Mortensen, eds., *Scattering in Polymeric and Colloidal Systems*. Amsterdam: Gordon & Breach, 2000, pp. 1–71. ISBN 90-5699-260-0.
48. E. Akkermans, P. E. Wolf, and R. Maynard, Coherent backscattering of light by disordered media—Analysis of the peak line-shape. *Phys. Rev. Lett.* 198656:1471–1474.
49. E. Hecht, *Optics*, 4th ed. Reading, MA: Addison-Wesley, 2002.
50. J. C. Ramella-Roman, S. A. Prahl, and S. L. Jacques, Three Monte Carlo programs of polarized light transport into scattering media: Part I. *Opt. Exp.* 2005;13:4420–4438.
51. C. F. Bohren and D. R. Huffman, *Absorption and Scattering of Light by Small Particles*. New York, NY: Wiley-VCH, 1998.
52. H. C. van de Hulst, *Light Scattering by Small Particles*. New York, NY: Wiley, 1957.
53. A. S. Martinez and R. Maynard, Faraday-effect and multiple-scattering of light. *Phys. Rev. B* 1994;50:3714–3732.
54. R. Lenke and G. Maret, Magnetic field effects on coherent backscattering of light. *Eur. Phys. J. B* 2000;17:171–185.
55. M. Born and E. Wolf, *Principles of Optics*, 7th (expanded) ed. Cambridge, U.K.: Cambridge University Press, 1999.
56. V. Turzhitsky, J. D. Rogers, N. N. Mutyal, H. K. Roy, and V. Backman, Characterization of light transport in scattering media at subdiffusion length scales with low-coherence enhanced backscattering. *IEEE J. Sel. Top. Quantum Electron.* 2010;16:619–626.

57. L. H. Wang, S. L. Jacques, and L. Q. Zheng, CONV—Convolution for responses to a finite diameter photon beam incident on multi-layered tissues. *Comput. Methods and Programs in Biomed.* 1997;54:141–150.

58. V. Turzhitsky, N. N. Mutyal, A. J. Radosevich, and V. Backman, Multiple scattering model for the penetration depth of low-coherence enhanced backscattering. *J. Biomed. Opt.* 2011;16:097006.

59. V. Turzhitsky, A. J. Radosevich, J. D. Rogers, N. N. Mutyal, and V. Backman, Measurement of optical scattering properties with low-coherence enhanced backscattering spectroscopy. *J. Biomed. Opt.* 2011;16:067007.

60. J. D. Rogers, I. R. Capoglu, and V. Backman, Nonscalar elastic light scattering from continuous random media in the Born approximation. *Opt. Lett.* 15 2009;34:1891–1893 .

61. V. M. Turzhitsky, A. J. Gomes, Y. L. Kim, Y. Liu, A. Kromine, J. D. Rogers, M. Jameel, H. K. Roy, and V. Backman, Measuring mucosal blood supply *in vivo* with a polarization-gating probe. *Appl. Opt.* 2008;47:6046–6057.

62. V. Backman, R. Gurjar, K. Badizadegan, I. Itzkan, R. R. Dasari, L. T. Perelman, and M. S. Feld, Polarized light scattering spectroscopy for quantitative measurement of epithelial cellular structures *in situ*. *IEEE J Select Top Quant Electron*. 1999;5:1019–1026.

63. R. S. Gurjar, V. Backman, L. T. Perelman, I. Georgakoudi, K. Badizadegan, I. Itzkan, R. R. Dasari, and M. S. Feld, Imaging human epithelial properties with polarized light-scattering spectroscopy. *Nat. Med.* 2001;7:1245–1248.

64. J. R. Mourant, T. M. Johnson, and J. P. Freyer, Characterizing mammalian cells and cell phantoms by polarized backscattering fiber-optic measurements. *Appl. Opt.* 2001;40:5114–5123.

65. A. Myakov, L. Nieman, L. Wicky, U. Utzinger, R. Richards-Kortum, and K. Sokolov, Fiber optic probe for polarized reflectance spectroscopy *in vivo*: Design and performance. *J. Biomed. Opt.* 2002;7:388–397.

66. L. Qiu, D. K. Pleskow, R. Chuttani, E. Vitkin, J. Leyden, N. Ozden, S. Itani et al. Multispectral scanning during endoscopy guides biopsy of dysplasia in Barrett's esophagus. *Nat. Med.* 2010;16:603–606, 1p following 606.

67. H. K. Roy, Y. Liu, R. K. Wali, Y. L. Kim, A. K. Kromine, M. J. Goldberg, and V. Backman, Four-dimensional elastic light-scattering fingerprints as preneoplastic markers in the rat model of colon carcinogenesis. *Gastroenterology* 2004;126:1071–1081; discussion 1948.

68. A. J. Gomes, H. K. Roy, V. Turzhitsky, Y. Kim, J. D. Rogers, S. Ruderman, V. Stoyneva et al. Rectal mucosal microvascular blood supply increase is associated with colonic neoplasia. *Clin. Cancer Res.* 2009;15:3110–3117.

69. H. K. Roy, A. Gomes, V. Turzhitsky, M. J. Goldberg, J. Rogers, S. Ruderman, K. L. Young et al. Spectroscopic microvascular blood detection from the endoscopically normal colonic mucosa: Biomarker for neoplasia risk. *Gastroenterology* 2008;135:1069–1078.

70. H. K. Roy, A. J. Gomes, S. Ruderman, L. K. Bianchi, M. J. Goldberg, V. Stoyneva, J. D. Rogers et al. Optical measurement of rectal microvasculature as an adjunct to flexible sigmoidoscopy: Gender-specific implications. *Cancer Prev. Res. (Phila)* 2010;3:844–851.

71. M. P. Siegel, Y. L. Kim, H. K. Roy, R. K. Wali, and V. Backman, Assessment of blood supply in superficial tissue by polarization-gated elastic light-scattering spectroscopy. *Appl. Opt.* 2006;45:335–342.

72. G. H. Weiss, R. Nossal, and R. F. Bonner, Statistics of penetration depth of photons re-emitted from irradiated tissue. *J. Mod. Opt.* 1989;36:349–359.

73. A. Amelink and H. J. Sterenborg, Measurement of the local optical properties of turbid media by differential path-length spectroscopy. *Appl. Opt.* 2004;43:3048–3054.

74. S. C. Kanick, H. J. Sterenborg, and A. Amelink, Empirical model of the photon path length for a single fiber reflectance spectroscopy device. *Opt. Express* 2009;17:860–871.

75. R. Reif, M. S. Amorosino, K. W. Calabro, O. A'Amar, S. K. Singh, and I. J. Bigio, Analysis of changes in reflectance measurements on biological tissues subjected to different probe pressures. *J. Biomed. Opt.* 2008;13:010502.

76. H. Subramanian, P. Pradhan, Y. L. Kim, and V. Backman, Penetration depth of low-coherence enhanced backscattered light in subdiffusion regime. *Phys. Rev. E Stat. Nonlin. Soft Matter Phys.* 2007;75:041914.
77. V. Turzhitsky, N. N. Mutyal, A. J. Radosevich, and V. Backman, Multiple scattering model for the penetration depth of low-coherence enhanced backscattering. *J. Biomed. Opt.* 2011;16:097006.
78. Y. Liu, Y. Kim, X. Li, and V. Backman, Investigation of depth selectivity of polarization gating for tissue characterization. *Opt. Express* 2005;13:601–611.
79. Y. L. Kim, L. Yang, R. K. Wali, H. K. Roy, M. J. Goldberg, A. K. Kromin, C. Kun, and V. Backman, Simultaneous measurement of angular and spectral properties of light scattering for characterization of tissue microarchitecture and its alteration in early precancer. *IEEE J Select Top Quant Electron* 2003;9:243–256.

6

Computational Biophotonics

The previous chapters dealt with a wide range of experimental techniques that are used in biophotonics research. In only a few applications, however, are these experimental methods sufficient without appropriately chosen analysis techniques. For instance, consider fluorescence spectroscopy that acquires a spectrum of endogenous fluorescence in tissue. This spectrum depends on the spatial distribution of fluorophores in tissue as well as the transport of both excitation and emission light through the tissue. Unless one can model the outcome of the experiment, either analytically or numerically, it would be essentially impossible to understand the origins of fluorescence and to characterize the fluorophores—the technique would only be able to generate purely empirical data with no clear connection to tissue properties.

As a second example, let us consider acquisition of diffuse reflectance signal from tissue. What does it tell us? What kind of structural or morphological properties of tissue can we infer from the signal? Again, we would never know unless we were able to accurately model light diffusion through tissue. In yet another example, we recall that most macroscopic imaging techniques discussed in this chapter (e.g., diffuse optical tomography [DOT] and photoacoustic tomography [PAT]) are based and, in fact, are completely dependent on the reconstruction of tissue absorption or fluorescence properties in three dimensions based on recorded signals.

In all the previously mentioned applications, as well as a great number of other applications, it is crucial to be able to model light interaction with tissue. Ideally, we would have an analytical theory capable of describing these processes, including light scattering, light transport, fluorescence, and inelastic interactions. Given the complexity of biological tissue, this is not a plausible option. Indeed, analytical solutions of light–matter interactions are essentially restricted to fairly simple systems with well-defined geometries or statistical properties. Tissue is much more complex than that. Even if a fully analytical theory for a particular type of light–tissue interaction were developed, its validation would be at best difficult if one relied on experimental studies only, while being able to compare the predictions of an analytical model with a numerical simulation would be highly advantageous.

A solution is frequently found in computational (numerical) modeling. Thus, it is imperative for a biomedical optics student to be proficient in not only experimental but also computational techniques. This chapter provides an introduction to the most popular computational techniques that are frequently used to address biomedical optics problems.

Overview of Computational Biophotonics Methods

Just as biological tissue is immensely complex with various tissue components spanning length scales from nanometers to centimeters, computational biophotonics techniques have been developed to describe light interaction at different levels of tissue complexity.

Ideally, of course, one would be able to model light–tissue interaction from the first principles with no simplifying assumptions. This would require including treatment of light–tissue interactions from nanometer-to centimeter-length scales and all elastic and inelastic phenomena. This is far beyond contemporary computational capabilities. Therefore, different computational techniques were designed to address specific light–tissue interactions for a certain range of length scales and approximations. In this section, we summarize the most frequently used techniques. In the following section, several exercises are presented. Each of these exercises is focused on a particular technique, and each lab presents a detailed description of the technique.

Most computational biophotonics methods can be conveniently categorized based on three aspects: (i) microscopic versus macroscopic interactions, (ii) field-based versus intensity-based, and (iii) length scales of details.

The main objective of microscopic computational techniques is to model light interaction—typically elastic scattering and absorption—with tissue structures with sizes comparable to the wavelength of light. For example, modeling of single-light scattering by a biological cell or light absorption by a red blood cell would fall in to the domain of microscopic techniques. Macroscopic methods, on the other hand, deal with multiple-light scattering and *light transport* in tissue over length scales from hundreds of microns to, in principle, centimeters. As an example, modeling of light diffusion in breast tissue for the purposes of DOT imaging of the three-dimensional distribution of blood and fat would be considered a microscopic method.

Considering that tissue is a medium with a spatially continuously fluctuating refractive index—as opposed to a suspension of isolated discrete particles—the distinction between single and multiple scattering is, of course, an approximation. However, it is a very convenient approximation that does not distract from being able to accurately model most light transport phenomena. In this concept, microscopic methods can be used to model single scattering, with their output becoming an input into macroscopic techniques. Thus, in literature we frequently see a distinction between light scattering (referring to single scattering) and light transport (referring to multiple scattering). The latter is sometimes incorrectly equated with light diffusion. Diffusion (see Chapter 5) is certainly a multiple-scattering phenomenon, but it does not describe the entire light transport process, and, technically, diffusion is possible only for a weakly absorbing medium where one can ignore short light path lengths.

Most rigorous computational methods treat light as an electromagnetic wave and model light interaction with tissue through electromagnetic fields. These methods are used when field-based phenomena such as interference and coherence are of interest. Intensity-based methods are, by definition, approximations. They ignore the wave nature of light and describe the propagation of light intensity. The intensity-based techniques can be further subdivided into vectorial and scalar depending on whether they keep track of polarization of light.

The basis for intensity-based computations is the equation of transport discussed in Chapter 5. As its name implies, this equation, whether in its scalar or vectorial form, treats light propagation as a process of energy transport due to scattering. Single scattering is described through its differential cross-section and absorption by an absorption coefficient. Terms that describe light emission as a result of fluorescence or inelastic scattering can be added. Single-scattering properties can be modeled by use of more accurate field-based techniques. The main reason intensity-based methods are used to address a myriad of light transport problems is because they are less computationally intense compared to field-based approaches. Although many intensity-based methods are designed to model macroscopic multiple-scattering processes, these two aspects should not be equated; some macroscopic computational techniques are actually field based.

The third classification aspect deals with the size of structural features of tissue a computational method is able to take into account. As discussed in Chapter 4, elastic scattering may depend on tissue components as small as tens of nanometers. Inelastic interactions occur predominately on molecular (nano) scales. Modeling of light scattering with this level of detail and in three dimensions is, although rigorous, highly computationally complex in terms of memory and time required given existing computational resources.

As a rule, the larger the overall length scale of the problem, the coarser the level of detail that are considered by a simulation. For example, while finite-difference time-domain method can be and is widely used to model single scattering with the spatial distribution of refractive index being modeled with nanoscale precision, in most cases, its computational complexity allows FDTD to be used with microscopic objects only. The pseudospectral time-domain method, on the other hand, can model light transport with macroscopic volumes of tissue (e.g., hundreds of microns), but at the cost of losing nanoscale sensitivity of FDTD; PSTD models spatial distribution of refractive index with details on the order of half wavelengths, that is, hundreds of nanometers.

With this classification in mind, and for each practical application, one can choose an appropriate computational method. We now briefly overview some of the most frequently used techniques. Mie theory, discussed in Chapter 4, and its extensions might be considered a computational method. Mie theory solves Maxwell's equations for light scattering and absorption by a uniform spherical particle of arbitrary size. The input parameters are the real and imaginary parts of the refractive index of a particle, the refractive index of the surrounding medium, the wavelength of light, and the radius of the particle. Although precise, Mie theory provides a solution in the form of an infinite series, which can only be evaluated numerically. This is a microscopic field-based technique.

Well-tested Mie theory codes are widely available for free download online. Mie theory has been widely used in tissue optics to model light scattering in tissue. For example, Mie theory can be used to estimate the size distributions of tissue scatterers based on angular or spectral-scattering patterns. Because realistic scattering particles in tissue are, generally, neither spherical nor homogeneous and are not located in the far field of each other (Mie theory is, strictly speaking, valid only for isolated scatterers located in the far field of each other), strictly speaking, this kind of analysis can be used only for qualitative assessment.

Exact solutions of Maxwell's equations for light scattering and absorption by particles of other specific shapes have been developed, too. Examples include spheroids, cubes, and cylinders. The solutions are in the form of infinite series. Codes are available online. Some refer to these algorithms as "Mie theory for spheroids." Although convenient, this terminology is incorrect, as Mie theory applies to spheres and spheres only. An extension of Mie theory has also been developed for multilayered spheres consisting of an arbitrary number of concentric layers with different dielectric properties. It must be pointed out that many of these codes have poor convergence, and when implemented, the user has to check whether the obtained solution is well convergent.

T-matrix is another popular computational method of modeling light scattering by nonspherical particles [1]. The term *T-matrix* stands for transition matrix. The method finds an asymptotically exact solution to Maxwell's equations for light scattering by, in principle, an arbitrary complex scattering particle. For example, in biomedical optics, the T-matrix method has been used to model light scattering from cellular organelles [2].

At the core of this formalism, the incident and scattered fields are expressed as superpositions of discrete basis sets of functions $\psi_n^{(i)}$ and $\psi_n^{(s)}$, respectively, with each of these functions being a solution of the Helmholtz equation $E_i = \sum_n^\infty a_n \psi_n^{(i)}$ and $E_s = \sum_n^\infty b_n \psi_n^{(s)}$. A T-matrix $\mathbf{T}$ relates vectors $\mathbf{a}$ and $\mathbf{b}$: $\mathbf{b} = \mathbf{Ta}$. The elements of the T-matrix are numbers rather than functions of angles, as would be the case with Muller matrices or scattering matrices. Conveniently, the T-matrix elements are independent of the direction of propagation and polarization of the incident light.

The key feature of the T-matrix method is that it separates all properties of the incident field and those of the scattering particle. In other words, the scattering properties of the object are completely determined by the T-matrix, while the incident field is fully described by vector $\mathbf{a}$. Therefore, one needs to calculate the T-matrix for a given scattering particle only once. After that, the matrix can be used to predict the scattered field for any configuration of the illumination as long as illumination vector $\mathbf{a}$ is known.

The T-matrix method can be compared to a vessel that needs to be filled; that is, the elements of the matrix must be calculated by one means or the other. When first introduced by Waterman, the elements of the T-matrix were calculated by using the so-called extended boundary condition method (EBCM), also known as the null-field method. Since then, some have equated T-matrix formalism with EBCM. Although EBCM is still the most popular approach used to calculate the T-matrices, it is by no means the only one; thus, EBCM should not be equated with the T-matrix method. It is more correct to talk about the T-matrix formalism—as a way to completely separate the properties of illumination and those of a scattering particle—and a method that is used to calculate the T-matrix itself. Besides EBCM, other computational approaches to calculating the T-matrix elements may include essentially any microscopic field-based simulations such as discrete dipole approximation (DDA) and the finite-difference time-domain simulations. Both DDA and FDTD are discussed later in this section, with laboratory exercises included in the following section.

A powerful extension of Mie theory is generalized multiparticle Mie (GMM) theory [3]. The theory provides an exact solution of Maxwell's equations for an ensemble of an arbitrary number of uniform spheres of arbitrary sizes. The spheres do not have to have the same size. They do not have to be in the far field of each other. For example, touching shears can be modeled. Because Maxwell's equations are solved, near-field interactions are accounted for. GMM provides a good compromise between the level of detail, the size of a problem, and the computational complexity, as long as refractive index distribution in tissue is modeled is a superposition of spheres. Fairly complex tissue structures can be modeled by GMM. A computer workstation is typically able to handle GMM simulations involving thousands of spheres approximating tissue within a 3-D volume on the order of $10^6 \, \mu m^3$. GMM is an example of a macroscopic field-based technique.

In the GMM theory, the internal electric fields of each sphere in the primary coordinate system are given by

$$\mathbf{E}_{int}(j,j_0) = -\sum_{n=1}^{\infty}\sum_{m=-n}^{n} i\mathbf{E}_{mn}\left[d_{mn}^{jj_0}\mathbf{N}_{mn}^{(1)} + c_{mn}^{jj_0}\mathbf{M}_{mn}^{(1)}\right],$$

where $\mathbf{M}_{mn}^{(1)}$ and $\mathbf{N}_{mn}^{(1)}$ are the vector spherical wave functions (the superscripts denote the kind of spherical Bessel function), $E_{mn} = i^n((2n+1)(n-m)!/(n+m)!)$, $c_{mn}^{jj_0}$ and $d_{mn}^{jj_0}$ are the internal coefficients, and (j,j_0) represents the internal field of the jth sphere expanded in the primary j_0th coordinate system. The internal intensity of each sphere is defined by $I(j,j_0) = |\mathbf{E}_{int}(j,j_0)|^2$.

A more rigorous representation of tissue is through continuous refractive index variations. Several microscopic field-based techniques have been designed to solve light-scattering problems where a high level of detail of refractive index distribution is taken into account. The starting point is typically not Maxwell's equations but the integral equation representation of the scattering problem. As discussed in Chapter 4, any scattering problem can be expressed as an integral equation. The scattering amplitude **f** can be expressed through the field **E** that exists inside a scattering region of volume V [4]

$$\mathbf{f}(\mathbf{s},\mathbf{s}_0) = -\frac{k^2}{4\pi} \int_V \mathbf{s} \times (\mathbf{s} \times \mathbf{E}(\mathbf{r}'))\left(n_1^2(\mathbf{r}')-1\right)e^{-ik\mathbf{s}\mathbf{r}'}d\mathbf{r}', \tag{6.1}$$

where **s** and $\mathbf{s}_0$ are the directions of propagation of the scattered and incident waves, k is the wave number in the surrounding medium ($k = k_{vacuum}n_0$), $n_1 = n/n_0$ is the relative refractive index within the scattering volume V, and the integration is taken over the volume of the particle. The scattered field can then be found as $\mathbf{E}_s(\mathbf{r}) = \mathbf{f}(\mathbf{s},\mathbf{s}_0)(e^{ikr}/r)$.

The physical interpretation of Equation 6.1 is that the field that exists inside a scattering potential as a result of the incident field as well as all scattered waves creates oscillating electric polarization vectors at each point in the scattering volume. This, in turn, generates scattered spherical waves originating from each point. All these waves interfere in the far field at the point of observation.

Of course, the field inside the scattering potential is not *a priori* known. Thus, various approximations, analytical and numerical, can be constructed by substituting field **E** by a defined function. In the simplest case—and "simplest" is a relative term—the field inside the scattering potential is approximated as the incident field. This is the first-order or single-scattering Born approximation. It is valid for weakly scattering media, including most biological tissue.

By definition, single-scattering approximation cannot model multiple scattering. To model light propagation in tissue, multiple scattering must be considered. There are two approaches to do so by using the Born approximation. First, for any given spatial distribution of refractive index, the integral in Equation 6.1 can be taken numerically or, in some cases, analytically. Thus, found differential cross-section can then be used in other macroscopic simulations that require differential cross-section as input information, as discussed later.

Another approach involves calculating higher-order terms in the Born approximation expansion. The first-order scattered field that exists with scattering potential can be added up to the incident wave and the resulting field substituted back into the integral instead of the incident field only. This is the *second-order Born approximation*. The iterative process can be continued for higher-order terms until convergence is achieved.

Computationally, a solution can be found by approximating a medium (or a scattering object) as a collection of dipoles, each interacting with incident and scattered waves. This is the *discrete dipole approximation* method. The first-order Born approximation is equivalent to its simpler version, uncoupled-dipole approximation, where the effect of scattered waves on the dipoles is neglected. A numerical implementation of the higher-order Born approximation is the *coupled-dipole approximation* (CDA), in which the scattered waves are traced and their contribution to the total field inside the particle is determined. CDA is a field-based numerical method that can easily handle microscopic dimensions and may, in principle, be applied to macroscopic problems, although practically it is more appropriate to talk about CDA as a mesoscopic technique. Typically, it is difficult to use CDA to model light propagation over millimeter-length scales.

The best-studied intensity-based macroscopic computational technique is the Monte Carlo method. This is a numerical equivalent of the radiative transport equation [5]. In its simplest form, Monte Carlo tracks only one property of light, its intensity, while polarization, phase, and other characteristics are neglected. Just as the transport equation is constructed based on an *a priori* known differential cross-section of single scattering and the absorption cross-section, Monte Carlo method uses the same single-scattering characteristics (typically in the form of the scattering and absorption coefficients and the phase function) as input. Light propagation is then modeled as a random process in which the direction of propagation of each photon after each scattering event is based on the phase function. Light intensity as a function of space is then obtained after billions or sometimes a greater number of photons are required to arrive at a convergent solution.

More complex versions of Monte Carlo track not only intensity but also polarization of light by means of Muller matrix, where the fields are described by their Stokes vectors. The Monte Carlo method can be easily adapted to model not only elastic processes and absorption but also fluorescence, inelastic scattering, etc. In these cases, new photons with a different wavelength are randomly generated throughout tissue. Vectorial Monte Carlo simulations can also be applied to model birefringent media.

Although most simulations are run for spatially statistically homogeneous media, the method can also be used with spatially heterogeneous media where the optical properties are allowed to vary depending on location [6]. Monte Carlo simulations have been a staple of understanding the reflectance spectroscopy of tissue [7]. In these applications, the goal is to noninvasively measure optical properties of tissue via the analysis of light reflected from the tissue surface by means of the process of multiple scattering—the process sometimes referred to as the diffuse reflectance. (We point out that the term "diffuse," although widely used, is generally not entirely correct, since a measurable component of the reflectance is not due to the diffusion but rather subdiffusive single and multiple scattering events that correspond to path lengths shorter than the transport mean free path length.) Another important application of Monte Carlo is that inverse calculations are used in diffuse optical tomography where an experimentally recorded distribution of light emerging from tissue is used and input into the simulations, and the endpoint is the spatial distribution of optical properties of tissue such as the absorption coefficient.

Not all Monte Carlo simulations are intensity based. A so-called partial-photon electric field Monte Carlo method was developed to model enhanced backscattering in 1980s. Another version of the field Monte Carlo was introduced in 2009 [8], which combined fully vectorial diffraction theory with electric field Monte Carlo simulations and enabled one to determine the amplitude and the phase of focused fields in turbid media. In field-based Monte Carlo, light is described as a propagating wave front that is characterized by tracking the direction of the wave vector k, its location within a medium, and the path length traveled between successive scattering events. Field-based Monte Carlo techniques have been applied to solve a few specialized problems only, and it is the intensity-based simulations that rule the field. However, it is expected that these advanced Monte Carlo techniques will continue finding more and more applications in the field of tissue optics, in particular when coherence properties of light are critical, such as optical coherence tomography, interferometry, and enhanced backscattering.

How does one decide which single-scattering characteristics to use as the input into Monte Carlo simulations? These can be obtained from microscopic simulations, such as the first-order Born approximation, DDA, or, in a more rigorous way, from finite-difference time-domain simulations. FDTD is currently the most robust and rigorous field-based method of computational electromagnetics. FDTD enables modeling of light–tissue interactions with an essentially arbitrary level of detail limited only to available computational resources.

Although FDTD was originally developed for military applications, the method was adapted for tissue optics applications by R. Richards-Kortum and colleagues [9] and has now established itself as having critical importance in tissue optics research. The goal of FDTD is to numerically solve Maxwell's equations. As opposed to CDA, which is an integral equation solver, FDTD is a differential equation solver. With currently available computers, the FDTD method allows modeling of full-vector field interactions with inhomogeneous tissue structures spanning as much as 100 wavelengths in three dimensions on a grid as fine as 1/20th of a wavelength, thus taking into account nanoscale detail of refractive index distribution. FDTD has been used to calculate differential scattering cross-section of internally heterogeneous and nonspherical scattering particles [10], the scattering coefficient of media with spatially continuous refractive index variations [11], and in many other applications.

Although in principle, FDTD could be used to model light interaction with tissue that spans macroscopic dimensions, this has not been feasible because even the most powerful existing computers lack the capabilities to deal with the enormous database of electromagnetic-field vector components mandated for FDTD. Thus, at least for the time being, FDTD remains a microscopic technique.

Recently, a variation of FDTD has been developed as a macroscopic field-based modeling approach, the pseudospectral time-domain method. Similar to FDTD, PSTD was first developed for military applications and introduced to solve tissue optics problems by A. Taflove, S. Tseng, and colleagues [12]. Similar to FDTD, PSTD solves full-vector Maxwell's equations. The difference, however, is that in the PSTD, the spatial derivatives in Maxwell's curl operators are implemented using fast Fourier transforms. This yields infinite-order accuracy, and essentially exact results for electromagnetic-field spatial modes are sampled according to the Nyquist criterion. Whereas FDTD requires a simulation grid that is 1/20th of the wavelength of light, the PSTD meshing density can approach two samples per wavelength, which yields a three-orders-of-magnitude decrease in computational resources relative to FDTD.

As a result, PSTD can model light propagation in realistic macroscopic three-dimensional tissue volumes that may span from hundreds of microns to millimeters. Within such a volume, PSTD can account for the near- and far-field interactions with no simplifications other than the half-wavelength discretization of the refractive-index distribution. The same aspect of PSTD that makes it so computationally efficient, that is, half-wavelength discretization, is perhaps the major weakness of the technique compared to FDTD. Indeed, while the fine grain used in FDTD makes it ideal to study subwavelength, nanoscale features of refractive index distribution, PSTD cannot model light scattering in tissue with a nanoscale level of detail.

Currently, 100 micron–scale tissue volumes can be modeled by use of PSTD on a fairly conventional workstation. With the help of state-of-the-art supercomputers, PSTD modeling of millimeter-sized media is possible. Beyond that, PSTD is just too computational complex, and Monte Carlo simulations are essentially the only viable option. However, with more powerful computer clusters made available in the future, progress in FDTD and PSTD and their application in tissue optics are only expected to increase.

Laboratory 1: Designing a Spherical Particle Suspension to Mimic Scattering Properties

Discussion

Development of instrumentation in biomedical optics often requires validation by measuring a sample with known optical properties. It is, therefore, important to have a means of physically mimicking a scattering medium that closely resembles the scattering properties of tissue. A material used to mimic tissue properties is usually referred to as a tissue phantom. The optical properties of a tissue phantom should be controlled by the design of the phantom.

A common example of such a phantom is the use of a suspension of polystyrene microspheres of known size and concentration. The optical properties, including the scattering coefficient μ_s and anisotropy coefficient g, can be calculated using Mie's solution to Maxwell's equations. Derivations of the Mie solution are found in numerous textbooks, for example, Bohren and Huffman [13], and the details will not be presented here. Briefly, the Mie solution is found by solving Maxwell's equations using boundary conditions for spheres (or infinite cylinders). For such geometries, it is convenient to use a spherical coordinate system. This choice of coordinate system allows the equations to be separated into radial and angular components. Application of the boundary condition requiring that the tangential field be continuous across the sphere's boundary leads to solutions involving a series of spherical Bessel functions.

These functions are easily computed, so programs can be written to calculate the internal and scattered electric field and hence the scattering function and scattering properties. Although the Mie solution applies to a single spherical particle, the result can still be used for particle suspensions provided that the particles are of uniform size and composition, randomly distributed, and separated on average by many wavelengths. When these conditions are satisfied, the scattering from each sphere can be treated as independent (no fixed phase relationship), and the total scattering function is simply taken as the Mie solution multiplied by the number of particles. For example, the scattering cross-section per unit volume (μ_s) is simply the scattering cross-section per sphere times the number of spheres per unit volume.

The ability to accurately compute scattering properties of spheres, along with the commercial availability of precise sphere sizes, makes microsphere suspensions a very useful model for tissue scattering. However, one major difference is that tissue is not composed of discrete scatterers. Since tissue is made of materials with a range of refractive index values that vary continuously with position, much like atmospheric turbulence, tissue can be described as a continuous random medium [14].

In a continuous random medium, there is no representative particle that can be analyzed, so instead, a representative volume must be examined. Any particular volume will have a specific effect on the scattered wave, but when average or expected values of the scattering process are of interest, the sum of interactions from uncorrelated volumes can be treated as an intensity sum. In this case, when the refractive index variations are small and the scattering is weak, the differential scattering cross-section per unit volume can be calculated using the refractive index correlation function $B_n(r)$. See Chapter 5 for a more complete discussion of modeling scattering in continuous random media. To summarize, the spectral density $\Phi(k_s)$ of the material is computed by taking the Fourier transform of $B_n(r)$. The differential scattering cross-section per unit volume, which describes the scattered intensity as a function of angle, is then computed by evaluating

$$\sigma(\theta, \varphi) = 2\pi k^4 (1 - \sin^2(\theta)\cos^2(\varphi))\Phi(2k\sin(\theta/2)) \tag{6.2}$$

where k is the wave number. This method of calculating scattering from refractive index correlation functions makes use of the Born approximation for weak scattering where the incident field is assumed to be constant throughout the volume of interest and requires that the scattered field be much weaker than the incident field (which is the case when the refractive index fluctuations are small). Interestingly, this method can also be applied to microsphere suspensions using the appropriately scaled autocorrelation function for a sphere.

Although the scattering properties of a microsphere suspension can be accurately computed using the Mie solution, the wavelength dependence of scattering properties for a single-sized sphere suspension will typically not match those observed in tissue. When spectral measurements of the scattering are made in tissue, a power–law dependence on wavelength is often observed. This can be modeled by a mass fractal representation of the medium, as discussed previously. In this case, a suspension of a single microsphere size is not sufficient to accurately mimic tissue. However, by using a mixture of microsphere sizes, the spectral scattering properties can be tuned to match a wide range of scattering models, including mass fractals, as discussed previously. In this lab, MATLAB-based Mie code will be used to design phantoms to mimic specific scattering properties.

Materials

MATLAB – Examples in this section will use MATLAB, but other software platforms may be adopted.

Mie code – The Mie code used here is based on Maetzler's code available at http://omlc. ogi.edu/software/mie/.

Exercises

Use Mie Code to Calculate q$_{sca}$ *and g of a Microsphere*

Use Mie() to calculate the total scattering efficiency q_{sca} and g for a 1.0-μm-diameter sphere with refractive index 1.59 surrounded by water with refractive index 1.33 for a range of wavelengths from 0.4 to 1.0 μm, and plot the results. Try varying the particle size (Figure 6.1).

Calculate Scattering Coefficient μ_s *and Reduced Scattering Coefficient* μ_s'

Scattering efficiency is the ratio of scattering cross-section to the geometrical cross-section. The code returns the scattering efficiency for a single particle. To calculate the scattering coefficient μ_s, multiply q_{sca} by the geometric cross-section to get the scattering cross-section, and then multiply by the particle concentration to get the scattering cross-section per volume or scattering coefficient. For example, given a polystyrene density of 1.05 and a suspension of 10% by weight, the particle density would be $\rho = 0.10/1.05/(4/3\pi(D/2)^3)$, where D is the sphere diameter. Calculate the reduced scattering coefficient

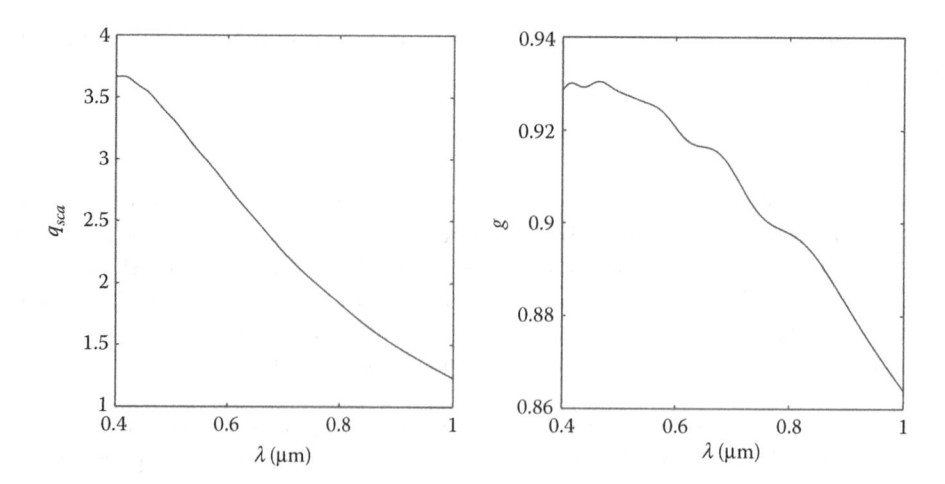

FIGURE 6.1
Scattering efficiency and anisotropy as a function of wavelength for a 1.0-μm sphere.

using $\mu_s' = \mu_s(1 - g)$. Plot these scattering properties over the wavelength range using a linear and log–log scale for a 10% suspension of 100-nm particles. In a log–log scale, a power law appears as a straight line. Do the scattering properties appear to follow a power law? Repeat the plot for a 1.0-μm sphere suspension (Figure 6.2).

Fit a Power Law for μ_s'

The best-fit power law can be determined by fitting a line in log–log space. Use polyfit() to determine the best-fit line for $\log(\mu_s')$ vs. $\log(\lambda)$. Repeat the plots above and include the power–law fit. Calculate the normalized error from the power–law fit. How well does a power–law fit the case when the wavelength is much larger or much smaller than the sphere? What about sphere diameters close to the wavelength?

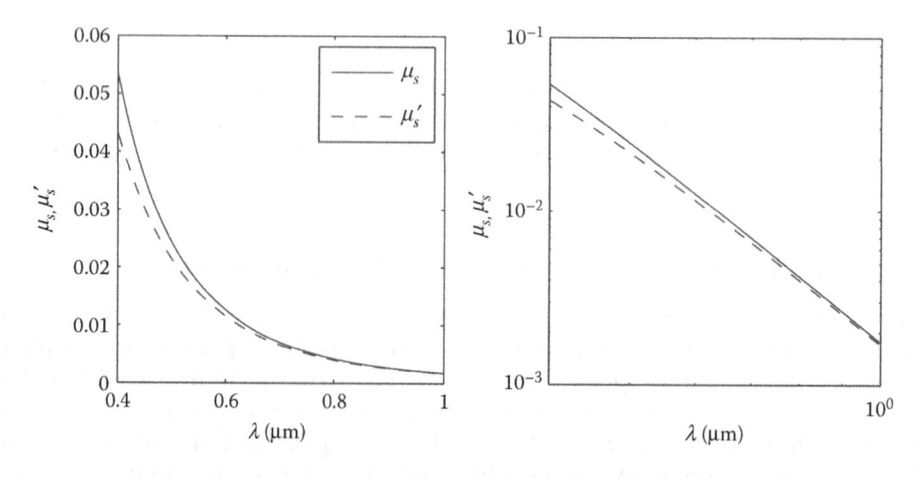

FIGURE 6.2
Scattering coefficient and reduced scattering coefficient for a 100-nm sphere suspension.

Local Fit for the Power Law

Increase the range of wavelengths to span several orders of magnitude. Since a power law can easily be seen as a line in a log–log plot, it is convenient to use a log scale for wavelengths. For example, *lambdas = logspace*(log 10(0.1), log 10(10), 128) generates an array of wavelength values equally spaced in a log scale spanning 0.1–10 in 128 steps. Use this range to plot μ_s' for a 0.3-μm-diameter sphere, and estimate the wavelength at which the power law changes. Most experimental measurements do not span orders of magnitude in wavelength, so measurement of the wavelength dependence will be a local measurement. Assuming μ_s' has a power–law dependence on wavelength of the form $b\lambda^\alpha$, the instantaneous power can be calculated by taking the derivative in log-space. Write an equation for the instantaneous value of α and plot (Figure 6.3).

Refractive Index and Dispersion

The refractive index contrast of the particle plays a large role in scattering. For a fixed wavelength, plot μ_s' as a function of the relative refractive index of the particle. For simplicity, the refractive index has been assumed to be constant over the wavelength ranges of interest; however, dispersion can play a significant role in the refractive index even over the range of visible wavelengths. Use the following formulas to approximate the refractive index of polystyrene and water over the visible range of wavelengths

```
npolystyrene = 1.5663 + 7.85e-3*lambdas.^-2 + 3.34e-4*lambdas.^-4;
nwater = 1.3231 + 3.3e-3*lambdas.^-2 - 3.2e-5*lambdas.^-4;
```

Plot the refractive indices over the visible range. Compare the plot of μ_s' with and without dispersion. How significant is the effect of dispersion in scattering? Repeat for large or small sphere diameters (Figure 6.4).

Multiple Particle Sizes

Scattering properties such as μ_s', μ_s, and g represent statistical average properties of a medium. As such, these properties can be treated as a linear superposition, allowing

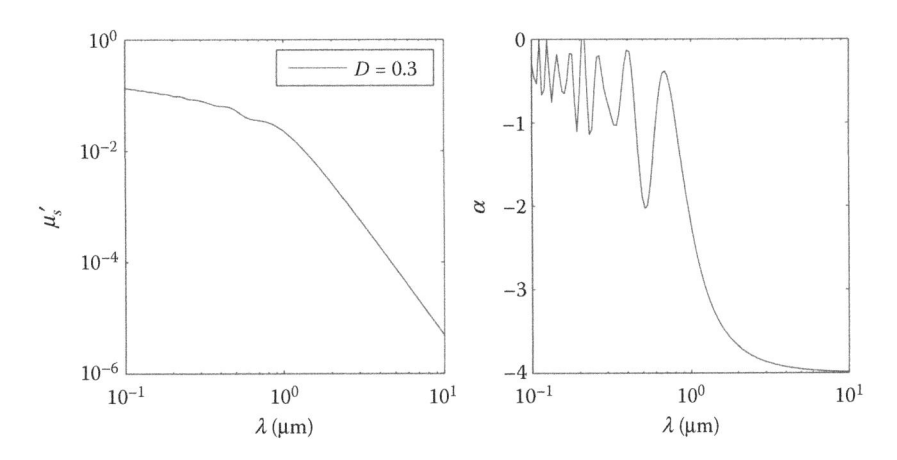

FIGURE 6.3
Instantaneous power law fit of μ_s'.

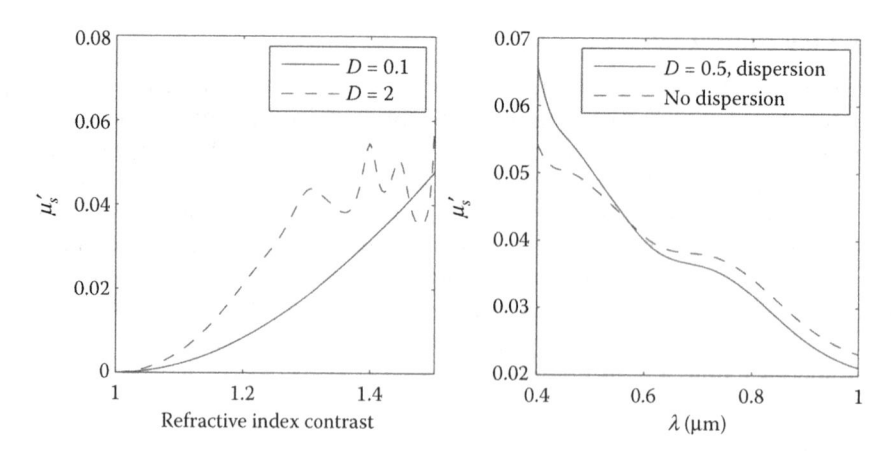

FIGURE 6.4
Effect of refractive index contrast and dispersion on reduced scattering coefficient.

average properties of a medium made of multiple particle sizes to be computed as a weighted average of the properties from the component suspensions. In the previous exercises, it was observed that the reduced scattering coefficient for a suspension of spheres is a power law with an exponent around −1 for particles much larger than the wavelength and rapidly transitions to an exponent of −4 for particles much smaller than the wavelength. However, measurements of tissue often result in intermediate values of the exponent. Calculate and plot μ_s' resulting from a combination of two sphere sizes by adding the fractional contribution from each (Figure 6.5).

Mimicking a Mass Fractal

Although the mixture of two sphere sizes achieves an intermediate power law, the value of the exponent is still highly dependent on the spectral range over which the power law is fit. In a previous section, tissue was modeled as a mass fractal, giving rise to a smooth power law over a wide range of wavelengths. This can be achieved by adding more spherical

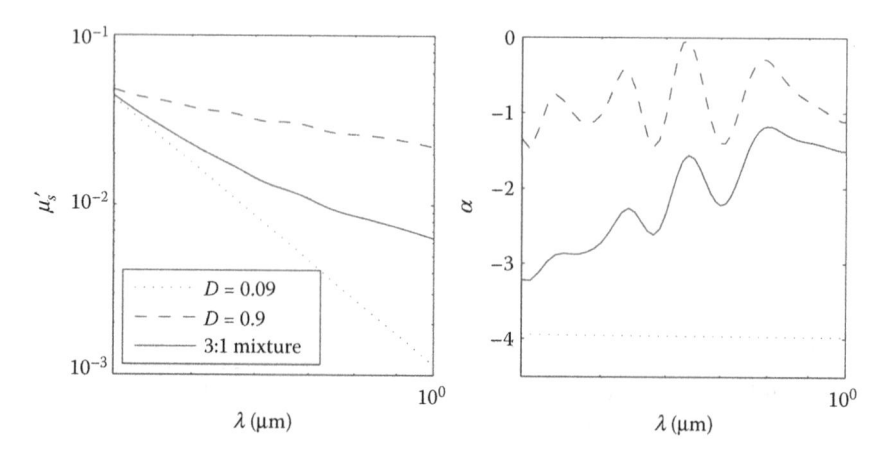

FIGURE 6.5
Mixture of two sphere sizes.

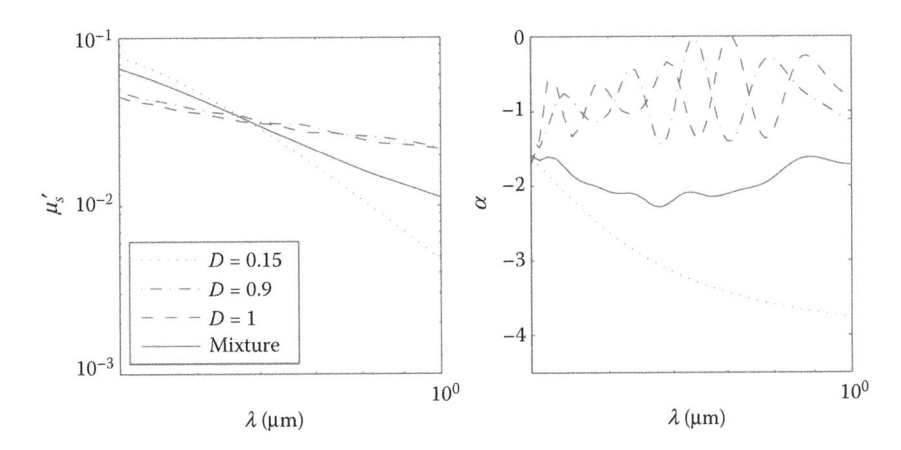

FIGURE 6.6
A mixture of three sphere sizes, 0.15, 0.9, and 1.0, in a ratio of 15:5:4 produces a more stable value of α.

particle sizes to the mixture. Figure 6.6 shows the result of mixing three sphere sizes in different ratios to produce a more stable value of the power–law exponent. Try other sizes and concentrations to create recipes for values of α ranging from –1 to –4. How could this process be automated? How does α relate to mass fractal dimension?

Plot the Refractive Index Correlation Function of a Sphere Suspension

The correlation function of a suspension of spheres is proportional to the autocorrelation of a sphere. The sphere autocorrelation function has an analytical form and can be written as an anonymous MATLAB function:

```
acfs = @(r,D) (((D-r).^2.*(r+2*D))./(2*D.^3)).*rectpuls(r/D-.5);
```

To calculate scattering using the first-Born approximation, the excess relative refractive index n_1 is used where $n_1 = n/<n> - 1$, where $<n>$ is the mean refractive index. Write a formula for the mean refractive index as a function of volume fraction Cv and refractive index of the medium nm and sphere ns. The correlation function of the medium is proportional to the variance of n_1. Write a formula for the variance of n_1. Plot the correlation function for a medium of spheres with diameter 1.0 μm and 0.1 concentration by weight. Assume the density of polystyrene is 1.05.

Use the Born Approximation to Calculate the Scattering Properties of a Sphere Suspension

Under the Born approximation, the differential scattering cross-section is calculated by taking the Fourier transform of the index correlation function to get spectral density $\Phi(k_s) = FT(B_{n1}(r))$ and then evaluating $2\pi(nk)^4(1 - \sin(\theta)\cos(\varphi))^2 \Phi(2nk\sin(\theta/2))$

```
dscs = @(vn,mn,D,k,theta,phi)
(2*pi*(k*mn).^4*(1-(sin(theta).*cos(phi)).^2))...
.*vn.*(3*((2*k*mn*sin(theta/2)).*D/2.*cos((2*k*mn*sin(theta/2))*D/2)-...

sin((2*k*mn*sin(theta/2))*D/2)).^2)./(2*(2*k*mn*sin(theta/2)).^6*pi^2
*(D/2)^3)
```

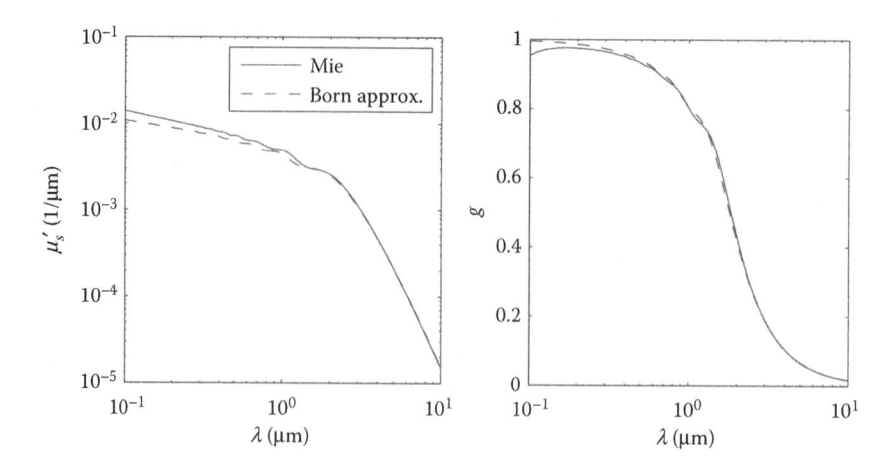

FIGURE 6.7

Reduced scattering coefficient and anisotropy factor calculated using Mie theory and using the first Born approximation for a 1-μm sphere with 1.1 refractive index contrast.

where vn = variance of n_1, mn is the mean index, D is the sphere diameter, k is the wave number, theta is the polar angle, and phi is the azimuthal angle.

Plot the differential scattering cross-section in spherical coordinates for a small and a large sphere size. The scattering coefficient and anisotropy can be calculated by numerically integrating $\mu_s = \int dscs \, \sin(\theta) \, d\theta \, d\varphi$ and $g = 1/\mu_s \int dscs \, \cos(\theta) \, \sin(\theta) \, d\theta \, d\varphi$.

Plot the reduced scattering coefficient and anisotropy factor calculated using Mie code and the previous method for several sphere sizes. Start with a small index contrast (nm = 1.0, ns = 1.1), and then increase the contrast in small steps. Where does the error occur? Can the error be explained in terms of phase and wavelength? (Figure 6.7.)

Calculate the Refractive Index Correlation Function of a Sphere Mixture

The autocorrelation of a medium composed of multiple sphere sizes can be calculated as the superposition of the component correlations. Plot the autocorrelation for five sphere sizes with radii 0.01, 0.0316, 0.10, 0.316, and 1.0 μm in a log–log scale. Note that these radii are equally spaced in a logarithmic scale. Next, plot the sum of these correlation functions. To approximate a fractal correlation function, the ACF should follow a power law. Adjust the variance (or relative weight of these correlation functions) in the sum to approximate a power law of the form r^-1.5, and plot this along with the resulting sum. What functional form do the coefficients take? How well does this superposition represent a power law? (Figure 6.8.)

Write a Script to Optimize a Sphere Mixture to Mimic a Mass Fractal Dimension D_f over the Visible Spectrum

A fractal is defined as something that exhibits self-similarity over a range of length scales. A random scattering medium can be considered a mass fractal when this self-similarity results in a correlation function that follows a power law. For example, a suspension of spheres spanning a large range of diameters "looks" similar regardless of the scale of observation. An ideal fractal is self-similar at *all* length scales, but real media will have some upper and lower limits to the length scales that follow a power–law correlation. Calculate and plot the reduced scattering coefficient for the recipe of spheres from the previous exercise. How well does the recipe mimic scattering from a mass fractal?

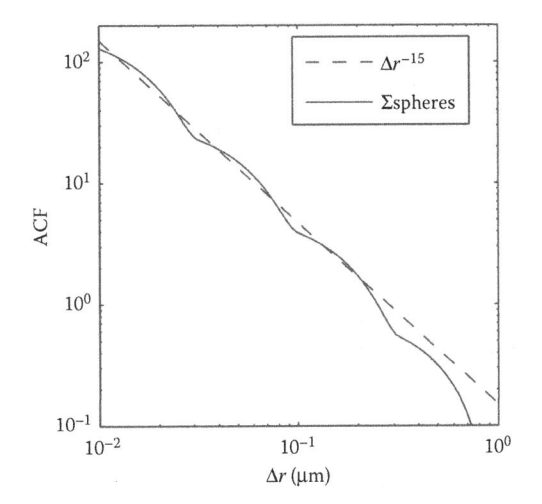

FIGURE 6.8
Autocorrelation of a superposition of spheres.

To design a recipe that mimics a mass fractal scattering medium, some practical limits must be considered. For example, precision microsphere suspensions do not come in arbitrary sizes, so a recipe must be optimized to make use of what is available. In general, the use of more sphere sizes introduces more degrees of freedom and allows better fitting to the desired properties, but pipetting errors and cost limit the practicality of using a large number of spheres. To optimize a fractal phantom, choose a practical number of sphere sizes, and calculate the scattering properties for each. Next, write a merit function that represents the error in a superposition of the suspensions from the desired power law. For example, minimizing the sum squared error is a good approach. Use the function fminsearch() to find volumes that minimize the error (Figure 6.9).

Suppose a phantom is to be designed to test an optical coherence tomography instrument that is sensitive to backscattering as opposed to reduced scattering coefficient. How would the method need to be modified to design a phantom for this case?

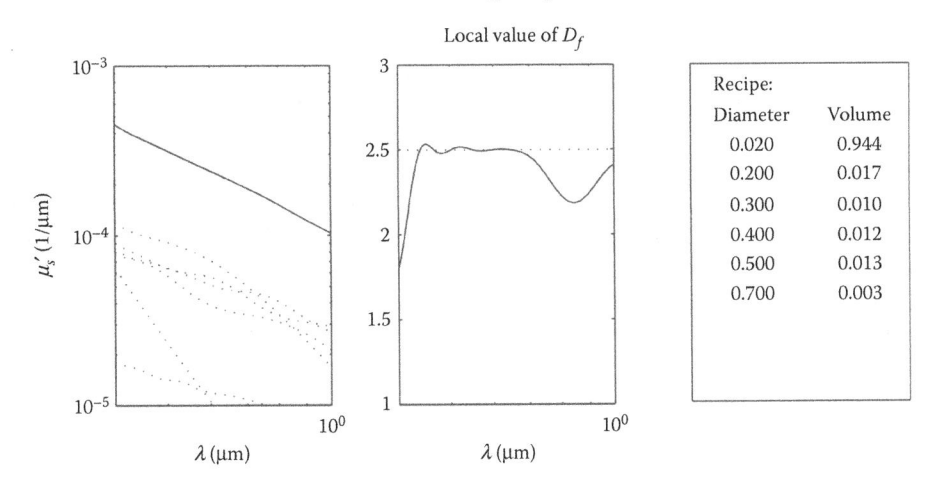

FIGURE 6.9
An example of a recipe for a fractal phantom designed to mimic a mass fractal dimension of 2.5 over the wavelength range 450–650 nm.

Laboratory 2: Generalized Multiparticle Mie Algorithm

As introduced previously, Mie theory provides the exact solution of the scattering field from a spherical particle by solving Maxwell's equation on the sphere boundary. However, a more realistic scenario often involves a cluster of particles that can interact with each other and create a coherent total scattering field that is beyond Mie theory. As discussed earlier in this chapter, one of the computational techniques that may be used to model more complex structures is the generalized multiparticle Mie solution. The objective of this laboratory is to learn the underlying principles of GMM and how it can be used to model light interaction with tissue-like media.

Background

In 1995, Yu-lin Xu reported a comprehensive solution that simulates the scattering field from an arbitrary number of isotropic and homogeneous spheres called GMM [15]. As a rigorous simulation tool, GMM serves as an extension to Mie theory thanks to its ability to simulate interparticle interactions. It has been used to analyze light transport through a chain of dielectric microspheres [16] and the near-field interaction between closely spaced metal nanoparticles in different geometries [17], and a similar study was reported to explore potential surface-enhanced Raman scattering due to an alignment of metallic nanoparticles [18]. Rigorous scattering calculation facilitates a better understanding of scattering from biological tissue in spectroscopy or microscopy [19,20]. It is also intended to simulate DNA structures by a chain of spheres to study the physical process of chromatin decondensation during gene transcription [21] to provide an understanding of optical contrast due to the different chromatin structures.

The central problem is to solve Maxwell's equations under the boundary condition on the spherical surface

$$\left(E_i^j + E_s^j - E_I^j\right)\times n^j = \left(H_i^j + H_s^j - H_I^j\right)\times n^j = 0, \tag{6.3}$$

where the superscript j stands for the jth sphere; the subscripts I, i, and s represent, respectively, the field inside the sphere (internal field) and the incident field on the sphere from both original illumination and the secondary scattering field from other spheres. The total incident field in Mie theory is the original illumination; in GMM, it is decomposed into different partial scattering fields from other spheres added to the original illumination.

Due to the radial symmetry of the sphere, the solution of the vector wave function in a spherical coordinate system (vector spherical functions) is applicable, which is essentially used to express every scattering field. Although a plane-wave illumination is considered throughout this discussion, the illumination can be arbitrary as long as it can be decomposed into the spherical functions. With that, the partial scattering field from ith sphere propagating to jth sphere is also expressed by a linear combination of the spherical functions with different coefficients, called extension coefficients [15]. Solving the extension coefficients for each sphere is a key step in GMM. Since the sphere location is arbitrary, but the scattering field for individual sphere is calculated as if the center of the sphere is the origin, a translation of the scattered field from ith sphere as the incident field to jth sphere with a different coordinate system originating at the center of jth sphere is required. It is realized in GMM by applying additional theorems for the spherical vector wave functions [22,23].

Without going through the detailed algorithms of GMM, the main purpose herein is to introduce a basic usage of the public GMM fortune code provided by Yu-lin Xu. The following introduction is according to the description of GMM code published online at http://www.scattport.org/index.php/programs-menu/multiple-particle-scattering-menu/356-gmm.

The incident electromagnetic field adopted is a plane wave propagating along the $+z$ direction. The incidence can be an arbitrary wave form in addition to a plane wave as long as it can be expressed as a superposition of the elementary spherical waves. Conventionally, the scattering function is described by the scattering matrix, often called S matrix

$$\begin{pmatrix} E_{\parallel s} \\ E_{\perp s} \end{pmatrix} = \frac{\exp(ikr)}{-ikr} \begin{bmatrix} S_2 & S_3 \\ S_4 & S_1 \end{bmatrix} \begin{pmatrix} E_{\parallel i} \\ E_{\perp i} \end{pmatrix}, \tag{6.4}$$

where S_j ($j = 1,2,3,4$) is the amplitude of the scattering matrix, which is a function of wavelength and the scattering direction defined by the scattering angle θ and the azimuth angle ϕ in a spherical coordinate as shown in Figure 6.10; k is the wave number defined as $2\pi/\lambda$; and r is the propagating distance from origin in the far field. The subscripts s and i stand for scattering and incidence. The electric field for both incidence and scattering has two polarizations, parallel and perpendicular to the scattering plane, which are defined by the directions of the incident and scattering field as illustrated in Figure 6.10. Often, an optical system has a particular polarization state so that the polarization of the incident field is decomposed in the x,y directions for convenience. Thus, the scattering matrix that GMM code outputs is modified

$$\begin{pmatrix} E_{\parallel s} \\ E_{\perp s} \end{pmatrix} = \begin{pmatrix} E_{\theta s} \\ -E_{\phi s} \end{pmatrix} = \frac{\exp(ikr)}{-ikr} \begin{bmatrix} S_2^x & S_3^y \\ S_4^x & S_1^y \end{bmatrix} \begin{pmatrix} E_{xi} \\ -E_{yi} \end{pmatrix}, \tag{6.5}$$

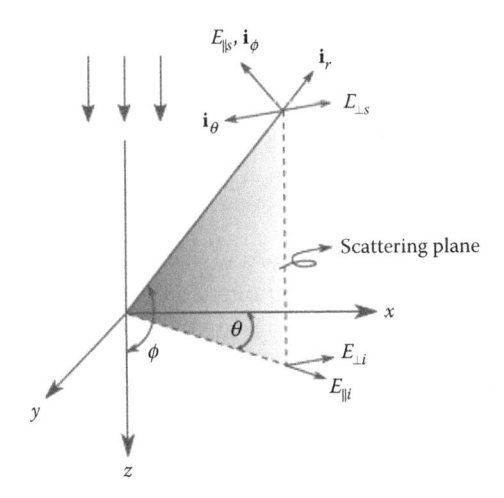

FIGURE 6.10
The coordinate system and the vector definitions. Both the polarization of scattering and the incidence field are conventionally defined with respect to the scattering plane set by the scattering direction $\mathbf{i}_r$ and the illumination wave vector, which is along the positive z axis in GMM. The parallel polarization $E_\parallel$ is inside the plane. The vector direction ($\mathbf{i}_r, \mathbf{i}_\theta, \mathbf{i}_\phi$) of the spherical coordinate is shown as well. The scattering and azimuth angles (θ, φ) determine the scattering direction. The incidence polarization here is defined to the scattering plane as well. The $E_{\perp i}, E_{\parallel s}$, and $+z$ direction form a right-handed triad.

such that S_j^p ($p = x,y$) is the new scattering matrix with respect to the incidence polarization in either the x or y direction. E_{xi} and E_{yi} are the amplitude of the x-polarized and y-polarized incident fields. For convenience, it is desired to transform the scattering polarization into x, y, and z directions so that the angle-dependent scattered field can be coherently integrated when it is used to synthesize numerical microscopic images. The transformation can be formulated

$$\begin{cases} E_{xs} = E_{\theta s} \cos\theta \cos\phi - E_{\phi s} \sin\phi \\ E_{ys} = E_{\theta s} \cos\theta \sin\phi + E_{\phi s} \cos\phi. \\ E_{zs} = -E_{\phi s} \sin\phi \end{cases} \tag{6.6}$$

With the complete scattering matrix, the scattering properties such as the scattering cross-section and the extinction cross-section are also calculated in GMM. In the case of plane-wave illumination, the backscattering cross-section is defined as

$$C_{bak} = \frac{4\pi}{k} \mid S(\theta = 180°) \mid^2 . \tag{6.7}$$

This is particularly useful because a back-reflectance detection scheme is used in the majority of noninvasive and *in vivo* biophotonics applications.

Laboratory Protocol

Objective

The purpose of the assignment in this lab section is to introduce the GMM Fortran code provided by Yu-lin Xu. Students should learn to use the downloadable code from an online resource, compile it to an executable program, and control the input and output suiting different application purposes.

Suggested Lab Components

MATLAB, Fortran code compiler (cygwin, g95 Fortran compiler).

Lab Preparation

1. Visit the ScattPort website at http://www.scattport.org/files/xu/codes.htm, and download the code gmm01f.f for homogeneous sphere clusters. A Fortran code compiler is required to compile the GMM code into an executable program. For Windows system users, we recommend running g95 compiler on Cygwin interface, which are both free and available online. Download Cygwin, and install it on a Windows system according to the instructions given at http://cygwin.com/install.html. Download g95 binary, and install it under Cygwin according to the instructions on page 4 of the g95 user's manual at http://www.g95.org/downloads.shtml#manual.

   After the installation, execute Cygwin and use basic Linux command lines to navigate to the directory of the downloaded GMM code. Use g95 compile commands to compile the code into an executable program. If the code is modified, the program needs to be recompiled for the update. The previous procedure is verified on a Windows 7 64-bit computer.

For users who are not familiar with Linux, g95 compiler can be directly installed on a Windows system, and the compilation commands can be executed by the Windows command line. Execute cmd.exe and navigate to the directory of the downloaded GMM code. Use g95 compile commands to compile the code into an executable program. Refer to the g95 installation manual when needed.

2. The executable requires two input files according to the most updated code (Dec. 2011). The input files specify the physical and optical parameters of the sphere clusters, such as the size, locations, and refractive indices of the spheres, as well as the computational parameters, such as the wavelength of the incidence, far-field scattering direction, and options for averaging geometry rotated with respect to the x, y, or z axis. An example of the two input files is shown here with the parameters explained by comments, notated by the double backslash symbol.

```
gmm01f.in:
----------
//The name of the input file 2 for physical and optical parameters
is specified.

s1323.k

//The GMM code is capable of calculating and output the averaging
scattering field from different rotational states of the same
cluster. The rotation is determined by the Euler angles, and the
following line defines the numbers of the Euler angle being
averaged. (nbeta - # of rotations around the z-axis, nthet - # of
rotations by which z-axis deviates from the original direction,
nphai - # of rotations by the Euler angle of "gamma"). For a
fixed geometry without rotational averaging, specify three numbers
to be 1:

1 1 1

//The range of the three Euler angles (min beta, max beta, min thet,
max thet, min gamma, max gamma) is then specified. For a fixed
geometry without rotational averaging, specify the minimum and
maximum number to be equal for all three angles. The unit is in
degrees:

0.  0.  0.  0.  0.  0.

// The indicator for Mie theory calculation. If it is specified to
be 1, GMM only calculates the coherent summation of all the spheres
without any interparticle interaction:

0

//The rotational averaging doubling option. When calculating a fixed
geometry, specify the option to be 0. When it is 1, the number of
orientations to be calculated is doubled.

Each orientation is coupled with the orientation that the aggregate
is rotated by 90° around the z axis. This ensures that the averaged
polarizations are zero when the scattering angle is either 0° or
```

```
180°, as it is supposed to be for an average over random
orientations:
```

```
0
```

```
//The rotational averaging scheme option. The above Euler angle
theta will be equally divided on the value of cosine(theta) when it
is specified to be 1, or divided on the value of the degree when it
is specified to be 0, or divided randomly over the angular range
when it is -1. The second parameter specifies the seed for a random
number generator, which is only valid when the scheme option is
-1. It needs to be changed when a different random rotation is
desired:
```

```
0 200
```

```
//The first two parameters are the computational options for
calculating the extension coefficients for the interparticle
interaction. In most cases, it has the best performance with numbers
[0 0]. Please refer to the code description if more detailed
information is needed. GMM uses an iterative method to solve the
optimal extension coefficients, and the last parameter specifies the
maximum number of iterations it allows:
```

```
0. 0. 100
```

```
//The number of terms to add to the scattering orders required by the
Wiscombe's criterion when solving the extension coefficients.
Normally, specify this to be 0:
```

```
0
```

```
//The computational error tolerances for calculating Mie coefficient
for single spheres and the extension coefficients for interparticle
interaction. In most cases, the following error can yield reliable
results:
```

```
1.d-20 1.d-6
```

```
//The interaction factor. It is defined as fint = [r_i + r_j]/d_ij,
where r_i and r_j are the radius of two spheres that can interact
with each other, and d_ij is the center to center distance. If it is
set to be 1, there will be no interaction as long as all spheres are
separated:
```

```
0.01
```

```
//The angular interval for the scattering angle(sang) and the
azimuth angle(pang). Unit: degrees:
```

```
1.0.
```

```
//The internal field options:
 0 0 0 0 0 0
```

```
s1323.k:
--------
//The wavelength of the incidence. Unit: microns:

3.9972

//The total number of the spheres:2

//The parameters for each sphere in the order of the x, y, z
coordinate, the radium, the relative real and imaginary refractive
indexes:

-6.4    0.0    6.4    3.2    1.615    0.008
 0.0    0.0    6.4    3.2    1.615    0.008
```

3. The parameters that GMM outputs are fairly self-explanatory, including the scattering matrix, the Mueller matrix, the cross-section of scattering, absorption, and extinction, etc. Please refer to the example and the code description files provided online. The primary parameter this lab involves is the scattering amplitude matrix. One can calculate the scattering field at any point in the far-field space with the scattering matrix.

4. As discussed above, conventionally, the scattering matrix **S** is based on the incidence polarization decomposed on the scattering plane (parallel and perpendicular to the scattering plane) including Mie theory. On the other hand, the scattering matrix output from GMM is based on the x,y polarized incidence. To compare the Mie theory and GMM in the following lab assignments, we need a transformation between two polarization definitions. For a general case, this transformation does not have a universal form. However, a relationship can be drawn if we consider a linearly polarized plane-wave incidence propagating along the z axis of the coordinate system.

$$\begin{cases} E_{\|i} = E_0(\cos\phi\cos\beta + \sin\phi\sin\beta) \\ E_{\perp i} = E_0(\sin\phi\cos\beta - \cos\phi\sin\beta)' \end{cases} \tag{6.8}$$

where E_0 and β are the amplitude and the polarization angle of incidence, respectively, and ϕ is the azimuth angle of the scattering plane. Therefore, any given S matrix calculated from Mie theory can be rewritten in terms of x or y polarized incidence

$$\begin{bmatrix} S_2' & S_3' \\ S_4' & S_1' \end{bmatrix} = \begin{bmatrix} S_2 & S_3 \\ S_4 & S_1 \end{bmatrix} \begin{bmatrix} \cos(\phi - \beta) \\ \sin(\phi - \beta) \end{bmatrix} E_0. \tag{6.9}$$

The value of β is 0 or 90° for the x or y polarized incidence. S′ is the matrix transformed from Mie theory for an arbitrary linear polarized illumination propagating along $+z$ direction.

Exercise

1. Compare the phase function from a cluster of spheres calculated by GMM with Mie theory.

   Calculate the S matrix from a fixed geometry without any rotational average. Input three dielectric spheres with the radius of 0.5 microns, and situate them at the coordinates (0, 0, 0), (2, 0, 0), (−2, 0, 0). Specify the relative refractive index as 1.19, the incidence wavelength as 0.5, and the intervals of the scattering and azimuth angles as 5 and 360. Specify the interaction factor to be 1. Plot the absolute value of S_2^x versus the scattering angle. Change the interaction factor to be 0.75, 0.5, 0.25, and 0.1, and repeat the experiments. Compare the S_2^x curves.

Calculate $|S_2(\theta, \phi = 0)|$ in terms of the scattering angle from a single sphere with the same properties using Mie theory from the previous lab section. Compare $|S_2^x(\theta, \phi = 0)|$ with GMM results when the interaction factors are equal to 1.

2. Compare the backscattering spectrum from a cluster of spheres calculated by GMM with Mie theory.

Program a MATLAB script to update GMM input files over a wavelength range from 0.5 to 0.7 microns with 0.01-micron increment; execute the GMM program, and load S matrix from the output files. *Hint*: Include a loop in the MATLAB script through the wavelength range. Inside the loop, update the GMM input files with the new wavelength; execute the GMM program and load S matrix. Each component of S matrix needs three dimensions (θ, ϕ, λ). Useful MATLAB commands in terms of input and output files: fopen, fclose, fgetl, fread, fwrite, fscanf, fprintf. To run an executable program in a MATLAB script, use the command "dos(program name.exe)."

Input three dielectric spheres with a radius of 0.025 microns, and situate them at the coordinates (0, 0, −2), (0, 0, 0), and (0, 0, 2). Specify the relative refractive index as 1.19 and the intervals of the scattering and azimuth angles as 180 and 360. Perform the calculation sequentially with the interaction factor equal to 1, 0.75, 0.5, 0.25, and 0.1. Plot the spectra of $|S_2^x(\theta = 180, \phi = 0, \lambda)|$ after each calculation, and compare with the Mie spectrum $|S_2(\theta = 180, \phi = 0, \lambda)|$ from a single sphere with the same properties.

Discussion and Questions

1. The interaction factor defines the range in which a sphere can interact with other spheres. Should the results be the same between two calculations with the interaction factor equal to 1 and the *Mie theory calculation option* equal to 1? Is there a slight discrepancy when all the spheres are behaving independently compared to the single-sphere Mie solution? Why?

2. Why do we translate the polarization of output in terms of the scattering plane to the *x*, *y* axis if we want to synthesize numerical images?

Laboratory 3: Monte Carlo Simulations of Radiative Transport: Modeling Radial Probability Distribution from a Semi-Infinite Medium and Comparison with the Diffusion Approximation

As discussed in the Chapter 5 section on diffusion approximation, propagation of light in random media is described according to the radiative transport equation. Under simplifying approximations, this equation can be solved analytically to yield a result that is accurate in specific limiting cases. While these simplifications provide a useful testing ground to understand the general trends expected for various tissue compositions, a full solution is still highly desirable. Unfortunately, for most cases, a full analytical solution of

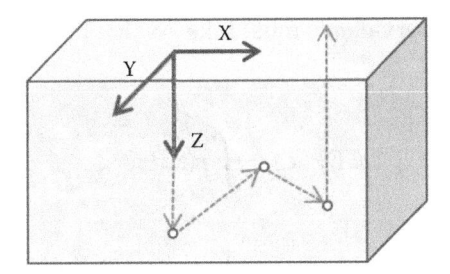

FIGURE 6.11
General diagram of scattering medium.

the RTE is not possible. Still, a full solution result can be obtained using a variety of different numerical methods. In this section, students will gain exposure to an extremely useful and widely applied alternative to solving the RTE, the Monte Carlo method.

The Monte Carlo method refers to a statistical sampling technique used to numerically solve complex stochastic equations that are otherwise difficult or impossible to solve analytically. This method was first developed by Metropolis, Ulam, and von Neumann in the 1940s while working on the Manhattan Project to study how neutrons diffuse through various materials [24,25]. Since then, MC algorithms have been developed for a wide variety of disciplines, including finance and business, computational biology, weather forecasting, and, importantly for us, biophotonics. The first application of the MC method for studying the propagation of light in biological tissue was performed by Wilson and Adam in 1983 [26], after which an almost uncountable number of codes have been developed, making MC the "gold standard" for studying light transport in tissue. Of particular note are codes that model layered tissue [27], polarization dependencies [28–30], coherent phenomena [30–33], deterministic geometries [34], and continuous media [30,35]. Although MC codes have been developed to simulate many different electromagnetic phenomena, each of these codes is in essence a simple photon trajectory-tracing algorithm. In this algorithm, a large number of photons are first directed into a medium composed of scattering and absorbing particles. Each of these photons is then tracked as it is scattered through the medium until eventually being either absorbed or transmitted/reflected out of the medium (Figure 6.11).

Monte Carlo Simulation Algorithm

Light propagation in random turbid media is a stochastic process. This means that (1) the propagation distance between scattering events and (2) the direction of propagation for a single photon of light are dependent on probability density functions (PDFs). The goal of MC simulations is to randomly sample these PDFs with a large number of photons in order to achieve a stable reflectance measurement. It should be noted that in this section (and in MC literature in general) the use of the term *photon* carries little quantum mechanical significance. In fact, in MC simulations, the photon is even allowed to be divided into smaller fractional components. For clarity here, the use of the term *photon* describes a single realization of an infinite number of possible paths in which light can propagate through a scattering medium.

The core of MC simulations is the ability to sample the PDFs that govern light propagation in a scattering medium. An arbitrary PDF $p(x)$ defined on some interval (a,b) is normalized such that $\int_a^b p(x)dx = 1$. As such, the cumulative distribution function (CDF) at some point x' within the interval (a,b) must take a value between 0 and 1 (represented by the symbol ζ) [36]

$$CDF(x') = \int_a^{x'} p(x)dx = \zeta \tag{6.10}$$

After integration, the resulting equation can be solved to find a single observation point x' on $p(x)$ as a function of ζ

$$x' = f(\zeta) \tag{6.11}$$

If we use a random number generator to create a value of ζ that is uniformly distributed between 0 and 1, we can, therefore, randomly determine a value for x' that satisfies the probability distribution of $p(x)$. Depending on the functional form of $p(x)$, an analytical solution for Equation 6.11 may or may not be possible. In the event that no analytical solution exists, a numerical integration can be performed to compute Equation 6.10.

As with any model of a physical system, it is necessary to define the relevant system parameters that affect that process. For light MC simulations, these include the physical dimensions of the bulk material along with the optical scattering and absorption properties. The scattering properties are specified by the scattering phase function $F(\theta)$ and scattering mean free path l_s, while the absorption properties are specified by the absorption mean free path l_a. $F(\theta)$ is a PDF that determines the probability of light being scattered in a particular angular direction, while the combined contribution of l_s and l_a determines how far the light will travel before encountering another optical event (i.e., either scattering or absorbing). Alternatively, l_s and l_a are often represented as the scattering coefficient ($\mu_s = 1/l_s$) and absorption coefficient ($\mu_a = 1/l_a$). In this case, the conceptual understanding of these terms is total scattered (μ_s) or absorbed (μ_a) power per unit volume. As discussed in the section on "Light Transport in a Medium with a Continuously Varying Refractive Index" in Chapter 5, *a priori* knowledge of the physical composition (i.e., discrete scatterers, continuous media, etc.) is needed to accurately define these optical properties.

One popular functional form for $F(\theta)$ is the Henyey–Greenstein phase function, which was originally developed to study the light scattering from interstellar dust. This one-parameter phase function is commonly normalized such that $\int_{-1}^{1} F_{HG}(\cos\theta)\,d\cos\theta = 1$ and takes the form [37]

$$F_{HG}(\cos\theta) = \frac{1}{2}\frac{1-g^2}{(1+g^2-2g\cos\theta)^{3/2}} \tag{6.12}$$

where θ is the polar angle defined on $[0, \pi]$, and g is the anisotropy factor. The parameter g is defined as the average $\cos\theta$ for a given phase function

$$g = \int_{-1}^{1} \cos\theta\, F_{HG}(\cos\theta)d\cos\theta \tag{6.13}$$

The shape of F_{HG} for different values of g can be seen in Figure 6.12. When $g = 0$, the phase function is isotropic, and there is an equal probability to scatter light in any direction.

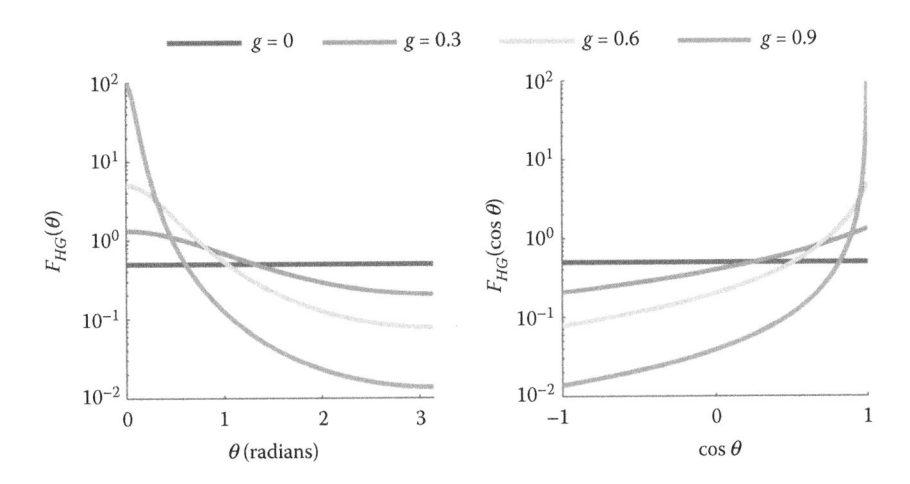

FIGURE 6.12
Illustration of the Henyey–Greenstein phase function F_{HG} for different values of g. The left panel shows F_{HG} as a function of the polar angle θ, while the right panel shows F_{HG} as a function of $\cos \theta$.

As g increases, the phase function becomes more forward-directed until it becomes an impulse in the forward direction when $g = 1$.

For this functional form, the CDF can be found analytically as

$$CDF_{HG}(\cos\theta) = \begin{cases} \dfrac{1-g^2}{2g} \cdot [(1+g^2 - 2g\cos\theta)^{-1/2} - (1+g)^{-1}], & g > 0 \\[2ex] 1/2(\cos\theta + 1), & g = 0 \end{cases} \tag{6.14}$$

Finally, in order to randomly generate a scattering angle, we invert Equation 6.14 as

$$\cos\theta = \begin{cases} \dfrac{1}{2g} \cdot \left[1 + g^2 - \left(\dfrac{1-g^2}{1+g(2\zeta - 1)}\right)^2\right], & g > 0 \\[2ex] 2\zeta - 1, & g = 0 \end{cases} \tag{6.15}$$

In this model, the phase function is rotationally invariant over the azimuthal angle and can therefore be found as $\varphi = 2\pi\zeta$.

In addition to specifying the medium optical properties, it is also necessary to initiate a collection grid to track the positions of photons as they exit the scattering medium. In the simplest geometry, we can track all photons that exit the medium within a radial annuli of radius ρ and width $\Delta\rho$ (Figure 6.13). This quantity is known as the reflectance $R_d(\rho)$. Since any computer will have limited memory capacity, the grid size will have a limited extent. Additionally, since MC simulations track a discrete number of photons, the noise level of the simulation will be inversely related to the grid resolution. Because of these limitations, the grid size and the resolution must be carefully chosen so that the simulation accurately represents the experimentally measurable reflectance with an acceptable level of noise.

After the system parameters are specified, the MC simulation is performed according the flow chart in Figure 6.14. The first step is to initiate a photon with a weight value W, position vector $[x, y, z]$, and trajectory $[\mu_x, \mu_y, \mu_z]$ before launching it into the medium. W represents

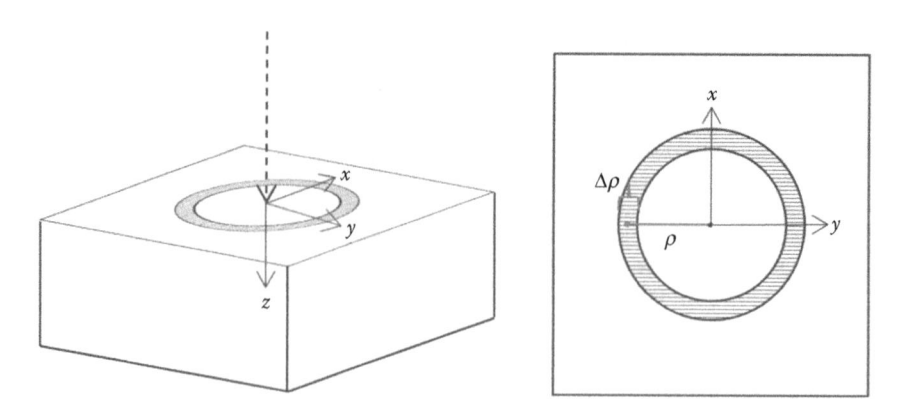

FIGURE 6.13
Schematic of MC geometry with coordinate system.

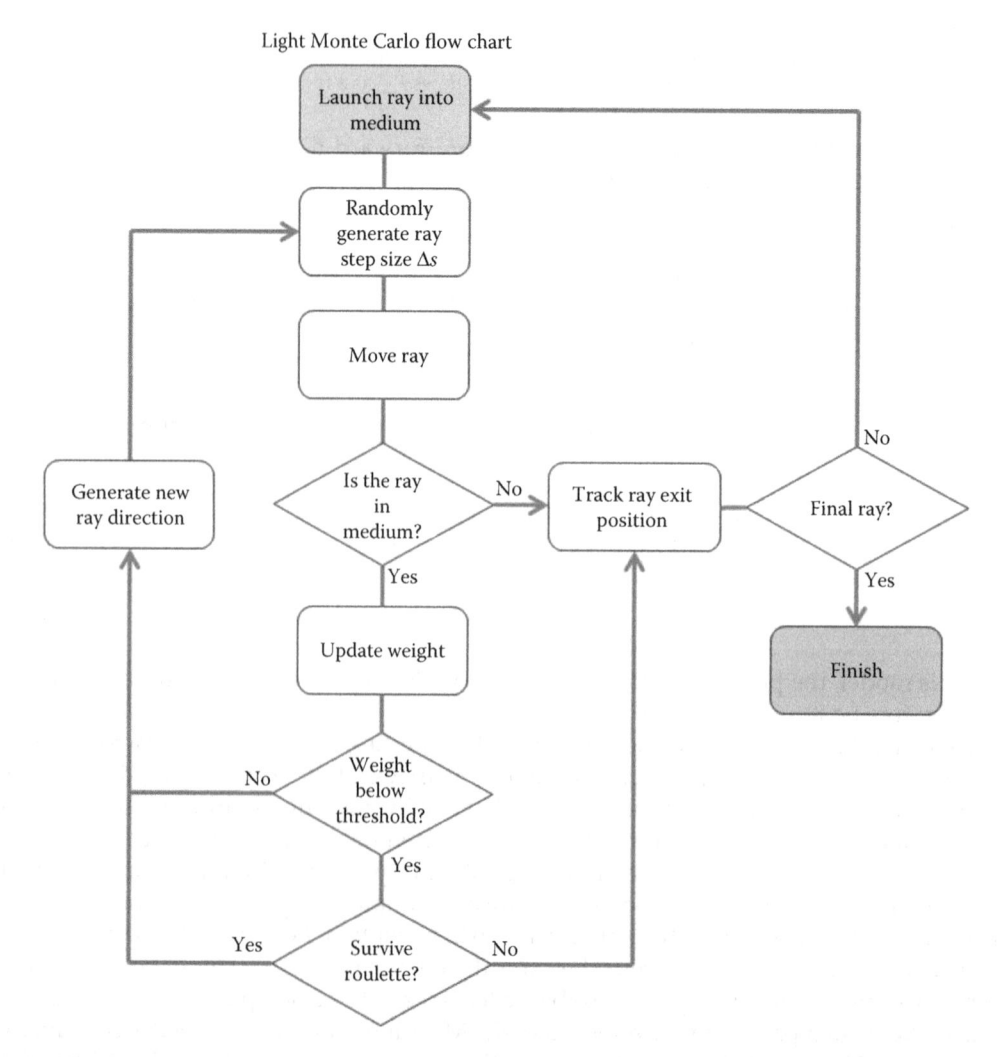

FIGURE 6.14
Flow chart for light MC simulation.

the fractional amount of a photon that survives after encountering an optical event and is defined on the interval [0 1]. The position vector records the physical position of the photon, typically in units of centimeters. The trajectory of the photon is specified by the direction cosines $[\mu_x, \mu_y, \mu_z]$, which are the projection of the photon's direction vector onto the $x/y/z$ basis axis. Typically, the photon is injected into the medium orthogonal to the surface at the origin. As such, the initial conditions are $W = 1$, $[x, y, z] = [0, 0, 0]$, and $[\mu_x, \mu_y, \mu_z] = [0, 0, 1]$.

After being launched into the medium, the photon is allowed to propagate until it reaches an optical event. The probability of a photon traveling to a distance x before encountering an optical event is called the "survival" and is governed by the Beer–Lambert law

$$Survival = e^{-(\mu_s + \mu_a) \cdot x} \tag{6.16}$$

Normalizing this equation as a PDF, we have

$$p(x) = (\mu_s + \mu_a) \cdot e^{-(\mu_s + \mu_a) \cdot x} \tag{6.17}$$

Using Equations 6.10 and 6.11, we can therefore randomly generate the photon step size as

$$s = -\frac{\ln \zeta}{(\mu_s + \mu_a)} \tag{6.18}$$

After determining s, the position vector is updated as

$$x' = x + \mu_x s \tag{6.19}$$

$$y' = y + \mu_y s \tag{6.20}$$

$$z' = z + \mu_z s \tag{6.21}$$

The photon then encounters the next optical event, which can be a combination of both scattering and absorbing. In the presence of absorption, the photon is split into two parts: (1) an absorbed portion that is terminated at the optical event and (2) a scattered portion that continues propagation with reduced weight. The fractional amount of scattered light that continues propagation after encountering an optical event is called the *albedo* $= \mu_s/(\mu_s + \mu_a)$. The scattered portion, therefore, continues propagation with a reduced weight of $W = (albedo)^n$ after n scattering events. Because of this loss, the weight will be reduced incrementally by some fraction at each optical event, and the value will never reach 0. Since photons with a W of nearly 0 contribute very little to the final result, it is very inefficient to continue tracking them. It is, therefore, desirable to terminate a photon when its weight falls below a preset threshold value (typically 0.001). However, if we simply terminate every photon whose weight falls below this threshold, the law of conservation of energy will be violated. Instead, in keeping with the stochastic nature of MC, the photon can be terminated by undergoing a random Russian roulette routine. This works by giving each photon with W lower than the threshold a one in m (typically 10) chance of surviving with a weight of $m \cdot W$ or otherwise being terminated. In this way, photons can be terminated in an efficient manner that is also probabilistically accurate. In the trivial case where there is no absorption, *albedo* $= 0$ and the photon's weight will never decrease below the threshold value.

Following the steps outlined previously, each photon will continue its propagation until it is either fully absorbed or it leaves the medium. Once a photon leaves the medium, its weight is recorded in the appropriate grid location, and a new photon is injected into the medium in the same fashion. This process continues until the completed trajectories of a preset number of s have been calculated. At this point, the ensemble diffuse reflectance profile $R_d(\rho)$ for all photon realization can be observed. Since numbers of photons have an arbitrary meaning, a proper normalization (of which there exist many) must be found. Typically, $R_d(\rho)$ is normalized relative to the total incident intensity. This can be done by dividing the output of the MC simulation by the total number of photons $\cdot \Delta\rho$.

The simulation described previously represents the spatial reflectance impulse-response for a homogeneous slab medium. That is, we measure the spatial reflectance response of a scattering medium to an infinitely narrow pencil beam impulse. However, for real-world application, a true impulse can never be formed. Because of this, a simulation for a beam of some finite size is needed to match with experimentally observable results. In principle, this can be done by adding an additional stochastic event where the location in which the impulse is sent into the medium is randomly chosen according to some desired illumination beam spot size. However, this would drastically increase the simulation time since many photon trajectories would have to be traced for each illumination position in order to guarantee a sufficiently low noise level. Instead, we can use linear system theory to greatly reduce computation time for finite beam spot sizes. If we know the impulse-response for any linear time-invariant system, then the output signal will simply be the convolution of the input signal with the impulse-response, as shown in Figure 6.15. In our case, the input signal is the illumination beam profile $I_{incident}(x,y)$ and the impulse-response is $R_d(x, y)$, where the x,y coordinate is the exit position of the photon. As a result, we can obtain the experimentally measurable output signal $I_{measured}(x,y)$ as

$$I_{measured}(x,y) = \int\int_{-\infty}^{\infty} I_{incident}(x-\alpha, y-\beta) \cdot R_d(\alpha,\beta)\,d\alpha\,d\beta \qquad (6.22)$$

$$= I_{incident} \circledast R_d$$

where $\circledast$ indicates the convolution operations.

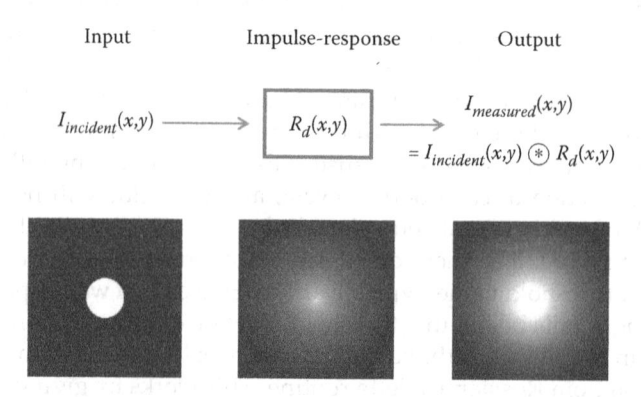

FIGURE 6.15

Demonstration of the properties of a linear time-invariant system. In our case, the illumination beam profile $I_{incident}(x,y)$ is the input, and $R_d(x,y)$ is the impulse-response. The output signal $I_{measured}(x,y)$ can therefore be found by convolving $I_{incident}(x,y)$ and $R_d(x,y)$.

Laboratory Exercises

The basic code for performing MC simulations with the scalar Henyey–Greenstein phase function for a homogenous scattering and absorbing slab can be found below in the MC.m file at http://ultraviolet.bme.northwestern.edu/~arad/#Code. Using this code as a starting point, perform the following studies:

1. The first step in working with any code is to validate that it is working properly. For a semi-infinite medium with $g = \mu_a = 0$, the diffusion approximation of Reference 38 is in almost complete agreement with Monte Carlo simulations. Validate your MC code by plotting the MC simulated reflectance versus exit position against the diffusion approximation. For the simulation, use $\mu_s = 100$ cm^{-1} and plot out to 0.10 cm. Note: The MC simulation is for reflectance in a particular radial annuli $2\pi\rho \cdot R_d(\rho)$, whereas the diffusion approximation equation is for reflectance at a particular source-detector separation $R_d(\rho)$.

2. Perform simulations with the optical properties in Exercise 1 using 10^2, 10^3, 10^4, 10^5, and 10^6 photons.

   a. Analyze the noise level of each simulation by subtracting the analytical diffusion approximation result from the MC simulated result. Plot the standard deviation of this noise versus the number of photons. This curve should follow a power law. Find the exponent of this power law, and discuss the implications for performing MC simulations.

   b. Make a plot of simulation versus number of photons. Describe the functional form of this plot, and discuss the implications for performing MC simulations.

3. Using the optical properties in Exercise 1 with 1e5 photons and maximum collection radius 0.10 cm, perform simulations with a number of bins of 200, 100, 50, 25, and 12. Observe the shape of the reflectance curve for these different simulations. Comment on the noise level and agreement with the diffusion approximation for a different number of bins. What is the tradeoff for performing simulations with different bin sizes?

4. Perform simulations with $g = \mu_a = 0$ and an infinite slab thickness for $\mu_s = 10, 50, 100$, and 200 cm^{-1}.

   a. Describe the difference in shape, and comment on the cause of these differences.

   b. Normalize $R_d(\rho)$ (i.e., y-axis) by μ_s^{-2} and ρ (i.e., x-axis) by μ_s. Comment on the shape of these new curves, and discuss the implications for performing MC simulations. What does this say about the units of $R_d(\rho)$?

5. Perform different simulations setting $\mu_a = 0$ cm^{-1} and $\mu_s = 100$ cm^{-1} and varying g from 0 to 0.9 in 0.1 step size.

   a. Plot simulation time versus g. Comment on the shape of this curve, and explain the trend.

   b. Describe and explain the shape $R_d(\rho)$ for increasing g. What happens to the agreement with the diffusion approximation as g is increased?

   c. As described in Exercise 4b, construct a plot of $R_d(\rho) \cdot \mu_s^{-2}$ vs. $\rho \cdot \mu_s$ for the different values of g. Comment on the difference in shape between curves.

   d. Now construct plots of $R_d(\rho) \cdot (\mu_s^*)^{-2}$ versus $\rho \cdot (\mu_s^*)$. Comparing with Exercise 5b, what does this tell you about dominating length-scale in a scattering medium?

6. Perform different simulations setting $g = 0$ and $\mu_s = 100$ cm^{-1} and vary μ_a to 0, 10, 50, 100, 200, and 500 cm^{-1}.

   a. Plot simulation time versus μ_a. Comment on the shape of this curve, and explain the trend.

   b. Describe and explain the shape $R_d(\rho)$ for increasing μ_a.

   c. Compare with the diffusion approximation, and comment on the agreement for increasing μ_a.

7. As described in other sections, the Whittle–Matérn phase function is a two-parameter $(\hat{g}, m)$ phase function that provides an added extra degree of flexibility for trying to understand light scattering in biological tissue. The functional form for the unnormalized scalar WM phase function is [35]

$$F_{WM}(\cos\theta) = \frac{\hat{g} \cdot (m-1)}{(1+\hat{g}^2 - 2\hat{g}\cos\theta)^m}$$

$$\hat{g} = 1 - \frac{\sqrt{1+4(kl_c)^2} - 1}{2(kl_c)^2}$$

   a. Normalize the above phase function such that $\int_{-1}^{1} F_{WM}(\cos\theta)d\cos\theta = 1$. Comment on the shape for different $\hat{g}$ and m. When $m = 1.5$, which functional form does this resemble?

   b. Find the CDF for the normalized phase function according to Equation 6.10.

   c. Find the equation to randomly generate $\cos\theta$, which follows the probability of $F_{WM}(\cos\theta)$ according to Equation 6.11.

   d. Modify the code developed for the Henyey–Greenstein phase function so that it now uses the Whittle–Matérn phase function. For a fixed $\hat{g} = 0.9$, perform simulations for $m = 1.0, 1.25, 1.5, 1.75$, and 2.00. Comment on the shape of $R_d(\rho)$ for increasing values of m.

   e. For a fixed $m = 1.0$, perform simulations for $\hat{g} = 0.1, 0.3, 0.5, 0.7$, and 0.9. Comment on the shape of $R_d(\rho)$ for increasing values of $\hat{g}$. What can you say about the shape of $R_d(\rho)$ for different values of $\hat{g}$ and m?

8. For many *in-vivo* applications, measurements of diffuse reflectance are carried out using optical fibers to both illuminate the tissue and detect the reflected light. Consider the optical fiber schematic shown in Figure 6.16. This geometry uses two 400-μm diameter optical fibers with a center-to-center distance of 800 μm. Fiber A is used as both an illumination and collection channel, while Fiber B is only for detection.

   a. Perform an MC simulation for tissue-relevant optical properties of $g = 0.9$, $\mu_s = 10$ cm^{-1}, and $\mu_a = 0.1$ cm^{-1}. Perform the convolution operation of Equation 6.22 to find the effective reflectance profile obtained by illuminating with Fiber A. Note: The convolution must be performed in (x,y) coordinates. You can either modify the MC code to track the (x,y) exit positions of photons or transfer $R_d(\rho)$ into (x,y) coordinates.

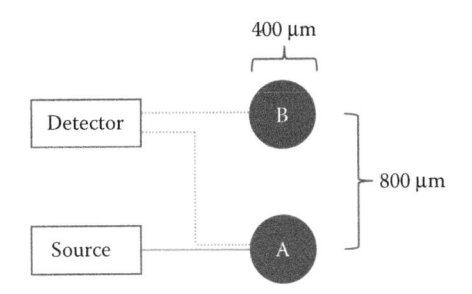

FIGURE 6.16
Schematic for Exercise 8.

    b. Determine the amount of light detected in Fiber A and Fiber B.

    c. Imagine that we are using this probe to detect the presence of cancer. If the optical properties change in cancer to $g = 0.92$, $\mu_s = 9$ cm^{-1}, and $\mu_a = 0.13$ cm^{-1}, what percent change would we detect in each of our fibers?

9. Modify the MC code to track maximum penetration depth of each photon. The output of the code should be the two-dimensional $R_d(\rho, z_{max})$.

    a. Run a simulation for the optical properties $g = 0.9$, $\mu_s = 10$ cm^{-1}, and $\mu_a = 0.1$ cm^{-1}. Using the two-dimensional $R_d(\rho, z_{max})$, make a plot of $R_d(\rho)$ versus ρ and $R_d(z_{max})$ versus z_{max}. Find the average penetration depth $\langle z \rangle$ of all photons. Hint: The average x value for any one-dimensional curve can be found as $\left(\int xf(x)dx \right) / \left(\int f(x)dx \right)$.

    b. Provide a plot of $\langle z \rangle$ vs. ρ. What does this tell you about the relationship between penetration depth and photon exit position?

    c. If the optical properties change in cancer to $g = 0.92$, $\mu_s = 9$ cm^{-1}, and $\mu_a = 0.13$ cm^{-1}, by how much does the penetration depth change for Fibers A and B?

Laboratory 4: The Finite-Difference Time-Domain Method in Computational Biophotonics

The behavior of macroscopic optical structures such as prisms, lenses, stops, and mirrors is successfully explained by a set of principles called *geometrical optics*. In this realm, the optical structures are much larger than the wavelength of light (400–700 nm), and the propagation of light can be understood using geometrical constructions such as *rays* and *wavefronts* [39,40]. Under the geometrical-optics approximation, the wavelength of light plays no significant role, except in higher-order corrections to this approximation (e.g., diffraction, chromatic aberrations). However, the analysis of light interactions with refractive index variations at micro and nanoscale structures, which, as we have seen in Chapters 2 and 4, is critically important when considering light propagation in biological tissue, requires more sophisticated approaches to the propagation and scattering of light beyond the ray-based small-wavelength approximations of geometrical optics. One of the most powerful and widely used computational biophotonics methods designed to solve exactly these kinds of problems is the finite-difference time-domain method. In this laboratory, we will learn how FDTD works and gain hands-on experience in how to use it.

Background

Introduction to the FDTD Method

Before we discuss how the FDTD method works, we should revisit the wave nature of light. Recall that light is an electromagnetic radiation—a form of energy transfer between electrically charged bodies. It is beyond the scope of this section to give a full account of Maxwell's equations of electrodynamics. For our purposes, a brief mention of the general characteristics of the equations will suffice. Maxwell's equations describe the evolution of two vector quantities, the electric (E) and magnetic (H) fields, in space and in matter. The amplitude of a field vector is commonly referred to as its *strength*, while its direction in space is referred to as its *polarization*. For example, one says that the electric field vector at a certain position is polarized in the x-direction (in a given Cartesian coordinate system) and has a strength of 10^{-4} volts/meter. The electric field vector quantifies the force per unit charge exerted on a charged particle at a given position in space and a given instant in time. The magnetic field vector quantifies the force per unit charge and velocity exerted on a *moving* particle at a given position and instant. When all currents are zero (the electrostatic regime), the magnetic field vanishes and does not enter the formulation. Maxwell's equations relate the electric and magnetic field vectors by expressing the spatial variation of one in terms of the time variation of the other. This interdependence creates a wave behavior that causes electromagnetic energy to be propagated in free space and matter.

The responses of a material to the electric and magnetic field vectors are determined by the *permittivity* (ε) and *permeability* (μ) of the material, respectively. These so-called *constitutive parameters* are macroscopic averages of the responses of individual atoms or molecules in the material. Before Maxwell, these two parameters were considered two very distinct properties of matter. When Maxwell unified the electric and magnetic forces in his electromagnetic theory, he showed that a combination of these two parameters determines the *velocity of electromagnetic wave propagation* in the material in the following way

$$v_p = c/n, \quad \text{where } n = \sqrt{\varepsilon\mu}/\sqrt{\varepsilon_0\mu_0},$$

This is called *Maxwell's relation*. Here, $c = 299{,}792{,}458$ m/s is the velocity of light in free space, and

$$\varepsilon_0 = 8.85418782 \times 10^{-12}\, \text{m}^{-3}\, \text{kg}^{-1}\, \text{s}^4\, \text{A}^2 \tag{6.23}$$

$$\mu_0 = 1.25663706 \times 10^{-6}\, \text{m}\, \text{kg}\, \text{s}^{-1} \tag{6.24}$$

are the constitutive parameters of free space, all of which are fundamental constants of nature. The reduction in the speed of light is governed by n, which is called the *refractive index* of the material. The preceding result due to Maxwell, reached by some brilliant physical insight and pure mathematical deduction, suggested strongly that light is an electromagnetic phenomenon. This fact was later confirmed experimentally by H. Hertz in 1887.

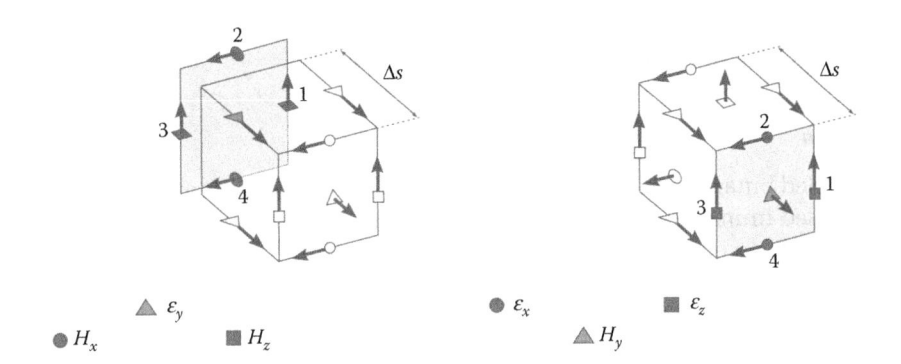

FIGURE 6.17
The discrete approximations to Ampere's Law (left) and Faraday's Law (right) in the FDTD scheme.

The analysis of the propagation and scattering of light using Maxwell's equations removes the approximations placed by geometrical optics and covers a very wide range of optical problems. However, exact solutions to Maxwell's equations are available only for simple problem geometries. Closed-form analytical solutions for complicated media such as biological cells can be obtained only under simplifying assumptions such as the weak-scattering, or Born, approximation [39,41,42]. More rigorous solutions to these sorts of problems require the use of numerical approximation methods for partial differential equations. In Project II.B, we will solve a nontrivial light-scattering problem numerically using the finite-difference time-domain method [43]. There are numerous other approaches to solving Maxwell's partial differential equations numerically. For more information on these approaches, see References 43–46.

The FDTD method provides an explicit solution algorithm for Maxwell's equations of electrodynamics. The simulation space is divided into rectangular cubes, and the electric/magnetic field components are defined on discrete positions on the faces and edges of these cubes. For example, the y-component of the electric field (E_y) is placed on the four edges of the cube that are parallel to the y axis, as seen on the left side of Figure 6.17. It is seen in the same figure that the z-component of the magnetic field (H_z) is placed on the two faces of the cube that are perpendicular to the z axis. At each discrete time instant, the electric field is updated to the next time instant using an approximate form of Ampere's Law in the shaded loop on the left of Figure 6.17. Similarly, the magnetic field is updated to the next time instant using Faraday's induction law on the shaded loop on the right side of the same figure. Once all six Cartesian field components (E_x, E_y, E_z, H_x, H_y, H_z) are updated, the cycle repeats, and the field components are again updated to the following time step.

Angora: An Open-Source FDTD Software Package

Angora is a software package for obtaining numerical solutions to complex electromagnetic problems using the FDTD method. It is freely available for download [47] and is licensed under the GNU Public License [48]. This license requires that the software can be freely copied, modified, and redistributed by everyone, and any software that derives from it is also released under the same license. Currently, *Angora* is available only for the GNU/Linux operating system. The only requirements for extracting and compiling *Angora* are the tar utility, GNU compiler collection (gcc), and the GNU make utility.

In Project II.B, this software package will be used for obtaining the scattering of an incident electromagnetic wave from a material sphere.

Binary Version

If a precompiled binary version of *Angora* is available for your system [47], it can be downloaded and used immediately. Otherwise, the source code for *Angora* must be downloaded and compiled. The compilation and installation instructions are given in the following section.

Compilation and Installation

Angora is dependent on the following libraries: *blitz++* [49], *libconfig* [50], *hdf5* [51], and *boost* [52]. These libraries should be installed on your system before *Angora* can be compiled. If possible, use the package manager for your specific GNU/Linux distribution (such as *Synaptic* in Ubuntu) to install the libraries directly from the package repository. Most major distributions provide these libraries in their package repositories. If you do not have root access to your system, you can install these libraries in your home directory. The installation instructions for the libraries usually provide detailed information on how to do this. The trick is to use the "--prefix=<local-path>" option when calling the configure script for the library package. After installing the dependency libraries, download the compressed package angora-<version>.tar.gz from the Web [47]. Replace <version> by the downloaded *Angora* version (e.g., 0.31). Extract the package using tar, and enter the created directory:

```
johndoe@mysystem:~$ tar xvf angora-<version>.tar.gz
johndoe@mysystem:~$ cd angora-<version>
```

Run the configure script in this directory to create the makefiles required to build the package:

```
johndoe@mysystem:~/angora-<version>$ ./configure
```

If any of the dependency libraries were installed in a local directory, then add the option "--with-<library>=<local-path-to-library>" to the above command line. For example, if the *blitz++* library was installed in "/home/johndoe/blitz-0.9," then the option to add is "--with-blitz=/home/johndoe/blitz-0.9." Type "./configure --help" in the directory "angora-<version>" for information on specifying the paths to the other dependency libraries. After the configure script finishes execution, compile and install *Angora* using the make utility:

```
johndoe@mysystem:~/angora-<version>$ make
```

If your system has multiple cores, you can speed up the compilation by parallelizing make. For example, you can use all four cores of your system by typing, instead of the above line,

```
johndoe@mysystem:~/angora-<version>$ make -j 4
```

This may take a couple of minutes, depending on your system. After make finishes, the executable "angora" will be located in the directory angora-<version>.

Running Simulations with Angora

Angora operates by reading a text file, called the *configuration file*, that specifies the details of the simulation. Every aspect of the simulation is configured by a related *configuration option* (or *option* for short) in the configuration file, which comprises either a single line or a number of lines. In general, the *Angora* executable is run by putting the name of the configuration file as a command-line option:

```
johndoe@mysystem:~/angora-<version>$ ./angora <path-to-config-file>
```

If the *Angora* executable is run without any command-line options, it looks for the configuration file named "angora.cfg" in the same directory from which the executable is run. Details on the specific configuration options for certain aspects of the simulation will be described as needed in Project II.B.

Some features of *Angora* are listed below:

- Flexible and user-friendly configuration via configuration files
- An automatic build and install mechanism for the GNU/Linux operating system
- Support for 3-D parallelization based on the MPI standard
- Support for (possibly lossy) planar multilayered media
- Support for HDF5, a portable file storage format
- Support for several basic shapes and reading materials from files
- Incident-beam creation (plane wave, coherent pulse, partially coherent beam)
- Near-field-to-far-field transform (time-domain and phasor domain)
- Field-value recording (2-D, 1-D, single point) and movie generation
- Random medium and random surface generation
- Numerical optical imaging

See the *Angora* website for more details [47]. The *Angora* project is still under development, and the optics and electromagnetics community is strongly encouraged to contribute to it. There are other freely available FDTD packages such as *Meep* [49] and *Tessa* [50] for FDTD numerical computation. The scattering problem in Project II.B can also be solved using these packages, but we will provide specific instructions only for *Angora*.

One of the noteworthy features of *Angora* is its numerical optical imaging ability. *Angora* can calculate the image-field distribution (also called the *aerial image* in photolithography literature) that would result if the scatterer under analysis were placed under a microscope. The image is calculated for a given illumination scheme and a set of microscope parameters such as the collection numerical aperture, the image-space refractive index, and the magnification. In Figure 6.18, the field-intensity (absolute square of the electric field) distribution of an example numerical image is shown in grayscale [46]. The surface profile of a buccal (cheek) cell is measured using an atomic force microscope (AFM), and this profile is inserted into the FDTD grid with a homogeneous internal refractive index of 1.38. The entire structure is 80 μm $\times$ 80 μm $\times$ 2 μm in size and is modeled in 20-nm spatial accuracy. The cell is placed on a glass half space and illuminated with a plane wave from above. The scattered light is collected over a numerical aperture of 0.6, and the numerical microscope image is calculated by applying the principles of geometrical optics and vectorial diffraction on this scattered field (see Reference 46 for details.) Because of the

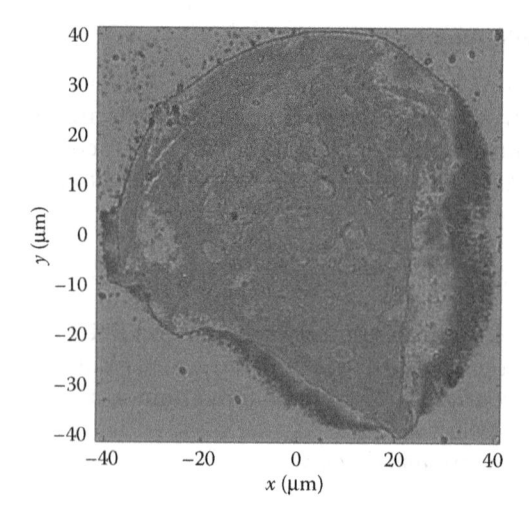

FIGURE 6.18
The numerical microscope image of a buccal (cheek) cell model with an AFM-measured surface profile. The numerical image was calculated entirely in the computer using *Angora*.

overwhelming size of this particular model, the simulation was parallelized and executed on hundreds of processors at the Quest high-performance computing facility [51].

Laboratory

One-Dimensional FDTD

The FDTD equations assume a very simple form when the problem is assumed to be *one-dimensional*, meaning that both the geometry of the structures under analysis and the illumination are assumed to vary only along a single dimension. Despite being fairly restrictive, this simplification allows the observation of the wave phenomenon in its most basic form. In this project, we will set up and numerically solve one-dimensional electromagnetic problems and observe the propagation and scattering of energy inside the solution space.

Let's assume that the problem is invariant in the x and y dimensions and dependent only on the z dimension. Because of this invariance, the partial derivatives in Maxwell's equations with respect to x and y vanish, yielding the following equations for the x and y components of the electric and magnetic field vectors [43]

$$\frac{\partial E_x}{\partial t} = -\frac{1}{\varepsilon_r \varepsilon_0} \frac{\partial H_y}{\partial z} \tag{6.25}$$

$$\frac{\partial H_y}{\partial t} = \frac{1}{\mu_r \mu_0} \frac{\partial E_x}{\partial z} \tag{6.26}$$

$$\frac{\partial E_y}{\partial t} = -\frac{1}{\varepsilon_r \varepsilon_0} \frac{\partial H_x}{\partial z} \tag{6.27}$$

$$\frac{\partial H_x}{\partial t} = \frac{1}{\mu_r \mu_0} \frac{\partial E_y}{\partial z} \tag{6.28}$$

Here, the *relative permittivity* and *permeability* ε_r, μ_r are defined as the actual constitutive parameters of the material (ε, μ) divided by the free-space constitutive parameters (ε_0, μ_0). These four equations can be grouped into two sets of equations, one involving the pair (E_x, H_y), and the other involving (E_y, H_x). These two sets of equations are completely decoupled from each other; therefore, they can be treated separately. In the following, we will assume that $E_y = H_x = 0$ and only consider the pair (E_x, H_y). For a homogeneous space with relative constitutive parameters ε_r, μ_r, it can easily be shown (see Problem 1) that Equations 6.25 and 6.26 allow the *propagating-wave* solutions

$$E_x(z,t) = Z \times H_y(z,t) = f(z \pm v_p t) \tag{6.29}$$

in which $f(\cdot)$ is an arbitrary function, and $Z = ((\mu_r \mu_0)/(\varepsilon_r \varepsilon_0))^{1/2}$ and $v_p = c/(\varepsilon_r \mu_r)^{1/2}$ are the wave impedance and the velocity of propagation in the material, respectively. The negative sign in Equation 6.29 corresponds to the upward-propagating solution (toward the $+z$ direction), while the positive sign corresponds to the downward-propagating one. In Figure 6.19, an upward-propagating pulse is shown in two different instances of time, separated by the interval τ. The profile of the pulse is unchanged in time, but its spatial position shifts in the $+z$ axis by an amount $\xi = v_p \tau$.

In the FDTD scheme, partial derivatives with respect to time and space are approximated by *finite central differences*

$$\frac{\partial f(z,t)}{\partial z} \cong \frac{f\left(z + \frac{\Delta z}{2}, t\right) - f\left(z - \frac{\Delta z}{2}, t\right)}{\Delta z} = \frac{f_{k+1/2}^n - f_{k-1/2}^n}{\Delta z}$$

$$\frac{\partial f(z,t)}{\partial t} \cong \frac{f\left(z, t + \frac{\Delta t}{2}\right) - f\left(z, t - \frac{\Delta t}{2}\right)}{\Delta t} = \frac{f_k^{n+1/2} - f_k^{n-1/2}}{\Delta t}$$

The discrete positions in space and time are indicated by subscripts and superscripts, respectively, with k and n assuming integer values. The *grid spacing* is denoted by Δz,

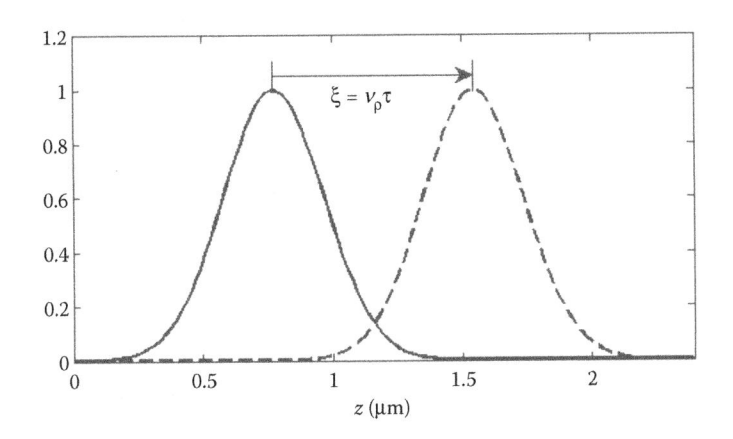

FIGURE 6.19
The temporal evolution of a one-dimensional pulse traveling in the $+z$ direction.

while the *time* step is denoted by Δt. With these approximations, the FDTD equations corresponding to Equations 6.25 and 6.26 for the electric and magnetic field components (E_x, H_y) become

$$E_{x_k}^{n+1} = E_{x_k}^{n} - \frac{\Delta t}{\varepsilon_r \varepsilon_0 \Delta z}\left[H_{y_{k+1/2}}^{n+1/2} - H_{y_{k-1/2}}^{n+1/2}\right] \tag{6.30}$$

$$H_{y_{k+1/2}}^{n+1/2} = H_{y_{k+1/2}}^{n-1/2} + \frac{\Delta t}{\mu_r \mu_0 \Delta z}\left[E_{x_{k+1}}^{n} - E_{x_k}^{n}\right] \tag{6.31}$$

These are in the form of *update equations* for E_x and H_y, which yield the field values at the next time step using the field information available in the current time step. These updates are done repeatedly over many time steps to evolve the fields in time and space. Note that the electric and magnetic fields are in staggered positions, 1/2 steps away from each other both in time and space. For this update scheme to be stable, the ratio of the time step Δt to the grid spacing Δz should be smaller than a certain value. The so-called *Courant–Friedrichs–Lewy stability condition* is stated as

$$S = c\Delta t / \Delta z < D \tag{6.32}$$

The parameter S is commonly referred to as the *Courant number*. Here, $c = (\mu_{min}\varepsilon_{min})^{-1/2}$ is the maximum velocity of propagation possible in the grid. The constant D depends on the dimensionality of the problem. In one dimension, $D = 1$; while in two and three dimensions, D is $1/\sqrt{2}$ and $1/\sqrt{3}$, respectively [43].

1. Confirm that Equation 6.29 is indeed a solution to Equation 6.25 and 6.26 by direct substitution.

2. Define two MATLAB arrays (say, Ex and Hy) for the electric and magnetic field components along the z axis, and assign Ex = Hy = 0. We will assume that the grid is terminated at both ends by an electric field component, so the Ex array will have one extra element in it. Let the length of the Ex array be N. For the purposes of this project, N = 200 will be enough. Next, define two MATLAB arrays (say, epsilon_r and mu_r) for the relative permittivities and permeabilities ε_r, μ_r along the z axis. Similarly to Ex, epsilon_r will have one extra element. Assign epsilon_r = mu_r = 1 to simulate free space. The free-space permittivities ε_0 and μ_0 are given by Equations 6.23 and 6.24.

   Assume a grid spacing of $\Delta z = 20$ nm (20×10^{-9} m). Assuming a Courant number of $S = 0.98$, calculate the time step Δt using Equation 6.32, and use this time step in the simulation. Let the excitation be a Gaussian waveform

$$f(t) = \exp(-(t - k\tau)^2/(2\tau^2)) \tag{6.33}$$

where τ is the *time constant* of the waveform, and k is the delay in terms of τ. Putting $k = 4$ or higher will ensure that the waveform builds up gradually starting from zero. To determine τ, it is necessary to look at the wavelength spectrum of the Gaussian waveform. In FDTD, the rule of thumb is that one should have ~20 grid spacings per minimum wavelength in the excitation waveform. Assuming that

$\lambda_{min} = (6.21 \times 10^8) \times \tau$ [in m] is the minimum wavelength in the waveform, determine the minimum τ admissible, and use this τ in the excitation waveform.

Set up a time loop with $N_t = 500$ time steps. At each time step, first excite the grid by assigning the above Gaussian expression to the first element of the Ex array. Do *not* update the last element of the Ex array. Then, update the electric and magnetic field components according to Equations 6.30 and 6.31.

Make a plot of Ex at each time step. Use the "DoubleBuffer" property of the MATLAB figures to update the plot more efficiently. You can use the following code snippet:

```
figure;
set(gcf,'DoubleBuffer','on');
myaxes = plot(1:N,zeros(1,N));
set(gca,'YLim',[-2 2]);
% time loop
for n=0:499
 % field updates, etc.
 % ...
 set(myaxes,'YData',Ex);
 pause(.03);
 drawnow;
end
```

Note that the simulation is slowed down by the pause command for easier observation.

Explain the behavior of the Gaussian pulse in the grid. Why is the pulse reflecting from both edges? (*Hint:* Think about the boundary conditions that are effectively enforced at the edges. The total solution is a combination of the downward- and upward-propagating solutions in Equation 6.29 that satisfies these boundary conditions.)

3. Implement an absorbing *boundary condition* (ABC) at the upper boundary of the grid (last element of the Ex array). This will prevent the reflection of the pulse from the top of the grid, simulating the extension of the grid toward positive infinity. It is sufficient for our purposes to use a first-order ABC, based on the upward-propagating solution in Equation 6.29. Let this solution be denoted as $E_x(z,t)$. Write down the first-order Taylor series expansions of E_x around (z,t) (which we assume correspond to the discrete position $E_{x_k}^n$) with respect to z and t, and obtain approximate expressions for $E_{x_k}^{n+1}$ and $E_{x_{k-1}}^n$. For more information on Taylor's series expansions, see Section 5.6 of Reference 52. Relate these two solutions and show that the updated value of Ex at time instant $n+1$ at the top of the grid can be approximated by

$$E_{x_N}^{n+1} \cong (1-S)E_{x_N}^n + SE_{x_{N-1}}^n \tag{6.34}$$

in which S is the Courant number, given by Equation 6.32. This equation expresses the updated value of Ex at the top of the grid in terms of the *present* values of Ex at the top of the grid and the nearest spatial position. Implement this ABC in your MATLAB code, and observe that the pulse is absorbed at the end of the grid. Record the value of Ex at the upper edge of the grid at each time step, and place these values in a MATLAB array. Do the field recording between the excitation, and update operations at each step. Show that the resulting array corresponds to the excitation waveform (6.33), except a propagation delay of $(N - 1)\Delta z/c$. (*Hint:* In the same figure, plot the excitation array with respect to the time array $\Delta t^*(0:N_t - 1)$ and the recorded Ex array with respect to the same time array plus the delay $(N - 1)\Delta z/c$.)

4. Now, we will investigate the behavior of the pulse when it encounters a discontinuity in the material properties. Place a half space made of glass into the grid by assigning the permittivity value 2.25 from the center of the grid upward (i.e., assign epsilon_r(N/2 + 1:N) = 2.25 in MATLAB). Assign the average permittivity to the interface, namely, epsilon_r(N/2) = (1 + 2.25)/2. Modify the ABC at the top of the grid such that the Courant number pertaining to the glass half space is used instead of the free-space one; in other words, substitute $S' = S/\sqrt{2.25} = S/1.5$ for S. Also, implement an ABC for the *bottom* of the grid where the wave is excited. This ABC will be based on the same approximation in Equation 6.34, except the present field values $E_{x_1}^n$ and $E_{x_2}^n$ will be used to update $E_{x_1}^{n+1}$. Turn off the excitation after 140 time steps, and apply the ABC to Ex(1) instead in the remaining time steps. Make a plot of Ex at each time step using the code snippet in Problem 2. Now, you should see a pulse originating from the bottom of the grid, which then splits into two pulses at the glass interface at the center. The pulse propagating upward is the *transmitted pulse*, whereas the one propagating downward is the *reflected pulse*. Both pulses should be absorbed by the ABCs at the respective edges. For this two-layered geometry, it is possible to obtain analytical solutions for both pulses. The amplitudes of the reflected and transmitted pulses are given by Reference 53

$$A_{refl} = \frac{1 - \sqrt{\varepsilon_r}}{1 + \sqrt{\varepsilon_r}} \tag{6.35}$$

$$A_{trans} = 1 + A_{refl} \tag{6.36}$$

where ε_r is the relative permittivity of the dielectric half space. Record the value of Ex at the top of the grid (Ex(N)), and show that it corresponds to the excitation pulse scaled by Equation 6.36 and delayed by $(N/2 - 1)\Delta z/c + (N/2)\Delta z/(c/1.5)$ (see the hint for the previous problem). The latter delay term is a result of the fact that the wave *slows down* in the dielectric by a factor $\sqrt{2.25} = 1.5$. Also, record the value of Ex at the origin of excitation (Ex(1)), and show that the latter-time response at that point (after 140 time steps) corresponds to the excitation pulse scaled by Equation 6.35 and delayed by $(N - 2)\Delta z/c$. It is a well-known fact that glass reflects 4% of the incident light intensity. Confirm this fact by checking the intensity reflection coefficient at the glass interface, which is equal to $|A_{refl}|^2$.

In Problems (3) and (4), there will be small discrepancies between the FDTD results and theory. These are due to several factors. First, the pulses are distorted as they propagate along the grid because of the numerical dispersion introduced by the discrete FDTD scheme. Second, the imperfect first-order ABCs cause a finite reflection. Finally, the theoretical reflection and transmission amplitudes (Equations 6.35 and 6.36) are only approximations for a discrete grid. Despite all this, you can expect that the pulses will agree within 1% of both their amplitudes and temporal displacements.

Scattering of Light from a Sphere: The Lorenz-Mie Solution

The exact solution for the scattering of light from a sphere has many interesting applications. It was first derived and used by Gustav Mie in 1908 to explain the optical properties of colloidal metallic solutions [54]. Since then, it has found use in many other applications. In astrophysics, chemistry, and meteorology, it is used to determine particle sizes and concentrations [55,56]. In biophotonics, it has been used to measure the average sizes of cell structures, with applications in cancer detection and screening [57]. The Mie solution can also help construct the computational models in ray-based Monte Carlo simulations of light propagation in biological tissue [58].

In this project, we will rigorously compute the scattering of light from a homogeneous dielectric sphere using two different methods and compare the results. The first method is based on the FDTD numerical approximation, and the second is the exact theoretical solution obtained by Lorenz [39,59]. The FDTD solution will be obtained using the *Angora* package described in Section B. The Mie solution will be computed using a standard MATLAB code, which can be downloaded from the Web [60]. The Mie code consists of three MATLAB scripts: Mie_S12.m, Mie_abcd.m, and Mie_pt.m. The file Mie_S12.m contains the main script for computing the scattered electric field, which in turn uses the other two scripts, Mie_abcd.m and Mie_pt.m.

The geometry of the scattering problem is depicted in Figure 6.20. The common origin of the Cartesian (x,y,z) and spherical (r,θ,ϕ) coordinates is at the center of the homogeneous sphere of radius a. A plane wave traveling in the $+z$ direction is assumed to illuminate the sphere from the bottom. Electromagnetic waves are *transverse* waves, meaning that their amplitude vectors are *perpendicular* to the direction of energy transfer. Without any loss in generality, we assume that the plane wave is polarized in the $+x$ direction.

1. Create an *Angora* script file (say, mie.cfg) for the problem.
   a. Use the Courant number $S = 0.98$ and the grid spacing $\Delta x = 20e-9$. Let the grid be of dimensions $1\,\mu m \times 1\,\mu m \times 1\,\mu m$ and the PML 200 nm thick. Let the simulation be run with 1500 time steps. For all this, put the following assignments in the script file:

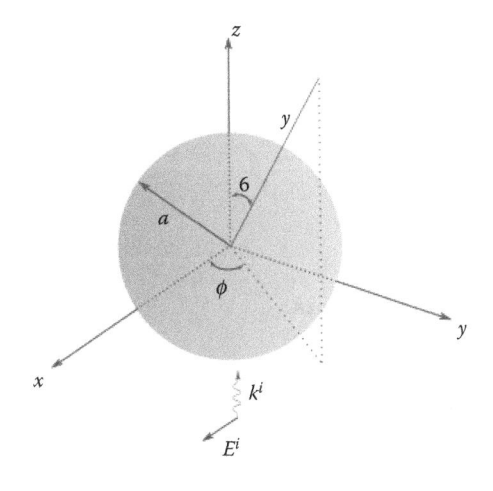

FIGURE 6.20
Geometry of the Mie-scattering problem.

```
courant = 0.98;
dx = 20e-9;
NCELLS_X = 50;
NCELLS_Y = 50;
NCELLS_Z = 50;
NPML = 10;
NSTEPS = 1500;
```

b. Create a spherical shape of radius 320 nm at the center of the grid using the "Spheres" subset of the "Shapes" configuration option:

```
Shapes:
{
    Spheres:
    (
        {
        shape_tag = "mysphere";
        center_x = 0;
        center_y = 0;
        center_z = 0;
        radius = 320e-9;
        }
    );
};
```

c. Create the glass material with relative permittivity 2.25 using the "Materials" configuration option:

```
Materials:
(
        {
        material_tag = "glass";
        rel_permittivity = 2.25;
        }
);
```

d. Using the shape and material tags you assigned to the spherical shape and the glass material as inputs to the "Objects" option, create a glass sphere of 320 nm at the center of the grid:

```
Objects:
(
        {
        material_tag = "glass";
        shape_tag = "mysphere";
        }
);
```

e. Create a sine-modulated Gaussian time waveform using the "ModulatedGaussianWaveforms" subset of the "Waveforms" option. Use the time constant $\tau = 2.127 \times 10^{-15}$ and center frequency $f_0 = 5.889 \times 10^{14}$. This corresponds to a center wavelength of 509 nm, which is at the lower green portion of the visible spectrum. Enter this into the script file as follows:

```
Waveforms:
{
    ModulatedGaussianWaveforms:
     (
      {
       tag = "mywaveform";
       modulation_type = "sine";
       tau = 2.12662e-15;
       f_0 = 5.88878e14;
      }
     );
};
```

f. Send a plane wave on the sphere using the "PlaneWaves" suboption. This is a subset of the more general "TFSF" option, which stands for "total-field/scattered-field." This is the technical name of the method by which *Angora* creates incident beams in the FDTD grid [43]. As explained above, the plane wave has an *x*-polarized electric field and is traveling in the +z direction. This is done in *Angora* by including the following in the configuration script file:

```
TFSF:
{
 PlaneWaves:
   (
     {
       THETA = 180;
       PHI = 0;
       PSI = 90;
       waveform_tag = "mywaveform";
     }
   );
};
```

Note that we are using the waveform with tag "mywaveform," which we defined earlier in (e).

g. Create a near-field-to-far-field transformer (NFFFT) using the "PhasorDomainNFFFT" option:

```
PhasorDomainNFFFT:
(
   {
    insert assignments here...
   }
);
```

The lines between the curly braces will be filled with various assignments in the following. The "PhasorDomainNFFFT" option calculates the electric field scattered by the sphere at "large" distances from the center of the sphere. More precisely, it calculates the asymptotic form of the electric field vector as r goes to infinity, with the r-dependent part taken out. For more details, refer to References 59 and 61. Record the far field only at $\lambda = 509$ nm by putting the following assignments in the "PhasorDomainNFFFT" option:

```
num_of_lambdas = 1;
lambda_min = 509.09e-9;
lambda_max = 509.1e-9;
```

Next, we must specify the directions at which the far field will be recorded. We will consider an array of directions on the positive-x portion of the xz plane. In spherical coordinates, these directions are specified by the spherical angles $\theta = 0$–$180°$ and $\phi = 0$. The theta angle is measured from the positive z axis downward, with $\theta = 0$ and 180 corresponding to the positive and negative z axes, respectively. These directions are entered into the "PhasorDomainNFFFT" option via the following assignments:

```
direction_spec = "theta-phi";
num_of_dirs_1 = 180;
dir1_min = 0;
dir1_max = 180;
num_of_dirs_2 = 1;
dir2_min = 0;
dir2_max = 0;
```

We now have 180 θ-angles between $0°$ and $180°$, and a single ϕ-angle, which is 0.

2. Run the *Angora* executable with the above script file as a command-line option (see Section B.3). By default, the far-field output file is written in the directory "output/nffft/pd" relative to the directory from which *Angora* is executed. The default name of the far-field file is FarField_pd_0_0.hd5. The format of this file is HDF5, which is a standardized portable data storage format [62]. There are freely available software tools and built-in MATLAB scripts for manipulating HDF5 files. Download the helper script "hdf5_read.m" [63] for reading the far-field information from FarField_pd_0_0.hd5.

a. Read the θ range from FarField_pd_0_0.hd5 using the hdf5_read script, and assign the result to the MATLAB array theta:

```
>> theta = hdf5_read('<full-path>/FarField_pd_0_0.hd5','theta');
```

Remember to enter the correct path to the file instead of *<full-path>*.

b. Next, read the θ component of the far field electric vector into the MATLAB array E_theta:

```
>> E_theta = hdf5_read('<full-path>/FarField_pd_0_0.
   hd5','E_theta');
```

c. Plot the absolute value of E_theta versus theta in a Cartesian plot. In the following, another plot will be placed in the same figure, and the two will be compared.

3. Compute the theoretical far field over the same θ range as above using the Lorenz–Mie solution provided by Mie_S12.m, which can be downloaded from the Web [60]. The function Mie_S12(m,x,u) in the script Mie_S12.m accepts three parameters:

m: The refractive index of the sphere, which is equal to the square root of the permittivity. It is equal to 1.5 for our problem.

x: 2*π*(radius of the sphere/wavelength). It is equal to 3.949 for our problem.

u: The cosine of the θ angle associated with the observation direction. Since we are scanning θ from 0 to 180°, this ranges from 1 to −1.

   a. Run the script Mie_S12(m,x,u) with the parameters above. Construct a loop over the θ values obtained in Problem 2.a, and use the cosine of the θ angle as the u parameter at every step. For every θ value, the function Mie_S12(m,x,u) will yield a vector with two values. Select the second value, and save the result in the MATLAB variable S2. Place all the S2 values in the MATLAB array S2_array.

   b. To get a far-field value that is comparable to the FDTD result in Problem 2.c, we must multiply S2 by some factors. The first factor, $1/(jk)$, is a result of the definition of S2 [55,64]. Here, j is the imaginary unit ($\sqrt{-1}$), and k is $2\pi/\lambda = 1.234 \times 10^7$. The second factor depends on the specific waveform that we employed in Problem 1.e. The script Mie_S12(m,x,u) assumes a monochromatic, sinusoidal unit-amplitude plane wave, while our waveform is a finite temporal pulse. This pulse can be decomposed into its sinusoidal components using the Fourier transform [65]. The weight of the sinusoidal component of our waveform at the wavelength 509 nm turns out to be

$$F(509\,\text{nm}) = -\frac{j}{4\pi}\sqrt{2\pi}\,\tau$$

Multiplying the result of the script Mie_S12(m,x,u) by this factor will yield the scattering response at 509 nm created by the modulated Gaussian pulse employed in Problem 1.e. Therefore, the total multiplicative factor in front of S2 becomes

$$K = -\frac{\tau}{k}\sqrt{\frac{1}{8\pi}}$$

Multiply the MATLAB array S2_array you obtained in Problem 3.a by this factor, and assign the result to the MATLAB array E_th_theta. (Make sure you always enter the imaginary unit as 1j or 1i in MATLAB. Entering simply j or i might cause a clash with another variable with the same name.)

   c. Plot the absolute value of E_th_theta versus θ in the same figure as the one you created in Problem 2.c. Show that the two plots are in good agreement.

Laboratory 5: Modeling Random Refractive Index Distribution in Tissue: Creating Random Volumes with Specified Covariance

As we learned in Chapters 2 and 4, local refractive index is a linear function of the local concentration of tissue's "solid" component such as protein, nucleic acids, and lipids. Spatially, both density and refractive index vary continuously. Although in some biophotonics techniques it is convenient (and customary) to approximate tissue as a suspension of isolated microscopic spherical particles (located in the far-field of each other) or slabs of uniform refractive index separated by well-defined boundaries that give rise to light reflection, we have to remember that these are approximations, and a continuously varying refractive index is a more realistic representation of how tissue is structured. Therefore, an initial step in modeling light–tissue interactions is to be able to model spatially random refractive index distribution in tissue. These distributions can then be used in analytical calculations of light scattering and transport properties (see Chapters 4 and 6) or as an input into computational methods such as GMM, T-matrix, DDA, and FDTD, all of which are discussed in this chapter. In this laboratory, we will learn how to model a spatially random refractive index.

Background

Random (or stochastic) models for the refractive index distribution of a medium can be useful when the exact properties of the medium are not known everywhere, but some average properties are available. Such models are used in microwave remote sensing applications [66], as well as for representing cells and tissue in biophotonics [58,67,68]. In this project, we will create samples from a random distribution of three-dimensional media with a specified mean and covariance function. We will calculate estimates for the statistical properties of the given sample and verify that the sample indeed conforms to the given random model.

The *mean refractive index* of a medium is defined as the expected value of the refractive index at a given point over all the possible realizations of the random medium

$$n_0(\boldsymbol{r}) = E\{n(\boldsymbol{r})\}$$

where $n(\boldsymbol{r})$ is a continuous sample of the random refractive index distribution. The operator $E\{\cdot\}$ denotes the statistical expected value, which is a weighted average over all the possible values of the random variable. We assume that the medium is *homogeneous*, meaning that the mean refractive index is the same everywhere along the medium (i.e., $n_0(\boldsymbol{r}) = n_0$). Another useful quantity associated with the random medium is the *two-point covariance function*, or simply the *covariance function*. It is defined as the expected value of the product of the mean-subtracted refractive indices at two points separated by the vector distance $\Delta \boldsymbol{r}$

$$B_n(\Delta r) = B_n(|\Delta \boldsymbol{r}|) = E\{\tilde{n}(\boldsymbol{r})\tilde{n}(\boldsymbol{r} - \Delta \boldsymbol{r})\} = E\{(n(\boldsymbol{r}) - n_0)(n(\boldsymbol{r} - \Delta \boldsymbol{r}) - n_0)\} \qquad (6.37)$$

Here, we define the mean-subtracted refractive index distribution $\tilde{n}(\boldsymbol{r})$. The distance over which the covariance function $B_n(\Delta r)$ retains an appreciable value is commonly referred to as the *correlation length* of the medium. It represents a distance scale over which the refractive index values "stay close to each other." Note that the covariance function depends only on Δr, which is the absolute value of $\Delta \boldsymbol{r}$. This applies only to a subset of random media with the property of being *shift invariant* and *isotropic*. The former means that the statistical relationship between two points of the medium does *not* change if they are both shifted in a certain direction by the same amount. The latter means that this relationship does not change when the medium is rotated as a whole.*

---

* To be more technical, shift invariance and isotropy are much more stringent conditions than required by Equation 15. They become equivalent only when the entire statistics of the medium are described by a jointly Gaussian random model. For more details, please consult texts on stochastic processes; for example, [71,75].

It should be stressed that the shift invariance and isotropy are only an approximation for a finite, discrete random sample generated in a computer. The statistics of a finite sample will evidently change near the edges but will approximate the real statistics near the middle.

Laboratory

1. As the simplest case, we will start with one-dimensional random media and see how independent samples from the same stochastic model can be created. This will serve to demonstrate the basic principle behind the creation of three-dimensional random media. To simplify the derivation, let's deal with mean-subtracted refractive index profiles; in other words, the fluctuations of the index around the mean. Let these fluctuations be represented by $\tilde{n}(x)$. Let's consider the following covariance function for the refractive index of a one-dimensional medium

$$B_n(\Delta x) = E\{\tilde{n}(x)\tilde{n}(x - \Delta x)\} \tag{6.38}$$

Any sample medium generated in the computer must be a discrete one; namely, it should be defined at a discrete set of spatial points. Let's consider the discrete version of Equation 6.38 and assume that a discrete refractive index profile $\tilde{n}[n]$ has the following discrete covariance function

$$B_n[s] = E\{\tilde{n}[n]\tilde{n}[n - s]\} \tag{6.39}$$

in which s is the (unitless) discrete separation between the two discrete points. Since the memory in any computer is limited, we also must limit the extent of the refractive index profile to a finite number. Let the truncated refractive index profile sample $\tilde{n}[n]$ be of length N. To generate a sample with the covariance function (6.39), we will make use of a powerful transformation between finite sequences: the discrete Fourier transform [69]. This transform yields a sequence of length N from a sequence of the same length and is fully reversible. In fact, the inversion operation is essentially the same as the one used for obtaining the transform in the first place. The availability of a highly efficient method called the fast Fourier transform for its computation has made the DFT very popular. The DFT of the sample mean-subtracted refractive index profile $\tilde{n}[n]$, denoted as $\tilde{N}[k]$, is defined as

$$\tilde{N}[k] = \sum_{n=0}^{N-1} \tilde{n}[n]e^{-j\frac{2\pi}{N}nk} \tag{6.40}$$

The inverse transform is given by [69]

$$\tilde{n}[n] = \frac{1}{N}\sum_{k=0}^{N-1} \tilde{N}[k]e^{j\frac{2\pi}{N}nk} \tag{6.41}$$

These two transforms are calculated in MATLAB using the built-in functions fft and ifft. The last expression clearly shows that the DFT is a decomposition of the sequence $\tilde{n}[n]$ into its complex harmonics $\exp(j(2\pi/N)nk)$. We will see shortly that the DFT sequence $\tilde{N}[k]$ holds the key to generating the original sequence $\tilde{n}[n]$. Note that $\tilde{N}[k]$ is also a random sequence, since $\tilde{n}[n]$ is random.

The *two-point correlation function* $B_N[s]$ of $\tilde{N}[k]$ is defined as*

$$B_N[s] = E\{\tilde{N}[k]\tilde{N}^*[k-s]\} \tag{6.42}$$

a. Substitute the definition of $\tilde{N}[k]$ in Equation 6.40 into Equation 6.42, employing the summation indices n and m in the first and second summations, respectively. Convert the product to a double summation over n and m, and carry the expected value inside the summation. Express the expected value in terms of the covariance $B_n[s]$ of $\tilde{n}[n]$ [see Equation 6.39].

b. Extend the summation over n from $-\infty$ to ∞. This approximation is justified if the length of $\tilde{n}[n]$ (N) is very large compared to the discrete correlation length n_c of $\tilde{n}[n]$.† Show that the summation over n is equal to the function $\Phi_n[k]$

$$\Phi_n[k] = \sum_{r=-\infty}^{\infty} B_n[r]e^{-j\frac{2\pi}{N}rk} \tag{6.43}$$

This function is known as the *power-spectral density* of the random sequence. It gives a measure of the powers carried by individual harmonic components. Finally, evaluate the summation over m, using the identity

$$\sum_{m=0}^{N-1} e^{-j\frac{2\pi}{N}ms} = N\delta[s] \tag{6.44}$$

where $\delta[s]$ is the Kronecker delta function (unity if $s = 0$, zero otherwise). Combine Equations 6.43 and 6.44, and write the final expression for $B_N[s]$. What does this result imply about the statistical correlation of $\tilde{N}[k]$ and $\tilde{N}^*[k-s]$ for $s \neq 0$? What about $E\{|\tilde{N}[k]|^2\}$? Based on the above derivations, describe a method for creating independent samples of the refractive index profile $\tilde{n}[n]$. (*Hint*: Is it easier to first generate the sequence $\tilde{N}[k]$ instead?)

2. Next, we will create three-dimensional samples of a random refractive index distribution with an exponential covariance function

$$B_n(\Delta r) = \sigma_n^2 \exp(-\Delta r / l_c) \tag{6.45}$$

Here, Δr is the vector distance between the two points, l_c is the correlation decay coefficient of the random medium, and σ_n^2 is the covariance, which quantifies the strength of the fluctuations of the refractive index around the mean value.

---

* Note that the second term in the curly brackets is $\tilde{N}^*[k-s]$, the complex conjugate of $\tilde{N}[k-s]$. This is the definition of the correlation function for complex random sequences.
† The discrete correlation length n_c is defined similarly to the continuous-domain correlation length l_c; namely, as the s value at which $B_n[s]$ vanishes to a negligible value.

The actual correlation length can be defined as several times l_c, depending on the convention.

a. Download the MATLAB script generate_corr_3d.m from the web [70]. Run the following MATLAB command to generate an exponentially correlated random volume sample of dimensions $200 \times 200 \times 200$, discrete correlation length nc = 10, and fluctuation strength sigma_n = 0.02:

```
>> medium = generate_corr_3D(0.02,10,200,200,200,2);
```

The last argument to generate_corr_3d indicates exponential covariance. Plot a cross-section of the array medium in a two-dimensional MATLAB image. Use the MATLAB command imagesc to scale the color range automatically. Generate three samples using the above command, and plot the cross-sections of these samples side by side.

b. We can approximate the covariance function $B_n(\Delta r)$ of the three-dimensional random array by the *normalized autocorrelation* of the sample array medium. The normalized autocorrelation of a three-dimensional array $x[m, n, p]$ is defined as

$$R_{xx}[i,j,k] = \frac{1}{MNP} \sum_{\substack{0 \le m \le M-1 \\ 0 \le n \le N-1 \\ 0 \le p \le P-1}} x[m,n,p]x[m-i,n-j,p-k] \tag{6.46}$$

This deterministic value, obtained from a single sample, is an estimate for the actual covariance function that is averaged over different samples. For a large class of random media (called *ergodic* media), the normalized autocorrelation for a single sample converges to the actual covariance function as M, N, P all tend to infinity [71]. The brute-force calculation of Equation 6.46 is computationally costly. It is more efficiently calculated using multidimensional fast Fourier transforms. Calculate the normalized autocorrelation of the array medium using the following command

```
>> corr_3D = ifftn(abs(fftn(medium)).^2)/(A*B*C);
```

Here, $A = B = C = 200$ are the dimensions of the array. Take one-dimensional strips of length 100 from corr_3D starting from corr_3D(1,1,1) in all three directions. Compare these three arrays to the discrete exponential covariance function

$$B_n[i,0,0] = \sigma_n^2 \exp(-|i|/n_c) \tag{6.47}$$

The value of i should range from 0 to 99. Show that medium has the desired exponential covariance.

c. Repeat (a) and (b) with $n_c = 60$. Observe that the match to the exponential covariance degrades for this n_c. This is a limitation of the Fourier-transform technique used by generate_corr_3D to generate the random samples. A good rule of thumb is that the extents of the array in every direction should be at least 10 times the discrete correlation length n_c. For our example, the maximum value of n_c for reliable statistical convergence is $n_c = 20$.

References

1. M. I. Mishchenko, L. D. Travis, and D. W. Mackowski, T-matrix computations of light scattering by nonspherical particles: A review. *J. Quant. Spectrosc. Radiat. Transf.* 1996;55(5):535–575.

2. M. G. Giacomelli, K. J. Chalut, J. H. Ostrander, and A. Wax, Application of the T-matrix method to determine the structure of spheroidal cell nuclei with angle-resolved light scattering. *Opt. Lett.* 2008;33(21):2452–2454.

3. Y. L. Xu, Electromagnetic scattering by an aggregate of spheres. *Appl. Opt.* 1995;34(21):4573–4588.

4. M. Born and E. Wolf, *Principles of Optics.* Cambridge, UK: Cambridge University Press, 1999.

5. B. C. Wilson and G. Adam, A Monte Carlo model for the absorption and flux distributions of light in tissue. *Med. Phys.* 1983;10(6):824–830.

6. L. H. Wang, S. L. Jacques, and L. Q. Zheng, MCML - Monte Carlo modeling of light transport in multilayered tissues. *Comput. Methods Programs Biomed.* 1995;47(2):131–146.

7. T. J. Farrell, M. S. Patterson, and B. Wilson, A diffusion-theory model of spatially resolved, steady-state diffuse reflectance for the noninvasive determination of tissue optical-properties invivo. *Med. Phys.* 1992;19(4):879–888.

8. C. K. Hayakawa, V. Venugopalan, V. V. Krishnamachari, and E. O. Potma, Amplitude and phase of tightly focused laser beams in turbid media. *Phys. Rev. Lett.* 2009;103(4):043903.

9. R. Drezek, A. Dunn, and R. Richards-Kortum, Light scattering from cells: Finite-difference time-domain simulations and goniometric measurements. *Appl. Opt.* 1999;38(16):3651–3661.

10. X. Li, A. Taflove, and V. Backman, Recent progress in exact and reduced-order modeling of light-scattering properties of complex structures. *IEEE J. Sel. Top. Quantum Electron.* 2005;11(4):759–765.

11. I. R. Capoglu, J. D. Rogers, A. Taflove, and V. Backman, Accuracy of the Born approximation in calculating the scattering coefficient of biological continuous random media. *Opt. Lett.* 2009;34(17):2679–2681.

12. S. H. Tseng, J. H. Greene, A. Taflove, D. Maitland, V. Backman, and J. Walsh, Exact solution of Maxwell's equations for optical interactions with a macroscopic random medium. *Opt. Lett.* 2004;29(12):1393–1395.

13. C. F. Bohren, D. R. Huffman, *Absorption and Scattering of Light by Small Particles.* New York: Wiley, 1983.

14. M. Moscoso, J. B. Keller, and G. Papanicolaou, Depolarization and blurring of optical images by biological tissue. *J. Opt. Soc. Am. A* 2001;18:948–960.

15. Y.-l. Xu, Electromagnetic scattering by an aggregate of spheres. *Appl. Opt.* 1995;34:4573–4588.

16. Z. Chen, A. Taflove, and V. Backman, Highly efficient optical coupling and transport phenomena in chains of dielectric microspheres. *Opt. Lett.* 2006;31:389–391.

17. B. Khlebtsov, V. Zharov, A. Melnikov, V. Tuchin, and N. Khlebtsov, Optical amplification of photothermal therapy with gold nanoparticles and nanoclusters. *Nanotechnology* 2006;17:5167.

18. A. Gopinath, S. V. Boriskina, B. M. Reinhard, and L. Dal Negro, Deterministic aperiodic arrays of metal nanoparticles for surface-enhanced Raman scattering (SERS). *Opt. Express* 2009;17:3741–3753.

19. J. Yi, J. Gong, and X. Li, Analyzing absorption and scattering spectra of micro-scale structures with spectroscopic optical coherence tomography. *Opt. Express* 2009;17:13157–13167.

20. W. Yip and X. Li, Multiple scattering effects on optical characterization of biological tissue using spectroscopic scattering parameters. *Opt. Lett.* 2008;33:2877–2879.

21. J. S. Kim, V. Backman, and I. Szleifer, Crowding-induced structural alterations of random-loop chromosome model. *Phys. Rev. Lett.* 2011;106(168102):1–4.

22. A. Stein, Addition theorems for spherical wave functions. *Q. Appl. Math.* 1961;19:15–24.

23. O. R. Cruzan, Translational addition theorems for spherical vector wave functions. *Q. Appl. Math.* 1962;20:33–40.

24. N. Metropolis and S. Ulam, The Monte Carlo method. *J. Am. Stat. Assoc.* 1949;44:335–341.

25. N. Metropolis, The beginning of the Monte Carlo method. *Los Alamos Sci.* 1987;15:125–130.

26. B. C. Wilson and G. Adam, A Monte Carlo model for the absorption and flux distributions of light in tissue. *Med. Phys.* 1983;10:824–830.

27. L. H. Wang, S. L. Jacques, and L. Q. Zheng, MCML - Monte Carlo modeling of light transport in multilayered tissues. *Comput. Methods Programs Biomed.* 1995;47:131–146.

28. J. C. Ramella-Roman, S. A. Prahl, and S. L. Jacques, Three Monte Carlo programs of polarized light transport into scattering media: Part II. *Opt. Exp.* 2005;13:10392–10405.

29. J. C. Ramella-Roman, S. A. Prahl, and S. L. Jacques, Three Monte Carlo programs of polarized light transport into scattering media: Part I. *Opt. Exp.* 2005;13:4420–4438.

30. A. J. Radosevich, J. D. Rogers, V. M. Turzhitsky, N. N. Mutyal, J. Yi, H. K. Roy, and V. Backman, Polarized enhanced backscattering spectroscopy for characterization of biological tissues at subdiffusion length-scales. *IEEE J. Sel. Top. Quantum Electron.* 2012;18(4):1313–1325 (invited).

31. M. Xu, Electric field Monte Carlo simulation of polarized light propagation in turbid media. *Opt. Express* 2004;12:6530–6539.

32. J. Sawicki, N. Kastor, and M. Xu, Electric field Monte Carlo simulation of coherent back-scattering of polarized light by a turbid medium containing Mie scatterers. *Opt. Exp.* 2008;16:5728–5738.

33. V. Turzhitsky, J. D. Rogers, N. N. Mutyal, H. K. Roy, and V. Backman, Characterization of light transport in scattering media at subdiffusion length scales with low-coherence enhanced backscattering. *IEEE J. Sel. Top. Quantum Electron.* 2010;16:619–626.

34. Y. C. Chen, J. L. Ferracane, and S. A. Prahl, A pilot study of a simple photon migration model for predicting depth of cure in dental composite. *Dent. Mater.* 2005;21:1075–1086.

35. V. Turzhitsky, A. Radosevich, J. D. Rogers, A. Taflove, and V. Backman, A predictive model of backscattering at subdiffusion length scales. *Biomed Opt. Express* 2010;1:1034–1046.

36. A. N. Witt, Multiple-scattering in reflection nebulae.1. Monte Carlo approach. *Astrophys. J. Suppl. Ser.* 1977;35:1–6.

37. L. Henyey and J. Greenstein, Diffuse radiation in the galaxy. *Astrophys. J.* 1941;93:70–83.

38. T. J. Farrell, M. S. Patterson, and B. Wilson, A diffusion-theory model of spatially resolved, steady-state diffuse reflectance for the noninvasive determination of tissue optical-properties *in vivo. Med. Phys.* 1992;19:879–888.

39. M. Born and E. Wolf, *Principles of Optics: Electromagnetic Theory of Propagation, Interference and Diffraction of Light*, 7th ed. Cambridge, U.K.: Cambridge University Press, 1999.

40. E. Hecht, *Optics*, 4th ed. New York: Addison-Wesley, 2002.

41. W. C. Chew, *Waves and Fields in Inhomogeneous Media*. New York, NY: Van Nostrand Reinhold, 1990.

42. A. Ishimaru, *Wave Propagation and Scattering in Random Media*. New York, NY: Wiley-IEEE Press, 1999.

43. A. Taflove and S. C. Hagness, *Computational Electrodynamics: The Finite-Difference Time-Domain Method*, 3rd ed. Boston, MA: Artech House, 2005.

44. A. F. Peterson, S. L. Ray, and R. Mittra, *Computational Methods for Electromagnetics*. New York, NY: IEEE Press, 1998.

45. J. Jin, *The Finite Element Method in Electromagnetics*. New York, NY: John Wiley & Sons, 2002.

46. I. R. Capoglu, J. D. Rogers, A. Taflove, and V. Backman, The microscope in a computer: Image synthesis from three-dimensional full-vector solutions of Maxwell's equations at the nanometer scale (accepted for publication in *Progress in Optics*; E. Wolf, ed.).

47. I. R. Capoglu, The Angora package. [Online]. Available: [Accessed 2012].

48. The GNU General Public License v3.0 - GNU Project - Free Software Foundation (FSF). [Online]. Available: http://www.gnu.org/licenses/gpl.html [Accessed Nov. 2011].

49. Meep - AbInitio. [Online]. Available: http://ab-initio.mit.edu/wiki/index.php/Meep [Accessed Nov. 2011].

50. Alioth: Tessa 3D-FDTD: Project Home. [Online]. Available: http://alioth.debian.org/projects/tessa/ [Accessed Nov. 2011].

51. High Performance Computing System - Quest. Northwestern University, IT Communications, [Online]. Available: http://www.it.northwestern.edu/research/adv-research/hpc/quest/index.html [Accessed Nov. 2011].
52. G. Arfken, *Mathematical Methods for Physicists*, 3rd ed. Orlando, FL: Academic Press, 1985.
53. C. A. Balanis, *Advanced Engineering Electromagnetics*. New York, NY: Wiley, 1989.
54. G. Mie, Beiträge zur Optik trüber Medien, speziell kolloidaler Metallösungen. *Ann. Phys.* 1908;330:377–445.
55. C. Bohren, E. Clothiaux, and D. Huffman, *Absorption and Scattering of Light by Small Particles*. Wiley-VCH, 2010.
56. H. Hulst, *Light Scattering by Small Particles*. New York: Courier Dover Publications, 1981.
57. J. W. Pyhtila and A. Wax, Polarization effects on scatterer sizing accuracy analyzed with frequency-domain angle-resolved low-coherence interferometry. *Appl. Opt.* 2007;46(10):1735–1741.
58. A. J. Radosevich, V. M. Turzhitsky, N. N. Mutyal, J. D. Rogers, A. Stoyneva, A. K. Tiwari, M. De La Cruz et al., Depth-resolved measurement of mucosal microvascular blood content using low-coherence enhanced backscattering spectroscopy. *Biomed. Opt. Express* 2010;1(4):1196–1208.
59. J. A. Stratton, *Electromagnetic Theory*. New York: McGraw-Hill, 1941.
60. I. R. Capoglu, Mie solutions scripts for MATLAB. [Online]. Available: [Accessed 2012].
61. R. F. Harrington, *Time-Harmonic Electromagnetic Fields*. New York, NY: Wiley-IEEE Press, 2001.
62. HDF Group - HDF5. [Online]. Available: http://www.hdfgroup.org/HDF5/. [Accessed Nov. 2011].
63. I. R. Capoglu, HDF5 file reading script for MATLAB. [Online]. Available: [Accessed 2012].
64. C. Mätzler, MATLAB functions for Mie scattering and absorption. [Online]. Available: http://omlc.ogi.edu/software/mie/maetzlermie/Maetzler2002.pdf [Accessed Nov. 2011].
65. A. V. Oppenheim, A. S. Willsky, and S. H. Nawab, *Signals and Systems*. Upper Saddle River, NJ: Prentice Hall, 1997.
66. C. D. Moss, F. L. Teixeira, Y. E. Yang, and J. A. Kong, Finite-difference time-domain simulation of scattering from objects in continuous random media. *IEEE Trans. Geosci. Remote Sens.* 2002;40(1):178–186.
67. J. D. Rogers, I. R. Capoglu, and V. Backman, Nonscalar elastic light scattering from continuous random media in the Born approximation. *Opt. Lett.* 2009;34(12):1891–1893.
68. I. R. Capoglu, J. D. Rogers, A. Taflove, and V. Backman, Accuracy of the Born approximation in calculating the scattering coefficient of biological continuous random media. *Opt. Lett.* 2009;34(17):2679–2681.
69. A. V. Oppenheim, R. W. Schafer, and J. R. Buck, *Discrete-Time Signal Processing*. 2nd ed. Upper Saddle River, NJ: Prentice Hall, 1999.
70. I. R. Capoglu, Random-medium generation script for MATLAB. [Online]. Available: [Accessed 2012].
71. A. Papoulis, *Probability, Random Variables, and Stochastic Processes*. New York, NY: McGraw-Hill, 1991.
72. libconfig - C/C++ configuration file library. [Online]. Available: http://www.hyperrealm.com/libconfig/. [Accessed Nov. 2011].
73. Blitz++ home page. [Online]. Available: http://www.oonumerics.org/blitz/. [Accessed Nov. 2011].
74. Boost C++ libraries. [Online]. Available: http://www.boost.org/. [Accessed Nov. 2011].
75. N. Kampen, *Stochastic Processes in Physics and Chemistry*. Amsterdam, The Netherlands: Elsevier, 2007.

Index